Joseph Mc Farland

Pathogenic Bacteria

Joseph Mc Farland

Pathogenic Bacteria

ISBN/EAN: 9783741103643

Manufactured in Europe, USA, Canada, Australia, Japa

Cover: Foto ©Klaus-Uwe Gerhardt /pixelio.de

Manufactured and distributed by brebook publishing software
(www.brebook.com)

Joseph Mc Farland

Pathogenic Bacteria

PREFACE TO THE SECOND EDITION.

In preparing this second edition of "Pathogenic Bacteria" I have endeavored to bring the work up to date in all departments of the subject by introducing brief mention of all recent work accomplished in bacteriology. In order to aid the student whose particular interest might make him desire to refer to the original papers, I have thought well to depart from the plan of the original work and give the references in the form of footnotes.

Without departing too much from the primary descriptive purpose of the book, I have made it a special point to add considerably to the amount of technique it contains, and so make it fulfill the double purpose of a systematic work upon bacteria and a laboratory guide.

New chapters have been added dealing with the bacteriology of Whooping-cough, Mumps, Yellow Fever, Hog-cholera, and Swine-plague ; describing the Bacillus aërogenes capsulatus and the Proteus vulgaris ; and describing the Methods of Determining the Value of Antiseptics and Germicides, and of Determining the Thermal Death-point.

To a number of friendly readers whose suggestions have been helpful in improving the work, I desire to extend my sincere thanks.

JOSEPH McFARLAND

PREFACE.

THE following pages are intended to convey to the reader a concise account of the technical procedures necessary in the study of bacteriology, a brief description of the life-history of the important pathogenic bacteria, and sufficient description of the pathological lesions accompanying the micro-organismal invasions to give an idea of the origin of symptoms and the causes of death.

The work being upon Pathogenic Bacteria, it does not cover the whole scope of parasitology, and the parasites of higher orders are all omitted. Malaria and amebic dysentery are omitted as logically as tape-worms and pediculi. The higher fungi are also omitted, both because they are not bacteria and because their proper consideration would make a small book in itself.

In leaving out the non-pathogenic bacteria of course a stumbling-block was encountered. The Sarcina ventriculi, for instance, may be a cause of dyspepsia, yet can scarcely be regarded as pathogenic, and, together with other similar bacteria of questionable deleterious operation, has been omitted ; on the other hand, it has been thought advisable to include and describe somewhat ·at length a long list of spirilla similar to, and probably closely allied with, the spirillum of cholera, yet not the cause of any particular diseased condition.

9

The aim has been to describe only such bacteria as can be proven pathogenic by the lesions or toxins which they engender, and, while considering them, to mention as fully as is necessary the species with which they may be confounded.

The book, of course, will find its proper sphere of usefulness in the hands of medical students; its pages, however, will be found to contain much that will be of interest and profit to those practitioners of medicine who graduated before modern science had thrown its light upon the etiology of disease.

In writing this work the popular text-books have been drawn upon. Hüppe, Flügge, Sternberg, Fränkel, Günther, Thoinot and Masselin, and others have been freely consulted.

The illustrations are mainly reproductions of the best the world affords, and, being taken from the great standards, are surely superior to anything new covering the same ground. Credit has carefully been given for each illustration.

J. McF.

PHILADELPHIA, Feb. 1, 1896.

CONTENTS.

PART I. GENERAL CONSIDERATIONS.

PART II. SPECIFIC DISEASES AND THEIR BACTERIA.

A. THE PHLOGISTIC DISEASES.

I. THE ACUTE INFLAMMATORY DISEASES.

II. THE CHRONIC INFLAMMATORY DISEASES.

CHAPTER IV.

CHAPTER V.

CHAPTER VI.

CHAPTER VII.

CHAPTER VIII.

B. THE TOXIC DISEASES.

CHAPTER I.

CHAPTER II.

CHAPTER III.

CHAPTER IV.

CHAPTER V.

C. THE SEPTIC DISEASES.

CHAPTER XIII.

CHAPTER XIV.

D. MISCELLANEOUS.

CHAPTER I.

CHAPTER II.

CHAPTER III.

CHAPTER IV.

CHAPTER V.

PATHOGENIC BACTERIA.

PART I. GENERAL CONSIDERATIONS.

INTRODUCTION.

IT is incorrect to begin the consideration of bacteriology, as is so often done, with the probable discoverer of bacteria, Leeuwenhoek, or with the so-called "Father of bacteriology," Henle. The controversies and ideas which stimulated the investigations and researches which have brought us to our present state of knowledge were begun hundreds of years before the beginning of the Christian era.

Excepting such as taught and believed that "in six days the Lord made heaven and earth, the sea and all that in them is," or a kindred theory of the origin of things, the thinkers of antiquity never seem to have doubted that under favorable conditions life, both animal and vegetable, might arise spontaneously.

Among the early Greeks we find that Anaximander (43d Olympiad, 610 B. C.) of Miletus held the theory that animals were formed from moisture. Empedocles of Agrigentum (450 B. C.) attributed to spontaneous generation all the living beings which he found peopling the earth. Aristotle (B. C. 384) is not so general in his view of the subject, but asserts that "*sometimes* animals are formed in putrefying soil, sometimes in plants, and sometimes in the fluids of other animals." He also formulated a principle that "every dry substance which becomes moist, and every moist body which becomes dried, produces living creatures, provided it is fit to nourish them."

Three centuries later, in his disquisition upon the Pythagorean philosophy, we find Ovid defending the same doctrine:[1]

> "By this sure experiment we know
> That living creatures from corruption grow:
> Hide in a hollow pit a slaughter'd steer,
> Bees from his putrid bowels will appear,
> Who, like their parents, haunt the fields and bring
> Their honey-harvest home, and hope another spring
> The warlike steed is multiplied, we find,
> To wasps and hornets of the warrior kind.
> Cut from a crab his crooked claws, and hide
> The rest in earth, a scorpion thence will glide,
> And shoot his sting; his tail in circles toss'd
> Refers the limbs his backward father lost;
> And worms that stretch on leaves their filmy loom
> Crawl from their bags and butterflies become.
> The slime begets the frog's loquacious race;
> Short of their feet at first, in little space,
> With arms and legs endued, long leaps they take,
> Raised on their hinder part, and swim the lake,
> And waves repel; for nature gives their kind,
> To that intent, a length of legs behind."

Not only was the doctrine of spontaneous generation of life current among the ancients, but we find it persisting through the Middle Ages, and descending to our own generation to be an accidental but important factor in the development of a new branch of science. In 1542, in his treatise called *De Subtilitate*, we find Cardan asserting that water engenders fishes, and that many animals spring from fermentation. Van Helmont gives special instructions for the artificial production of mice, and Kircher in his *Mundus Subterraneus* (chapter "De Panspermia Rerum") describes and *actually figures* certain animals which were produced under his own eyes by the transforming influence of water on fragments of stems from different plants.[2]

---

[1] Ovid's *Metamorphoses*, translated by Mr. Dryden, published by Sir Samuel Garth, London, 1794.

[2] See Tyndall: *Floating Matter in the Air.*

About 1668, Francesco Redi seems to have been the first to doubt that the maggots familiar in putrid meat arose *de novo:* "Watching meat in its passage from freshness to decay, prior to the appearance of maggots, he invariably observed flies buzzing around the meat and frequently alighting on it. The maggots, he thought, might be the half-developed progeny of these flies. Placing fresh meat in a jar covered with paper, he found that although the meat putrefied in the ordinary way, it never bred maggots, while meat in open jars soon swarmed with these organisms. For the paper he substituted fine wire gauze, through which the odor of the meat could rise. Over it the flies buzzed, and on it they laid their eggs, but the meshes being too small to permit the eggs to fall through, no maggots generated in the meat; they were, on the contrary, hatched on the gauze. By a series of such experiments Redi destroyed the belief in the spontaneous generation of maggots in meat, and with it many related beliefs."

It was not long before Leeuwenhoek, Vallismeri, Swammerdan, and others, following the trend of Redi's work, contributed additional facts in favor of his view, and it may safely be asserted that ever since the time of this eminent man the tide of scientific opinion has turned more and more strongly against the idea that life is spontaneously generated.

About this time (1675) one whose name has been already mentioned, Anthony van Leeuwenhoek, and who is justly called the "Father of microscopy," came into prominence. An optician by trade, Leeuwenhoek devoted much time to the perfection of the compound microscope, which was just coming into use. The science of optics, however, was not sufficiently developed to enable him to overcome the errors of refraction, and after the loss of much time he turned to the simple lens, using it in so careful and remarkable a manner as to be able to record his observations in one hundred and twelve contributions to the *Philosophical Transactions.* Leeuwenhoek, among other things, demonstrated the continuity of arteries and veins through intervening capillaries, thus affording ocular proof of Harvey's discovery of the circulation of the blood; discovered the rotifers, and also the bacteria, seeing them first in saliva. •

Although one of those who contributed to the support of Redi's arguments against the spontaneous generation of maggots, Leeuwenhoek involuntarily reopened the old controversy about spontaneous generation by bringing forward a new world, peopled by creatures of such extreme minuteness as to suggest not only a close relationship to the ultimate molecules of matter, but an easy transition from them. Interested in Leeuwenhoek's work, Plencig of Vienna became convinced that there was an undoubted connection between the microscopic animals exhibited by the microscope and the origin of disease, and advanced this opinion as early as 1762. Unfortunately, the opinions of Plencig seem not to have been accepted by others, and were soon forgotten.

In succeeding years the development of the compound microscope showed these minute organisms to exist in such numbers that putrescent infusions, both animal and vegetable, literally teemed with them, one drop of such a liquid furnishing a banquet for millions.

Much hostility arose in the scientific world as years went on until two schools attained prominence—one headed by Buffon, whose doctrine was that of "organic molecules;" the other championed by Needham, whose doctrine was the existence of a "vegetative force" which drew the molecules together.

Experimentation was begun and attracted much attention. Among the pioneers was Abbé Lazzaro Spallanzani (1777), who filled flasks with organic infusions, sealed their necks, and, after subjecting their contents to the temperature of boiling water, placed them under conditions favorable for the development of life, without, however, being able to produce it. Spallanzani's critics, however, objected to his experiment on the ground that air is essential to life, and that in his flasks the air was excluded by the hermetically-sealed necks.

Schulze (1836) set the objection aside by filling a flask only half full of distilled water, to which animal and vegetable matters were added, boiling the contents to destroy the vitality of any organisms which might already exist in them, then sucking daily into the flask a certain amount of air which had passed through a series of bulbs containing concentrated sulphuric acid, in which it was supposed that whatever germs of life the air might contain would be destroyed. This flask was kept from May to August; air was passed through it daily, yet without the development of any infusorial life.

The term "infusorial life" having been used, here it is well to observe that during all the early part of their recognized existence the bacteria were regarded as animal organisms and classed among the infusoria.

Cagniard Latour and Schwann in the year 1837 succeeded in proving that the minute oval bodies which had been observed in yeast since the the time of Leeuwenhoek were living organisms—vegetable forms—capable of growth; and when Boehm succeeded a year later in demonstrating their occurrence in the stools of cholera, and conjectured that the process of fermentation was concerned in the causation of that disease, the study of these low forms of life received an immense impetus from the important position which they began to assume in relation to medical science.

The experiments of Schwann, by proving that the free admission of calcined air to closed vessels containing putrescible infusions was without effect, while the admission of ordinary air brought about decomposition, suggested that the causes of putrefaction which were in the air were living entities.

In 1862, Pasteur published a paper "On the Organized Corpuscles existing in the Atmosphere," in which he showed that many of the floating particles which he had been able to collect from the atmosphere of his

laboratory were organized bodies.   If these were planted
in sterile infusions, abundant crops of micro-organisms
were obtainable.   By the use of more refined methods
he repeated the experiments of Schwann and others, and
showed clearly that "the cause which communicated life"
to his infusions came from the air, but was not evenly distributed through it."

Three years later he showed that the organized corpuscles which he had found in the air were the spores or
seeds of minute plants, and that many of them possessed
the property of withstanding the temperature of boiling
water—a property which explained the peculiar results
of many previous experimenters, who failed to prevent
the development of life in boiled liquids enclosed in
hermetically-sealed flasks.

Chevreul and Pasteur (about 1836) proved that animal
solids did not putrefy or decompose if kept free from
the access of germs, and thus suggested to surgeons that
the putrefaction which occurred in wounds was due rather
to the entrance of something from without than to some
change within.   The deadly nature of the discharges
from these wounds had been shown in a rough manner
by Gaspard as early as 1822 by injecting some of the
material into the veins of animals.

Examinations of the blood of diseased animals were
now begun, and Pollender (1849) and Davaine (1850)
succeeded in demonstrating the presence of the anthrax
bacillus in that disease.   Several years later (1863) Davaine, having made numerous inoculation-experiments,
demonstrated that this bacillus was the *materies morbi*
of the disease.

Tyndall enlarged upon the experiments of Pasteur,
and very conclusively proved that the micro-organismal
germs were in the dust suspended in the atmosphere, not
ubiquitous in their distribution.   His experiments were
very ingenious and are of interest to medical men.   First
preparing light wooden chambers, with one large glass
window in the front and one smaller window in each

side, he arranged a series of empty test-tubes in the bottom and a pipette in the top, so that when desired the tubes, one by one, could be filled through it. The chamber was first submitted to an optical test to determine the purity of its atmosphere, and was allowed to stand undisturbed and unused until a powerful ray of light passed through the side windows failed to reflect rays from suspended particles of dust when viewed from the front. When the dust had settled so as to allow the optical test of its purity, the tubes were filled with urine, beef-broth, and a variety of animal and vegetable broths, boiled by submergence in a pan of hot brine; the tubes were then allowed to remain undisturbed for days, weeks, or months. In nearly every case life failed to develop after the purity of the atmosphere was established.

In 1873, Obermeier observed that actively motile, flexible spiral organisms were present in large numbers in the blood of patients in the febrile stages of relapsing fever.

Thus evidence slowly accumulated to establish the theory for which Henle had labored as early as 1821, that for many diseases at least there was a distinct and specific *contagium vivum*, and the "GERM THEORY" was propounded.

Is it not strange that the very idea which was to be the outcome of all this investigation and discussion—an idea which would form a new era in scientific medicine and become a fundamental principle of pathology—was one which had been conceived and taught by a philosopher who lived nearly two thousand years ago? Among the numerous works of Varro[1] is one entitled *Rerum Rusticarum libri tres*, from which the following is quoted: "Animadvertendum etiam, si qua erunt loca palustria— quod crescunt animalia quaedam minuta, quae non possunt oculi consequi et per aëra intus in corpus per os ac nares perveniunt atque efficiunt difficilis morbus" (I., xii. 2).—"It is also to be noticed, if there be any marshy

[1] *Univ. Med. Mag.*, vol. iii., No. 3, Dec., 1890, p. 152.

places, that certain minute animals breed [there] which are invisible to the eye, and yet, getting into the system through mouth and nostrils, cause serious disorders (diseases which are difficult to treat)''—a doctrine which, as Prof. Lamberton, to whom the writer is in-debted for the extract, points out, is handed down to us from ''the days of Cicero and Cæsar,'' yet corresponds closely to the ideas of malaria which we entertain at present.

Pasteur had long before suggested that for the different kinds of fermentation there must be specific ferments, and by fractional cultures had succeeded in roughly separating them.

Klebs, who was one of the pioneers of the germ theory, published in 1872 his work upon septicemia and pyemia, in which he expressed himself convinced that the causes of these diseases must come from without the body. Billroth strongly opposed such an idea, asserting that fungi had no especial importance either in the processes of disease or in those of decomposition, but that, existing everywhere in the air, they rapidly developed in the body as soon as through putrefaction a ''Faulniss-zymoid,'' or through inflammation a ''phlogistische-zymoid,'' supplying the necessary feeding-grounds, was produced.

Klebs was not alone in the opposition aroused. Davaine no sooner announced the contagium of anthrax than critics declared that inasmuch as he introduced blood from the diseased animal into the other animal to whom the disease was to be communicated, it was altogether unreasonable to believe the bacilli which were in all probability accidentally present in that blood were the cause of the disease.

In 1875 the number of scientific men who had embraced the germ theory of disease was small, and most of those who accepted it were experimenters. A great majority of medical men either believed, like Billroth, that the presence of fungi where decomposition was in progress was an accidental result of their universal distribution, or, being still more conservative, retained the old un-questioning faith that the bacteria, whose presence in putrescent wounds as well as in artificially prepared media was unquestionable, were spontaneously generated there.

The following extracts from Tyndall's work[1] will illus-trate the slow growth of the germ theory even among men of eminence :

"At a meeting of the Pathological Society of London, held April 6, 1875, the 'germ theory' of disease was formally introduced as a subject for discussion, the debate being continued with great ability and earnestness at sub-sequent meetings. The conference was attended by many distinguished medical men, some of whom were profoundly influenced by the arguments, and none of whom disputed the facts brought forward against the theory on that occasion."

"The leader of the debate, and the most prominent speaker, was Dr. Bastian, to whom also fell the task of replying on all the questions raised."

"The coexistence of bacteria and contagious disease was admitted; but, instead of considering these organisms as probably the essence, or an inseparable part of the es-sence, of the contagium, Dr. Bastian contended that *they were pathological products spontaneously generated in the body after it had been rendered diseased by the real con-tagium.*"

"The grouping of the ultimate particles of matter to form living organisms Dr. Bastian considered to be an operation as little requiring the action of antecedent life as their grouping to form any of the less complex chem-ical compounds." "Such a position must, of course, stand or fall by the evidence which its supporter is able to produce, and accordingly Dr. Bastian appeals to the law and testimony of experiment as demonstrating the soundness of his view." "He seems quite aware of the

[1] *Op. cit.*

gravity of the matter at hand ; this is his deliberate and almost solemn appeal : 'With the view of settling these questions, therefore, we may carefully prepare an infusion from some animal tissue, be it muscle, kidney, or liver ; we may place it in a flask whose neck is drawn out' and narrowed in the blowpipe flame; we may boil the fluid, seal the vessel during ebullition, and, keeping it in a warm place, may await the result, as I have often done. . . . . After a variable time the previously heated fluid within the hermetically-sealed flask swarms more or less plentifully with bacteria and allied organisms, even though the fluids have been so much degraded in quality by exposure to the temperature of 212° F., and have in all probability been rendered far less prone to engender independent living units than the unheated fluids in the tissues would be.' "

These somewhat lengthy quotations are of great interest, for they show exactly the state of the scientific mind at a period as recent as twenty years ago.

In 1877 the introduction of the anilin dyes by Weigert made possible a much more thorough investigation of the bacteria by enabling the observers to color them intensely, and thus detect their presence in tissues and organs where their transparency had caused them to be overlooked.

Rapid strides were immediately made, and before another decade had passed discoveries were so numerous and convincing that it was impossible to doubt that bacteria were causes of disease.

Before the publication of the discoveries of which we speak, however, there was suggested a practical application of the little known about bacteria which produced greater agitation and incited more observation and experimentation than anything suggested in surgery since the introduction of anesthetics—namely, *antisepsis.*

"The seminal thought of antiseptic surgery may perhaps be traced to John Colbach, a member of the College of Physicians, England, whose collection of tracts, printed

1704, contained a description of a new and secret method of treating wounds, by which healing took place quickly without inflammation or suppuration; but it is to one of old Scotia's sons, Sir Joseph Lister, that the everlasting gratitude of the world is due for the knowledge we possess in regard to the relation existing between micro-organisms and inflammation and suppuration, and the power to render wounds aseptic through the action of germicidal substances." [1]

Lister was not the discoverer of carbolic acid nor of the fact that it would kill bacteria; but, convinced that inflammation and suppuration were due to the entrance of germs from the air, instruments, fingers, etc. into wounds, he suggested the antisepsis which would insist upon the use of sterile instruments and clean hands and towels; which would keep the surface of the wound moist with a germicidal solution to kill such germs as accidentally entered; and which would conclude an operation by a protective dressing to exclude the entrance of germs at a subsequent period.

Listerism, originated (1875) a few years before Koch published his famous work on the *Wundinfectionskrankheiten* (traumatic infectious diseases) (1878), spread slowly at first, but surely in the end, to all departments of surgery and obstetrics.

The discovery of the yeast-plant by Latour and Schwann as the cause of fermentation, and the later discovery by Bassi of the yeast-like plant causing the miasmatic contagious disease of silkworms, had led Henle (1840) to believe that the cause of miasmatic, infective, and contagious diseases must be looked for in fungi or in other minute living organisms. Unfortunately, the methods of study employed in Henle's time prevented him from demonstrating the accuracy of his belief.

" It would indeed have been difficult at that period to satisfy every condition that he required to be fulfilled: the methods now in use were then unknown, and have

[1] Agnew's *Surgery*, vol. i. chap. ii.

only been perfected by workers as it has been found necessary from time to time to comply in the most minute
detail with Henle's conditions, and as, one point being
carried, it was found necessary to advance on others.
The first of these was that a specific organism should
always be associated with the disease under consideration.
As such presence, however, might be accidental, these
organisms were not only to be found in pus, etc., but actually in the living body. As they might be, even then,
merely parasitic, and not associated directly with the
causation of the disease, it would be necessary to isolate
the germs, the contagium organisms, and the contagium
fluids, and to experiment with these separately with
special reference to their power of producing similar
diseases in other animals. We now know that it has
only been by strict compliance with all these conditions,
again postulated by Koch, that the most brilliant scientific observers and experimentalists in Germany, France,
England, [and America] have been able to determine
the causal connection between micro-organisms and
disease.'' [1]

The refined methods of Pasteur, but more especially
of Koch, by making possible the fulfilment of the postulates of Henle caused an enormous increase in the
rapidity with which data upon disease-germs were gath-
ered. Almost within a decade the causes of the most
important specific diseases were isolated and cultivated.

In 1879, Hausen announced the discovery of bacilli in
the cells of leprous nodules. The same year Neisser
discovered the gonococcus to be specific for gonorrhea.

In 1880 the bacillus of typhoid fever was first observed
by Eberth, and independently by Koch.

In 1880, Pasteur published his work upon ''chicken-
cholera.'' In the same year Sternberg described the
pneumococcus, calling it the *micrococcus Pasteuri.*

In 1882, Koch made himself immortal by his discovery of and work upon the tubercle bacillus. The same

---

[1] Woodhead: *Bacteria and their Products*, p. 65.

year Pasteur published a work upon *Rouget du Porc*, and Löffler and Schütz reported the discovery of the bacillus of glanders.

In 1884, Koch reported the discovery of the "comma bacillus," the cause of cholera, and in the same year Löffler discovered the diphtheria bacillus, and Nicolaier the tetanus bacillus.

In 1892, Canon and Pfeiffer discovered the bacillus of influenza.

In 1892, Canon and Pielicke first found the bacillus now thought to be specific for measles.

In 1894, Yersin and Kitasato independently isolated the bacillus causing the bubonic plague then prevalent at Hong-Kong.

A new era in bacteriology, and probably the most triumphant result of the modern scientific study of disease, was inaugurated in 1890 by Behring, who presented to the world the "Blood-serum therapy," and showed as the result of prolonged, elaborate, and profound study of the subject of immunity that in the blood of animals with acquired immunity to certain diseases (diphtheria and tetanus) a substance was held in solution which was potent to save the lives of other animals suffering from the same diseases.

CHAPTER I.

BACTERIA.

A BACTERIUM is a minute vegetable organism consisting of a single cell principally composed of an albuminous substance, which Nencki has called *mycoprotein*. Nencki found the chemical analysis of bacteria in the active state to consist of 82.42 per cent. of water. In 100 parts of the dried constituents he found 84.20 parts of mycoprotein; 6.04 of fat; 4.72 of ash; 5.04 of undetermined substances.

Mycoprotein, which has the composition C 52.32, H 7.55, N 14.75, is a perfectly transparent, generally homogeneous body, which probably varies somewhat according to the species from which it is obtained, the culture-medium in which it is grown, and the vital products which the organism produces by its growth. Sometimes the mycoprotein is granular, as in bacillus megatherium ; sometimes it contains fine granules of chlorophyl, sulphur, fat, or pigment. Each cell is surrounded by a cell-wall, which in some species shows the cellulose reaction with iodin.

When subjected to the influence of nuclear stains the bacteria not only take the stain faintly, but in such a manner as to show the existence of a large nucleus situated in the centre of the cell and constituting its great bulk. The cell-wall generally is not stained, but when it does tinge, a delicate line of unstained material can sometimes be made out between the nucleus and the cell-wall, showing the existence of a protoplasm.

The anilin dyes, which possess a great penetrating power, color the organisms so intensely as to preclude the differentiation of the cellular constituents. Under

30

these conditions the bacteria appear as solidly-colored spheres, rods, or spirals, as the case may be.

The cell-walls of some of the bacteria seem at times to undergo a peculiar gelatinous change or to allow the exudation of gelatinous material from the protoplasm, so that the individuals appear surrounded by a distinct halo or capsule. This is not only a peculiarity of certain individuals, but one which only takes place when they develop under certain conditions; thus, Friedländer points out that the capsule of his pneumonia bacillus, when it was found in the lung or in the "prune-juice" sputum, was very distinct, while it could not be demonstrated at all when the organisms grew in gelatin.

From the cell-walls of many bacteria numerous delicate straight or wavy filaments project. These are called *cilia* or *flagella*, and seem to be organs of locomotion. Sometimes they are only observed projecting from the ends or from one end; sometimes they are so numerous and so regular in their distribution as to give the organisms a woolly appearance.

Many of the bacteria which are thus supplied with flagella are actively motile and swim about like microscopic serpents. In all probability the locomotory powers of the bacteria are not entirely dependent upon the presence of the flagella, but may sometimes be due to contractility of the protoplasm within an elastic cell-wall. The micro-organisms most plentifully supplied with them are those of the rod and spiral shape. Only one of the spherical forms, Micrococcus agilis of Ali-Cohen, has been shown to have flagella. This and one other species are probably the only motile cocci. Observing that the organisms known to be most active are those best supplied with flagella, it is reasonable to conclude that the motility is dependent upon the flagella.

The presence of flagella, however, does not necessarily imply motility, for some of the bacilli amply provided with these appendages are not motile. The flagella may not only serve as organs of locomotion, and be of use to the organism by conveying it from an area where the nutrition is less to one where it is greater, but, as Woodhead points out, may, in the non-motile species, serve to stimulate the passage of currents of nutrient material past the organism, so as to increase the food-supply. The flagellate bacteria have a greater number of representatives among those whose lives are spent in water and in fermenting and decaying materials than among those inhabiting the bodies of animals. This is an additional fact in favor of the view that locomotion and flagella are provisions favorable to the maintenance of the species by keeping the individuals supplied with food.

It may be added that such parasitic disease-producing bacteria as do not habitually gain access to the tissues, but inhabit the intestine, as the bacillus of typhoid fever and the spirillum of cholera, are actively motile, like the saprophytes, while those habitually entering the tissues and multiplying there are motionless and without flagella. Of course this example is open to criticism, because the spirillum of relapsing fever, which has never been found elsewhere than in the blood and spleen of affected animals, is actively motile.

One of the linear organisms, known as the Bacillus megatherium, has a distinct but limited ameboid movement.

The commonly observed dancing movement of the spherical forms seems to be the well-known Brownian movement, which is simply a physical phenomenon. It is sometimes difficult to determine whether an organism is really motile or whether it is only vibrating. In the latter case it does not change its relative position to surrounding objects.

The bacteria are so minute that a special unit of measurement has been adopted by bacteriologists for their estimation. This is the *micro-millimeter* (μ), or one-thousandth part of a millimeter, and about equivalent to the one-twenty-five-thousandth of an inch.

As a rule, the spherical organisms are the smallest and the spiral organisms the longest, except the chains of bacilli called *leptothrix.* Their measurements vary from 0.15 μ (micrococcus of progressive abscess-formation in rabbits) to 2.8 μ (Diplococcus albicans amplus) for cocci, and from 1×0.2 μ (bacillus of mouse-septicemia) to 5×1.5 μ (anthrax bacillus) for bacilli. Some of the spirilla are very long, that of relapsing fever measuring 40 μ at times.

This estimation of size almost prepares one for the estimation of weight given by Nägeli, who found that an average bacterium under ordinary conditions weighed $\frac{1}{100000000000}$ of a milligram.

The bacteria multiply in two ways : by direct division (fission) and by the development of spores, seeds, or eggs (sporulation). The more common mode is by binary division. The bacterium which is about to divide appears a little larger than normal, and, if a spherical organism, more or less ovoid. No karyokinetic changes have been observed in the nuclei, though they may occur. When the conditions of nutrition are good, the process of fission progresses with astonishing rapidity. Buchner and others have determined the length of a generation to be from fifteen to forty minutes.

The results of binary division, if rapidly repeated, are almost appalling. "Cohn calculated that a single germ could produce by simple fission two of its kind in an hour ; in the second hour these would be multiplied to four ; and in three days they would, if their surroundings were ideally favorable, form a mass which can scarcely be reckoned in numbers, or, if reckoned, could scarcely be imagined—four thousand seven hundred and seventy-two billions. If we reduce this number to weight, we find that the mass arising from this single germ would in three days weigh no less than seventy-five hundred tons." "Fortunately for us," says Woodhead, "they can seldom get food enough to carry on this appalling rate of development, and a great number die both for

3

want of food and because of the presence of other conditions unfavorable to their existence.''

When the conditions for rapid multiplication are no longer good, the organism assumes a protective attitude and develops in its interior small oval eggs, seeds, or, as they are more correctly called, *spores* (Fig. 1). Such

FIG. 1.—Diagram illustrating sporulation : *a*, bacillus enclosing a small oval spore; *b*, drumstick bacillus, with the spore at the end ; *c*, clostridium ; *d*, free spores; *e* and *f*, bacilli escaping from spores.

spores developed within the bacteria are called *endospores.* When the formation of such a spore is about to commence, a small bright point appears in the protoplasm, and increases in size until its diameter is nearly or quite as great as that of the bacterium. As it nears perfection a dark, highly-refracting capsule is formed about it. As soon as the spore arrives at perfection the bacterium seems to die, as if its vitality were exhausted in the development of the permanent form.

Endospores are generally formed in the elongate bacteria—bacillus and spirillum—but Zopf has described similar bodies as occurring in micrococci. Escherich also claims to have found undoubted spores in a form of sarcina.

The spores found in the bacilli are either round or oval. As a rule, each bacillus produces a single spore, which is situated either at its centre or at its end. When, as sometimes happens, the diameter of the spore is greater than the diameter of the bacillus, it causes a bulging of the organism, with a peculiar appearance described as *clostridium.* When the distending spore is in the centre of the bacillus, it produces a barrel-shaped organism; when situated at the end, a "Trommelschläger," or drumstick-shaped one. As the degeneration of the protoplasm of the bacillus sets the spore free, it appears as a clear, highly-refracting sphere or ovoid situated in a little collection of granular matter.

Spores differ from the bacteria in that their capsules seem to prevent evaporation and to enable them to withstand drying and the application of a considerable amount of heat. Ordinarily, bacteria are unable to resist a temperature above 60° C. for any considerable length of time, only a few resistant forms tolerating a temperature of 70° C. The spores, however, are uninjured by such temperatures, and can even successfully resist that of boiling water (100° C.) for a short time. The extreme desiccation caused by a protracted exposure to a temperature of 150° C. will, however, destroy them. Not only can the spores resist a considerable degree of heat, but they are also unaffected by cold of almost any intensity.

While the cell-wall of the bacterium is easily penetrated by solutions of the anilin dyes, it is a matter of much difficulty to accomplish the staining of spores, so that we see they are probably more resistant to the action of chemical agents than the bacteria themselves.

When a spore is accidentally dropped into some nutrient medium a change is shortly observed. The protoplasm, which has been clear, becomes somewhat granular, the capsule a little less distinct; the body increases slightly in size, and in the course of time splits open to allow the escape of the young organism. The direction in which the escape of the young bacillus takes place is of interest, as varying in the different species. The Bacillus subtilis escapes from the end of the spore, where a longitudinal fissure occurs; the bacillus of anthrax escapes from the side, sometimes leaving the capsule of the spore in the shape of two small cups.

As soon as the young bacillus escapes it begins to increase in size, develops around its soft protoplasm a characteristic capsule, and, having once established itself, presently begins the propagation of its species by fission.

In addition to the endospores, of which we have just been speaking, there are *arthrospores*. The formation of these is much less clear. It seems to be an effort to convert the entire microbe into a permanent form. This process is observed particularly in the micrococci, where the substance of a cell is said to break up into segments, each of which becomes a resisting body fruitful in prop‐ agating its species. Of the arthrospores little has, so far, been learned. It is not improbable that among the micrococci, and also among some of the smaller bacilli in whom no spores have been observed, the maintenance of the species when conditions of life become unfavor‐ able is due to the assumption of a permanent form by some of the individuals, without the formation of any spore-like bodies. This is at present largely a matter of conjecture, but the indications pointing in that direction are numerous.

It is believed by Fränkel and others that sporulation in the bacteria is not a sign of the exhaustion of nutri‐ tion, but a sign of the vital perfection of the organism. These observers regard spore-formation as analogous to the flowering of higher plants, which takes place only when the conditions and development are best.

Morphology.—The morphology of the bacteria is quite varied. Three principal forms, however, exist, from which the others seem to be but variations.

The most simple appear as minute spheres, and from

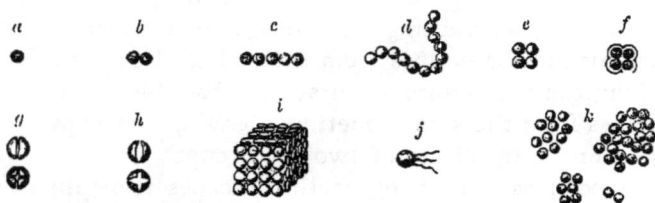

Fig. 2.—Diagram illustrating the morphology of the cocci: *a*, coccus or micrococcus; *b*, diplococcus; *c*, *d*, streptococci; *e*, *f*, tetragenococci or meris‐ mopedia; *g*, *h*, modes of division of cocci; *i*, sarcina; *j*, coccus with flagella; *k*, staphylococci.

their fancied resemblance to little berries are called *cocci* or *micrococci* (Fig. 2, *a*). When the bacteria of this form multiply by fission the resulting two organisms not infrequently remain attached to each other, producing what is called a *diplococcus* (Fig. 2, *b*). The diplococci sometimes consist of two perfect spheres, but more often show a flattening of the contiguous surfaces, which are not in absolute apposition (Fig. 2, *g*). In a few cases, as the gonococcus, the approximated surfaces are slightly concave, causing the organism to somewhat resemble the German biscuit called a "semmel," hence biscuit- or semmel-cocci (Fig. 2, *h*). Frequently a second binary division occurs, causing four individuals to remain closely approximated, without disturbing the arrangement of the first two. When division of this kind produces a distinct tetrad, the organism is described as a *tetragenococcus*, while to the entire class of cocci dividing so as to produce fours, eights, twelves, etc. on the same plane the name *merismopedia* is given (Fig. 2, *e* and *f*).

If, as sometimes happens, the divisions take place in three directions, so as to produce cubical masses or "packages" of cocci, the resulting aggregation is described as a *sarcina* (Fig. 2, *i*). This form slightly resembles a dice or a bale of cotton in miniature.

If the divisions always take place in the same direction, so as to produce a chain or string of beads, the organism is described as *streptococcus* (Fig. 2, *d*). When there are diplococci joined in this manner a *strepto-diplococcus* is of course formed.

More common than any of the forms already described is one in which, without any definite arrangement, the cocci occur in irregular groups having a fancied resemblance to bunches of grapes. These are called *staphylococci*, and, as it is very unusual to find cocci habitually occurring isolated, most cocci not classified under one of the above heads are called staphylococci.

When cocci are associated in globular or lobulated clusters encased in a resisting glutinous, homogeneous mass, the name *ascococcus* has been used in describing them. A modified form of this, in which the cocci are in chains or solitary and are surrounded by an encasement almost cartilaginous in consistence, has been called *leuconostoc.*

Certain bacteria, constituting a better-known if not more important group, are not spherical, but elongate or "rod-shaped," and bear the name *bacillus* (Fig. 3).

FIG. 3.—Diagram illustrating the morphology of the bacilli: *a, b, c,* various forms of bacilli; *d, e,* bacilli with flagella; *f,* chain of bacilli, individuals distinct; *g,* chain of bacilli, individuals not separated.

I would remark that the absence of a standard by which to separate the cocci from the bacilli is the cause of much confusion. In the judgment of the author, it would be well to place all individuals having one diameter greater than the other among the bacilli. This would prevent the error of describing one species as "oval cocci" and another as "nearly round bacilli," and by giving a definite standard would greatly aid in the formation of a differential table.

The bacilli present a considerable variety of forms. Some are quite short, with rounded ends, so as to appear elliptical; some are long and delicate. Some have rounded ends, as subtilis; others have square ends, as anthrax. Some are enormously large, some exceedingly small. Some are always isolated, never forming threads or chains; others nearly always occur in these forms.

The bacilli always divide by transverse fission, so that the only peculiarity of arrangement is the formation of threads or chains.

In the older writings the short, stout bacilli were all described under the generic term *bacterium.* This genus, like some of the species it comprehended, has now passed out of use. Some of the flexile bacilli, whose movements are sinuous, much resembling the swimming of a snake or an eel, were described as *vibrio*, but this name also has passed into disuse.

o The long filaments formed by the division of bacilli without their distinct separation are sometimes call'ed *leptothrix*, and when these long threads form distinct masses surrounded by a jelly-like material, the name *myconostoc* is sometimes applied to them.

Certain forms much resembling bacilli in their isolated state, characterized by the formation of long filaments with a peculiar grouping which gives the appearance of a false branching, are described as *cladothrix;* others in which true branchings are seen, as *streptothrix*. One other bacillus-like form, consisting of long, thick, not distinctly segmented, straight threads, is called *beggiatoa*. The only important difference between it and leptothrix is that its filaments are thick and coarse, while those of leptothrix are very delicate.

Some of the elongate bacteria have a remarkably twisted form and bear some resemblance to a cork-screw. These are called *spirilla* (Fig. 4). A subdivision

FIG. 4.—Diagram illustrating the morphology of the spirilla: *a, b, c,* spirilla; *d, e,* spirochæta.

of them, whose individuals are not only twisted but are also very flexible, is called *spirochæta*. Though not formerly differentiated from vibrio, these forms are quite distinct.

A spiral organism of a ribbon shape is called *spiro-*

monas, while a similar organism of spindle shape is called a *spirulina*. One species of spiral bacteria in whose protoplasm sulphur-grounds have been detected has been called *ophidiomonas*.

Some of the spirilla are exceedingly long and deli-. cate, as the spirochæta of relapsing fever; others which are stouter, like the spirillum of cholera, habitually occur in such short individuals as to be easily mistaken for slightly-bent bacilli.

Classification.—Leeuwenhoek, when he first saw the bacteria—and his successors even to so recent a date as to include Ehrenberg and Dujardin—did not doubt that they belonged to the infusoria.

It was not until biologists had concluded that organisms which take into their bodies particles of solid or semi-solid material, digest that which is useful, and extrude the remainder, are animals, and that those which live purely by osmosis and exosmosis are vegetables, that the bacteria, which we have seen provided with a resistant cell-wall, allowing of no possibility of nutrition except by osmosis and exosmosis, could be finally and correctly classed among the members of the vegetable kingdom.

The extremely simple organization of bacteria naturally places them among the lowest members of the vegetable kingdom, in that class of the Cryptogamia known as Thallophytæ, comprising the algæ, lichens, and fungi.

The algæ are mostly water-plants, containing chlorophyl and obtaining their nourishment from inorganic substances.

The lichens are plants, some of which contain chlorophyl. They live upon inorganic matter, which they generally absorb from the air. According to the new view of the subject, some, if not all, of these plants are regarded as fungi growing parasitically upon algæ.

The fungi, the lowest group of all, are minute or large plants, mostly devoid of chlorophyl, living upon organic matter, which they obtain as saprophytes from decomposing animal and vegetable matters, or as parasites upon the tissues or juices of living animals or plants. This lowest family, the fungi, are divisible into the—

> Hyphomycetes or Mucorini, or moulds;
> Saccharomycetes, or yeasts; and
> Schizomycetes, or bacteria.

Cohn divided the bacteria, according to their morphology, into—

> Sphero-bacteria, or cocci ;
> Micro-bacteria—short rods ;
> Desmo-bacteria—bacilli ;
> Spiro-bacteria—spirilla.

Davaine suggested a classification based upon motility, making four classes—Bacterium, Vibrio, Bacteridium, and Spirillum, neglecting to provide for the cocci.

Zopf arranged them, according to his theory of pleomorphism, into the COCCACEÆ, comprising those known only in the coccus form, and comprehending the *streptococci*, *merismopedia*, *sarcina*, *micrococcus*, and *ascococcus;* the BACTERIACEÆ, comprehending the genera *bacterium*, *spirillum*, *vibrio*, *leuconostoc*, *bacillus*, and *clostridium* (chiefly coccus, rod, and thread forms ; the former may be absent ; in the latter there is no distinction between base and apex ; threads straight or screw-like) ; and the LEPTOTHRICHEÆ, comprehending *crenothrix*, *beggiatoa*, *phragmidiothrix*, and *leptothrix* (coccus, rod, and thread forms ; the latter show a distinction between base and apex ; threads straight or screw-like ; spore-formation not demonstrated).

This classification is, however, based upon what is probably an erroneous principle, the pleomorphism of the bacteria.

Van Tieghem, DeBary, and Hüppe formed classifications the main feature of which was the formation of endospores or arthrospores, but, as the sporulation of many species is as yet unknown, they cannot be properly placed in it.

It has even been suggested to classify the bacteria by the size and number of their flagella, of which so little is known.

The most convenient classification, though it cannot be purely scientific, seems to be the morphological one. given by Cohn. Baumgarten, recognizing the relative pleomorphism of certain of the species, has modified it as follows, and thus made it answer all the needs of the pathologist at least:

I. Cocci,
II. Bacilli, } species relatively monomorphous.
III. Spirilla,

IV. Spirulina,
V. Leptothrix, } species relatively pleomorphous.
VI. Cladothrix,

The members of the first group, the cocci, bacilli, and spirilla, are practically the only ones which are of pathological significance.

CHAPTER II.

BIOLOGY OF BACTERIA.

THE distribution of bacteria is wellnigh universal. They and their spores float in the atmosphere we breathe, swim in the water we drink, grow upon the food we eat, and luxuriate in the soil beneath our feet. Nor is this all, for, entering the palpebral fissures, they develop upon the conjunctiva ; entering the nares, they establish themselves in the nose ; the mouth is always replete with them ; and, as many are swallowed, the digestive apparatus always contains them. The surface of the body never escapes their establishment, and so deeply are some individuals situated beneath the epithelial cells that the most careful washing and scrubbing and the use of the most powerful germicides are required to rid the surgeon's hands of what may prove to be dangerous hindrances to the healing of wounds. The ear is not without its microscopic flora ; special varieties live beneath the finger-nails, and especially the toe-nails, in the vagina, and beneath the prepuce.

While so general, however, they are not ubiquitous. Tyndall succeeded in proving that the atmosphere of high Alpine altitudes was free from them, and likewise that the glacier ice contained none. Wherever man, animals, or even plants, live, die, and decompose, bacteria are sure to be present.

Notwithstanding their extreme familiarity with the animal body, there are certain parts of it into which bacteria do not enter, or, entering, remain vital for a very short time, for the *body-juices and tissues of normal animals are free from them, and their occurrence there may almost always be accepted as a sign of disease.*

The presence of bacteria in the air is generally de-

43

pendent upon their previous existence in the soil, its pulverization, and its distribution by currents of the atmosphere. Koch has shown that the upper stratum of the soil is exceedingly rich in bacteria, but that their numbers decrease as the soil is penetrated, until below a depth of one meter there are very few. Remembering that bacteria can live only upon organic matter, this is readily understandable. Most of the organic matter is upon the surface of the soil. Where, as in the case of porous soil or the presence of cesspools and dung-heaps, the decomposing materials are allowed to penetrate to a considerable depth, the bacteria may occur much farther from the surface, yet they are rarely found at any great depth, because the majority of the known species require oxygen.

The water of stagnant pools always teems with bacteria, but that of deep wells rarely contains many unless it is polluted from the surface of the earth.

Being generally present in the soil, which the feet of men and animals grind to powder, the bacteria, together with the pulverized earth, are blown from place to place into every nook and cranny, until it is impossible to escape them. It has been suggested by Soyka that the currents of air passing over the surface of liquids might take up bacteria, but, although he seemed to show it experimentally, it is not generally believed. Where bacteria are growing in colonies they seem to remain undisturbed by currents of air unless the surface becomes roughened or broken.

Most of the bacteria which are carried about by the air are what are called saprophytes, and are perfectly harmless to the human being; but not all belong to this class, nor will they do so while tuberculous patients are allowed to expectorate upon the sidewalks, and typhoid patients' wash to dry upon the clothes-line, and their dejecta to be spread upon the ground.

The growth of bacteria is profoundly influenced by environment, so that a consideration of the conditions favorable or detrimental to their existence becomes a necessity.

Conditions influencing the Growth of Bacteria.— (*a*) *Oxygen.*—The majority of bacteria grow best when exposed to the air. Some develop better when the air is withheld; some will not grow at all where the least amount of oxygen is present. Because of these peculiarities bacteria are divisible into the

Aërobic bacteria, those growing in oxygen.

Anaërobic bacteria, those not growing in the presence of oxygen.

As, however, some of the aërobic forms will grow almost as well without as with oxygen, the term *optional* (facultative) *anaërobics* has been applied to the special class made to include them.

As examples of strictly aërobic bacteria the Bacillus subtilis and the Bacillus aërophilus may be given. These forms will not grow if oxygen is denied them. The staphylococci of suppuration and the bacilli of typhoid fever, pneumonia, and anthrax, as well as the spirillum of cholera, will grow almost equally well with or without oxygen, and hence belong to the optional anaërobics. The bacillus of tetanus and of malignant edema, and the non-pathogenic forms, the Bacillus butyricus, Bacillus muscoides, and Bacillus polypiformis, will not develop at all where any oxygen is present, and hence are strictly anaërobic.

(*b*) *Nutriment.*—The bacteria do not seem able to derive their nourishment from purely inorganic matter. Proskauer and Beck, however, have succeeded in growing the tubercle bacillus in a mixture containing ammonium carbonate 0.35 per cent., potassium phosphate 0.15 per cent., magnesium sulphate 0.25 per cent., glycerin 1.5 per cent. They grow best where diffusible albumins are present. The ammonium salts are rather less fitted to support them than their organic compounds. The individual bacterium varies very widely in the nutriment which it requires. Some of the water-microbes can live in distilled water to which the smallest amount of organic matter has been added; others require so concentrated a medium that only blood-serum can be used for their cultivation. Sometimes a species with a preference for a particular culture-medium can gradually be accustomed to another, though immediate transplantation causes the death of the transplanted organism. Sometimes the addition of such substances as glucose and glycerin has a peculiarly favorable influence upon bacteria, causing, for example, the tubercle bacillus to grow upon agar-agar.

(c) *Moisture.*—A certain amount of water is always necessary for the growth of bacteria. The amount can be exceedingly small, however, so that the Bacillus prodigiosus is able to develop successfully upon crackers and dried bread. Materials used as culture-media should not be too concentrated; at least 80 per cent. of water should be present. Most bacteria grow best in liquid media; that is, they form the longest threads, and diffuse themselves throughout the liquid so as to be present in far greater numbers than when on solid media.

The statement that certain forms of bacteria can flourish in clean distilled water seems to be untrue. When transferred to such a medium the organisms soon die and undergo a granular degeneration of their substance. If, however, in their introduction a good-sized drop of culture-material is carried with them, the distilled water ceases to be such, and becomes a dilute bouillon fitted to support life for a time.

(d) *Reaction.*—Should the pabulum supplied to bacteria contain an excess of either alkali or acid, the growth of the organisms is inhibited. Most true bacteria grow best in a neutral or feebly alkaline medium. There are exceptions to this rule, for the Bacillus butyricus and the Sarcina ventriculi can grow well in strong acids, and the Micrococcus urea can tolerate excessive alkalinity. Acid media are excellent for the cultivation of moulds.

(e) *Light.*—Most species of bacteria are not influenced in their growth by the presence or absence of light. The direct rays of the sun, and to a less degree the intense rays of the electric arc-light, retard and in numerous instances kill bacteria. Some colors are distinctly inhibitory to their growth, blue being especially prejudicial. ·Some of the chromogenic forms will only produce their colors when exposed to the ordinary light of the room. The Bacillus mycoides roseus will not produce its red pigment except in the absence of light. The pathogenic bacteria have their virulence gradually attenuated if grown in the light.

(*f*) *Electricity.*—Very little is known about the action of electric currents upon bacteria. Very powerful discharges of electricity through culture-media are said to kill the organisms, to change the reaction of the culture, and the rapidly reversed currents of high intensity to destroy the pathogenesis of the bacteria and change their toxic products into neutralizing protective (antitoxin?) bodies. Much attention has recently been devoted to this subject by Smirnow, Arsonval and Charin, Bolton and Pease, Bonome and Viola, and others.

(*g*) *Movement.*—When bacteria are growing in a liquid medium perfect rest seems to be the condition best adapted for their development. A slow-flowing movement does not have much inhibitory action, but violent agitation, as by shaking a culture in a machine, greatly hinders or prevents their growth. The practical application of this will show that rapidly flowing streams, whose currents are interrupted by falls and rapids, will, other things being equal, furnish a better drinking-water than a deep, still-flowing river.

· (*h*) *Association.*—It occasionally happens that bacteria grow better when associated with other species, or have their pathogenic powers augmented when grown in combination. Coley found the streptococcus toxin more active when combined with Bacillus prodigiosus.

Occasionally the reverse is true, and Pawlowski found that mixtures of anthrax and bacillus prodigiosus were less virulent than cultures of anthrax alone.

Rarely, the presence of one species of microörganism entirely eradicates another species. Hankin found that the Micrococcus Ghadialli destroyed the typhoid and colon bacilli, and suggested the use of this coccus to purify waters polluted with typhoid.[1]

(*i*) *Temperature.*—The question of temperature is of importance from its bearing upon sterilization. According to Fränkel, bacteria will scarcely grow at all below 16° and above 40° C.

· The researches of Flügge show that the Bacillus subtilis will grow very slowly at 6° C., and as the temperature is elevated it is said that until 12.5° C. is reached fission does not occur oftener than every four or five hours. When 25° C. is reached the fission occurs every three-quarters of an hour, and at 30° C. about every half hour.

Most bacteria die at a higher temperature than 60–75° C. The spores can resist boiling water, but are killed by dry heat if exposed to 150° C. for an hour or to 175° C. for five to ten minutes. Freezing kills many, but not all bacteria, but does not affect the spores at all.

Most bacteria grow best at the ordinary temperature of a comfortably heated room, and are not affected by its occasional slight changes. Some, chiefly the pathogenic forms, are not cultivable except at the temperature of the animal body (37° C.); others, like the tubercle bacillus, grow best at a temperature a little above that of the body—40° C.

Variations in the amount of oxygen, temperature, moisture, etc., beyond what have been described, are prejudicial to the growth and development of bacteria, first inhibiting their growth, thus tending toward their destruction. In the practical application of our knowledge of the biology of the bacteria we constantly make use of such precautions as removing from surgical dressings, sponges, etc., every substance that can possibly afford nutriment to bacteria, and heating such materials, as well

[1] *Brit. Med. Jour.*, Aug. 14, 1897, p. 418.

as culture-media and a variety of other substances, to a temperature beyond that known to be the extreme limit of bacterial endurance.

The presence of certain substances—especially some of the mineral salts—in an otherwise perfectly suitable medium will prevent the development of bacteria, and when added to grown cultures of bacteria will destroy them. Carbolic acid and bichlorid of mercury are the best known examples.

It is interesting to mention in this connection the results of the experiments of Trambusti, who found it possible to produce a tolerance to a certain amount of bichlorid of mercury by cultivating Friedländer's bacillus upon culture-media, containing gradually increasing amounts of the salt, until from 1-15,000, which inhibited ordinary cultures, it could accommodate itself to 1-2000.

(*j*) *x-Rays.*—The action of the *x*-rays upon bacteria has been investigated by Bonome and Gros and others. When the cultures are exposed to their action for prolonged periods, their vitality and virulence seem to be slightly diminished. They are not killed by the *x*-rays.

Some forms of the bacteria are never found except in the tissues of diseased animals. Such organisms are called *parasites.* The parasitic group really is divisible into the *purely parasitic* and the *occasionally parasitic* bacteria. Of the first division the tubercle bacillus may be used as an illustration, for, so far as is known, it is never found in other places than the bodies and dejecta of diseased animals. The cholera spirillum illustrates the second group, for, while it produces the disease known as Asiatic cholera when admitted to the digestive tract, it is a constant inhabitant of certain waters, where it multiplies with luxuriance.

Bacteria which do not enter the animal economy, or if accidentally admitted do no harm, but live upon decaying animal and vegetable materials, are called *saprophytes.*

According to their products of metabolism, bacteria are often described as—

4

Zymogenic, or bacteria of fermentation.
Saprogenic, or bacteria of putrefaction.
Chromogenic, or color producers.
Photogenic, or phosphorescent bacteria.
Aërogenic, or gas producers.
Pathogenic, or disease producers.

The parasitic organisms alone possess much interest to the physician, but as in their growth the saprophytes exhibit many interesting vital manifestations, it is not well to exclude them or their products from the following consideration of the

Results of Vital Activity in Bacteria.—1. *Fermentation.*—The alcoholic fermentation, which is a familiar phenomenon to the layman as well as to the brewer and the chemist, is not the work of a bacterium, but of a yeast-plant, one of the saccharomyces fungi. The acetic-acid, lactic-acid, and butyric-acid fermentations are, however, caused by bacilli. A considerable number of bacilli seem capable of converting milk-sugar into lactic acid, sometimes associating this with coagulation of milk, sometimes not. The production of coagulation in milk is not always associated with acid-production, but with the production of a curdling ferment similar to that belonging to the gastric juice. There seems to be no real specific micro-organism for the lactic-acid fermentation, although the Bacillus acidi lactici seems to be the most powerful generator of the acid. There may also be several bacteria which produce the acetic fermentation, though it is generally attributed to a special common form, the Mycoderma aceti or Bacillus aceticus. The butyric fermentation is generally due to the Bacillus butyricus, though it also may be caused by other bacilli, the one named simply being the most common. (For an exact description of the chemistry of the fermentations reference must be made to text-books upon that subject, as their consideration here would occupy too much space.)

2. *Putrefaction.*—This process is in many respects sim-

ilar to the preceding, except that instead of occurring in carbohydrates it takes place in nitrogenous bodies. The first step seems to be the transformation of the albumins into peptones, then the splitting up of the peptones into a large number of gases, acids, bases, and salts. In the process the innocuous albumins are frequently changed to toxalbumins, and sometimes to distinct animal alkaloids known as *ptomaïnes*. Vaughan and Novy declare the term "animal alkaloid" to be a misnomer, as ptomaïnes are sometimes produced from vegetable substances in the process of decomposition; they suggest the term "putrefactive alkaloids" as preferable. The definition of a ptomaïne given by these observers is "a chemical compound, basic in character, formed by the action of bacteria on organic matter." The chemistry of these bodies is very complex, and for a satisfactory description of them Vaughan and Novy's book[1] is brief and excellent. Among the ptomaïnes, the following appear to be important: Methylamin (CH_3NH_2), the simplest organic base formed in the process of putrefaction; dimethylamin (($CH_3)_2NH$); trimethylamin (C_3H_9N ($CH_3)_3N$); ethylamin ($C_2H_5.NH_2$); diethylamin ($C_4H_{11}N$ ($C_2H_5)_2$-NH); triethylamin ($C_6H_{15}N$ ($C_2H_5)_3N$); propylamin ($C_3H_7.NH_2$); butylamin $C_4H_{11}N$); iso-amylamin; caproylamin; tetanotoxin; spasmotoxin; dihydrolutidin; putrescin; cadaverin; neuridin; saprin; pyocyanin; and tyrotoxicon. It is supposed that the cases of ice-cream and cheese-poisoning that sometimes occur are due to tyrotoxicon produced by the putrefaction of the proteid substances of the milk before it is frozen into ice cream or made into cheese. The safeguard is to freeze the milk only when perfectly fresh and avoid adding the sugar and flavoring substances, allowing the whole to stand some time, and then freezing. Numerous others have been described, some toxic, others harmless.

It is to compounds of this kind that the occasional cases of "Fleishvergiftung" or "meat-poisoning" are

[1] *Ptomaïnes and Leucomaïnes.*

due, the growth of various bacteria in stale meat bringing about in its proteid substances the development of toxic ptomaïnes. Kaensche[1] carefully investigated the subject, and gives a synoptical table containing all the bacteria of this class described. His researches show that there are at least three different bacilli whose growth causes the development of poisonous ptomaïnes in meat.

Toxins and toxalbumins are also very common.

3. _Chromogenesis._—Those bacteria which produce colored colonies or impart color to the medium in which they grow are called _chromogenic;_ those with which no color is associated, _non-chromogenic._ Most chromogenic bacteria are saprophytic and non-pathogenic. Some of the pathogenic forms, as the Staphylococcus pyogenes aureus and citreus, are, however, color-producers. It seems likely that the bacteria do not form the actual pigments, but certain chromogenetic substances which, uniting with substances in the culture-medium, produce the colors.

Galleotti has described two kinds of pigment, one of which, being soluble, readily penetrates all neighboring portions of the culture-medium, like the colors of Bacillus pyocyaneus, and an insoluble pigment which does not tinge the solid culture-media at all, but is constantly found associated with the colonies, like the pigment of Bacillus prodigiosus. The pigments are found in their greatest intensity near the surface of the colony. The coloring matter never occupies the protoplasm of the bacteria (except the Bacillus prodigiosus, in whose cells occasional pigment-granules may be seen), but occurs in an intercellular excrementitious substance.

The pigments are so varied as to give almost every known color. It sometimes happens that a bacterium will elaborate two or more colors. The Bacillus pyocyaneus thus produces pyocyanin and fluorescin, both being soluble pigments—one blue, the other green. Gessard has shown that when the Bacillus pyocyaneus

[1] _Zeitschrift für Hygiene_, etc., Bd. xxii., Heft 1, June 25, 1896.

is cultivated upon white of egg, it produces only the green fluorescent pigment, while in pure peptone solution it grows with the production of blue pyocyanin alone. His experiments prove a very interesting fact, that for the production of fluorescein it is necessary that the culture-medium contain a definite amount of a phosphatic salt. Sometimes one pigment is soluble, the other insoluble, so that the colony will appear one color, the medium upon which it grows another. Some organisms will only produce their colors in the light; others, as the Bacillus mycoides roseus, only in the dark. Some produce them only at the room-temperature, but, though growing luxuriantly in the incubator, refuse to produce pigments at so high a temperature. Thus, Bacillus prodigiosus produces a brilliant red color when growing at the temperature of the room, but is colorless when grown in the incubator. Colored lights seem to have no modifying influence upon the pigment-production. Even if for successive generations the bacterium be grown so as to be colorless, it speedily recovers its primitive color when restored to its old environment, no matter what the color of the light thrown upon it. Bacteria which have been robbed of their color by incubation, when placed in the normal environment produce the original color, no matter what color the light they receive. Some of the pigments—perhaps most of them—are formed only in the presence of oxygen.

4. *Liquefaction of Gelatin.*—When certain forms of bacteria are grown in gelatin the culture-medium is partly or entirely liquefied. This characteristic is entirely independent of any other property of the bacterium, and is one manifested alike by pathogenic and non-pathogenic individuals. Sternberg and Bitter have shown that if from a culture in which liquefaction has taken place the bacteria be removed by filtration, the filtrate will retain the power of liquefying gelatin, showing that the property is not resident in the bacteria, but in some substance in solution in their excreted products.

These products are described as "tryptic enzymes" by Fermi, who found that heat destroyed them. Mineral acids seem to check their power to act upon gelatin. Formalin renders the gelatin insoluble. As some of the bacteria not only liquefy the gelatin, but do so in a peculiar and constantly similar manner, the presence or absence of the change becomes extremely useful for the separation of different species.

5. *Production of Acids and Alkalies.*—Under the head of "Fermentation" the formation of acetic, lactic, and butyric acids has been discussed. These, however, are by no means all the acids resulting from microbic metabolism. Ziegler mentions formic, propionic, baldrianic, palmitic, and margaric as being among those produced, and even this list may not comprehend them all. As the acidity due to the microbic metabolism progresses, it impedes, and ultimately completely inhibits, the development of the bacteria. The addition of phenolphthalein and litmus to the culture-medium is one of the best methods for detecting the acids. Milk, to which litmus is added, is particularly convenient. Rosalic acid may also be used, the acid converting its red into an orange color. The same tests will also determine the alkali-production, which occurs rather less frequently than acid-formation and depends chiefly upon the salts of ammonium.

6. *Production of Gases.*—This seems, in reality, to be a part of the process of decomposition and fermentation. Among the gases due to bacterial action, CO_2, H_2S, NH_4, CH_4, and others have been described. If the bacterium be anaërobic and develop at the lower part of a tube of gelatin, not infrequently a bubble of gas will be formed about the colonies. This is almost constant in tetanus and malignant edema. Ordinarily, the production or liberation of gases passes undetected, the vapors escaping from the surface of the culture-medium.

To determine the gas production where it is suspected but not apparent, the ordinary fermentation-tubes can be

employed. They are filled with glucose bouillon, sterilized as usual, inoculated and allowed to grow. If gases
are formed, the bubbles ascend and the
gas accumulates at the top of the tube.
In estimating quantitatively, one must
be careful that the tube is not so constructed as to allow the gas to escape as
well as to ascend in the main reservoir.

For the determination of the nature
of the gases ordinarily produced, some
of which are inflammable and some not,
Theobald Smith has recommended the
following methods:

Fig. 5.—Smith's fer-
mentation-tube.

"The bulb is completely filled with a 2 per cent. solution of sodium hydroxid (NaOH) and tightly closed
with the thumb. The fluid is shaken thoroughly with
the gas and allowed to flow back and forth from the bulb
to closed branch, and the reverse several times to insure
intimate contact of the CO_2 with the alkali. Lastly,
before removing the thumb all the gas is allowed to collect in the closed branch so that none may escape when
the thumb is removed. If CO_2 be present, a partial
vacuum in the closed branch causes the fluid to rise suddenly when the thumb is removed. After allowing the
layer of foam to subside somewhat the space occupied by
gas is again measured, and the difference between this
amount and that measured before shaking with the
sodium hydroxid solution gives the proportion of CO_2
absorbed. The explosive character of the residue is
determined as follows: "The cotton plug is replaced and
the gas from the closed branch is allowed to flow into the
bulb and mix with the air there present. The plug is
then removed and a lighted match inserted into the
mouth of the bulb. The intensity of the explosion varies
with the amount of air present in the bulb."

7. *Production of Odors.*—Of course, such gases as H_2S
and NH_3 are sufficiently characteristic to be described as
odors. There are, however, a considerable number of

pungent oders which seem dependent purely upon odoriferous principles dissociated from gases. Many of them are extremely unpleasant, as the onion-like odor of the tetanus bacillus. The odor does not have any direct relation to decomposition, but, like the colors and acids, seems to be a peculiar individual characteristic of the metabolism of the organism.

8. *Production of Phosphorescence.*—A Bacillus phosphorescens and numerous other organisms have a distinct phosphorescence associated with their growth. It is said that so much illumination is sometimes caused by a gelatin culture of some of these as to enable one to tell the time by a watch. Most of them are found in seawater, and are best grown in sea-water gelatin.

9. *Production of Aromatics.*—The most important of these is *indol*, which was at one time thought to be peculiar to the cholera spirillum. For the method of determining its presence, see "Dunham's Solution." At present we know that a variety of organisms produce it, and that it and phenol, kresol, hydrochinon, hydroparacumaric acid, and paroxy-phenylic-acetic acid are by no means uncommon.

10. *Reduction of Nitrites.*—A considerable number of bacteria are able to reduce nitrites present in the soil or in culture-media prepared for them into ammonia and nitrogen. To the horticulturist this is a matter of much interest. Winogradsky has found a specific nitrifying bacillus in soil, and asserts that the presence of ordinary bacteria in the soil causes the reduction of no nitrites so long as his special bacillus is withheld.

11. *Peptonization of Milk.*—Numerous bacteria possess the power of digesting—peptonizing—the casein of milk. The process differs with different bacteria, some digesting the casein without any apparent change in the milk, some producing coagulation, some gelatinization of the fluid. In some cases the digestion of the casein is so complete as to transform the milk into a transparent watery fluid.

Milk usually contains bacteria, entering it from the dust of the dairy, possessing this power. In the process of peptonization the milk may become bitter, but need not change its original reaction. As the peptonization progresses the milk very often becomes poisonous, especially to individuals under two years of age, and may bring about a fatal enterocolitis or "summer complaint." The disease does not only occur in consequence of toxic substances formed from the split-up albumins, or from the presence of metabolic products of the bacteria, but, as Lübbert has shown,[1] from the presence of the bacteria themselves. One reason that the enterocolitis caused in this way comes on in summer is that it is only in unusually warm weather that these bacteria are able to grow luxuriantly.

Sometimes the properties of coagulation and digestion of milk are valuable aids in the separation of different species of bacteria.

12. *Production of Disease.*—Bacteria which produce diseases are known as *pathogenic;* those which do not, as *non-pathogenic.* Between the two groups there is no sharp line of separation, for true pathogens may be cultivated under such adverse conditions that their virulence will be entirely lost, while at times bacteria ordinarily harmless may be made toxic by certain manipulations or by introducing them into animals in certain combinations. The diseases produced are the result of the sum of numerous activities exhibited by the bacteria. For example, it may be that a microbe, having effected its entrance into an animal, grows with great rapidity, completely blocking up the blood- and lymph-channels, so that the proper circulation of these fluids is stopped and disease and death must result. Perhaps more common than this is a local establishment of the organisms, with a resulting inflammation, due partly to the presence of the foreign organisms, and partly to their toxic metabolic products. More often, however, the pathogenic

---

[1] *Zeitschrift für Hygiene,* xxii., Heft 2, 1896, p. 1.

bacteria produce powerful metabolic poisons—toxins, ptomaïnes, etc.—which either cause widespread destruction of the tissues immediately acted upon, or, circulating throughout the organism, produce fever, nervous excitation, and a general overthrow of the normal physiological equilibrium. These peculiarities serve to divide the bacteria into

>Phlogistic bacteria,
>Toxic bacteria,
>Septic bacteria.

The bacteria of suppuration probably act in several ways. Their products may be of a violently chemotactic nature, or their virulence, exerted upon the surrounding tissue, may destroy large numbers of the cells, whose dead bodies may be chemotactic. When the suppuration is violent the toxic product of the bacterium is itself most probably strongly chemotactic.

The great majority of suppurations depend upon bacteria, but there are sterile suppurations which sometimes follow the use of croton oil, turpentine, etc. The difference between infectious and sterile pus is marked, for the former, containing the virulent germs, tends to invade new tissue or distribute its disease-producers to new parts of the body, while the latter remains local.

There are few purely toxic bacteria, the tetanus and diphtheria bacilli serving as typical examples. By septic bacteria, I mean those whose habitual tendency is to grow in the blood and lymph and distribute to all the organs. Anthrax is a type of the class.

How the disease-producing bacteria effect their entrance into the tissues is an interesting question. The channels naturally open to them are those leading into the interior of the organism, and must be separately considered.

(*a*) *The Digestive Tract.*—Attention has already been called to the facility with which the bacteria enter the digestive tract in foods and drinks. Once their metabolism is in active progress, the poisons which they produce are ready for absorption. It seems probable that the absorption of the toxic substances by reducing the vitality of the individual predisposes to the formation of local lesions through which the bacteria may enter the intestinal walls to continue their existence and produce greater damage than before. Some such theory may explain the activity of such organisms as those of typhoid, cholera, and meat-poisoning, but it is not true that all bacteria can be admitted into the intestinal structure in this way, for the experiments of Max Neisser,[1] who fed mice, guinea-pigs, and rabbits upon a variety of pathogenic and non-pathogenic bacteria, both before and after injuries to the intestine caused by the ingestion of powdered glass, chemical agents, and irritating bacteria, failed to show that with the exception of those bacteria whose particular tendency is toward the production of intestinal disease, none entered either the chyliferous system, the blood-vessels, or the organs.

The occurrence of the staphylococcus aureus and other bacteria in osteomyelitis, and of tubercle bacilli in deep-seated diseases of the bones and internal organs, has led many to believe that the intestine is a point of easy entrance. There is, however, no reason to believe that penetration of the digestive mucous membrane is any easier than that of the respiratory or other similarly delicate tissues.

On the other hand, Beco[2] is of the opinion that, without any apparent lesion of the intestine, bacteria—bacillus coli—escape from it into the blood during life. His experiments showed that immediately after death the colon bacillus could be found in small numbers in the spleen, in many cases. After twenty-four hours, in three cases, they were present in immense numbers. When, however, they were absent from the organ immediately after death, they were also absent after twenty-four hours.

[1] *Zeitschrift für Hygiene,* June 25, 1896, Bd. xxii., Heft 1.
[2] *Ann. de l'Inst. Pasteur,* 1895, No. 3.

Achard[1] studied 49 cases to determine whether or not the intestinal bacteria entered the organism during the death agony. In 14 bacteria were found *intra vitam* in the liver and in the blood. In 24 no bacteria were found during life, but after death. In 11 no bacteria were found either during life or after death—before twenty-two to twenty-seven hours, when his autopsies were made. The passage of bacteria into the blood during agony was unusual. The bacteria most commonly found during life were the streptococci and staphylococci. In the dead body the one most frequently encountered was the bacillus coli communis. Before reaching the intestine the bacteria pass through the stomach, and must resist the deleterious action of the acid gastric juice, which few are able to do.

(b) *The Respiratory Tract.*—Notwithstanding the moist interiors of the mouth and nose and the lashing cilia of the pharyngeal and tracheal mucous membrane, numbers of bacteria enter the smaller bronchioles, and occasionally penetrate as deeply as the air-cells. It is usual to find a few bacteria in a section of healthy lung.

Thomson and Hewlett[2] estimate that from 1500 to 14,000 bacteria are inspired every hour. As expired air is usually sterile, they sought to determine what became of these organisms, and agree with Lister and with Hildebrandt that the organisms are arrested before they reach the air-cells. They found by killing a number of animals and examining the tracheal surface that it was sterile, and conclude that the great majority of bacteria are stopped in the nose against the moist surfaces of its vestibules, where they found great numbers in the crusts. No doubt the ciliated cells of the nose have something to do with getting rid of the bacteria.

An ingenious experiment was performed by placing some bacilli prodigiosus upon the septum naris, and making a culture from the spot at intervals during two

---

[1] Archives de médecine expérimentale et d'antomie pathologique, 1895, No. 1, p. 25.

[2] *British Med. Jour.*, Jan. 18, 1896, p. 137.

hours. Cultures made within five minutes showed confluent colonies of the bacilli, which became fewer and fewer in number, until after two hours not a trace of a bacillus prodigiosus could be found.

Wurtz and Lermoyez assert that the nasal mucus exerts a germicidal action, but this is not substantiated. These writers conclude that the bacteria were carried away by the action of the cilia and trickling mucus.

It seems to have been proven by Buchner that microorganismal infection may take place through the lungs without definite breach of continuity of the alveolar walls. He mixed anthrax spores and lycopodium powder together, and caused mice and guinea-pigs to inhale them. Out of the 66 animals used in his experiments, 50 died of anthrax and 9 of pneumonia. Our knowledge of the disposition of foreign particles in the lung probably explains such infection by assuming that the presence of the lycopodium attracted numerous leucocytes to the affected aircells; that these took up the powder, and with it the spores; and that the leucocytes, being cells of very susceptible animals, were unable to resist the growth into bacilli of the spores which they had carried into the lymph-channels.

On the other hand, it has been shown that when the entering spores are unaccompanied by a mechanical irritant like the lycopodium powder, but are inspired in a pulverized liquid, infection takes place much less readily.

Tuberculosis and pneumonia are in all probability generally the result of the inspiration of the specific organisms.

(c) *The Skin and the Superficial Mucous Membranes.* — The entrance of bacteria into the tissues by way of the skin is probably extremely rare if the skin is sound. Numerous experimenters have caused infection by rubbing bacteria or their spores upon the skin. It would seem probable that in these cases there must have been some microscopic lesions into which the bacteria were forced. My own investigations have shown virulent staphylococci of suppuration upon the conjunctivæ in health. It is very improbable that the bacteria habitually present upon the skin, and ready to enter the least abrasion, can penetrate the outer coverings of the body, except when disease or accident · has rendered them abnormally thin or macerated.

Turro seems to have shown that the gonococcus can enter the tissues without any pre-existing lesion, for he asserts that if a virulent culture simply be touched to the meatus urinarius, the disease will be established.

(*d*) *Wounds.*—The results of the entrance of bacteria into unprotected wounds are now so familiar that no one deserving of the name of surgeon dares to allow a wound to go undressed.

(*e*) *The Placenta.*—Very frequently the occurrence of specific diseases during pregnancy causes abortion of the product of conception. In certain cases the specific contagion passes through the placenta and infects the fetus. This has been pretty clearly demonstrated for variola, malaria, syphilis, measles, anthrax, symptomatic anthrax, glanders, relapsing fever, typhoid, and in rare cases for tuberculosis.

Anche found streptococci and staphylococci in the tissues of aborted fœti in cases of variola.[1] Except in the case of wounds, it must be observed that, although the bacteria are in the body—*i. e.*, respiratory, digestive, or sexual apparatus, etc.—they are still not in the blood, and really not in, but only upon the surfaces of the tissues.

For their actual entrance into the circulation, Kruse[2] gives the following possible modes:

1. Passive entrance of the bacteria through the stomata of the vessels where the pressure of the inflammatory exudate is greater than the intravascular pressure.

2. Entrance of the bacteria into the vessel in the body of leucocytes that have incorporated them.

---

[1] *La Semaine Méd.*, 1892, No. 61.
[2] Flügge's *Mikroörganismen.*

3. Actual penetration of the vessel-wall by the growth of the microörganism.

4. Entrance into the vessels *via* the lymphatics, either passively or in leucocytes.

Seeing that the channels by which bacteria can enter the body are so numerous, and that there is scarce a moment when some part of us is not in contact with them, how is it that we are not constantly subject to disease? The consideration of this question, together with the closely related questions why we should be subject to certain diseases only, and to these diseases at certain times only, must be reserved for another chapter, in which the subjects *Immunity* and *Susceptibility* can be taken up at length. Before passing on to it, however, some attention must be paid to the subject of the

Elimination of Bacteria from the Body.—There is every reason to think that non-pathogenic bacteria entering the body ordinarily, or being experimentally injected into it, follow the same course as inert, non-vital particles; concerning which, the experiments of Siebel have shown that they accumulate in the finest capillaries, especially in the lung, liver, spleen, and bone-marrow, and are slowly transferred to the surrounding tissues, either to be collected in the connective-tissues, carried to the lymphatic nodes, or to be subsequently excreted with the bile, succus entericus, etc., or to be discharged from the surface of the mucous membranes, pulmonary alveoli, tonsils, etc. They also escape from suppurating wounds to which they may be carried by leucocytes. They are not excreted by the kidneys.

The experiments of Wyssokowitsch are in accord with the results of Siebel's work, and show that the kidney rarely eliminates bacteria. Cavazzani found that the kidney had the power to retain bacteria in the blood, unless the epithelium was injured.

The principal avenues of escape for the bacteria are, therefore, for the non-pathogenic forms, the mucous membranes, the bile, and the sweat. For the pathogenic forms, the mucous membranes, the intestine in particular in such diseases as anthrax, typhoid, and cholera; the bile almost always; the sweat generally; the kidney when damaged; the mammæ in tuberculosis and septicemia particularly, and, of course, such of the pathological products of the disease-process as pus from abscesses, dejecta of typhoid and cholera, expectoration in diphtheria and tuberculosis, etc.

The bacteria that are not excreted, but retained in such organs as the spleen, bone-marrow, and lymphatic nodes, are probably slowly devitalized and dissolved.

CHAPTER III.

IMMUNITY AND SUSCEPTIBILITY.

ONE of the most interesting things observed in physiology and pathology is the resistance which certain animals show to the invasion of their bodies by the germs of disease.

Thus, man suffers from typhoid fever, cholera, and other infectious diseases which are never observed in the domestic animals; cattle are subject to a pleuro-pneumonia which does not affect their attendants; man, the cow, and the guinea-pig are peculiarly susceptible to tuberculosis, which the cat, dog, and horse resist; yellow fever is a highly contagious, infectious disease which is almost certain to attack all new arrivals of the human species when epidemic, but which rarely, if ever, attacks animals.

The popular mind accepts the statement of such facts as these without any other explanation than that the animals are different, and so of course their diseases are different; but the more the scientific man contemplates them, the more complicated the matter becomes; for, while it might be admitted that a difference in the body-temperature and chemistry might explain why a frog will resist anthrax, which readily kills a white mouse, it will not explain why a house-mouse, whose chemistry must be almost identical with that of the white mouse, can successfully combat the disease. Nor is this all. That one attack of yellow fever, of typhoid fever, or of scarlet fever renders a second attack almost impossible is not the less interesting because of its every-day observation. The mouse that has recovered from tetanus will not take tetanus again, and most interesting and

5

extraordinary is the fact that a few drops of blood from the recovered mouse injected into another will protect it from tetanus.

Immunity is the condition in which the body of an animal resists the entrance of disease-producing germs, or, having been compelled to allow them to enter, resists their growth and pathogenesis. The resistance so manifested is a distinct, potential vital phenomenon.

Susceptibility is the opposite condition, in which, instead of resistance, there is a passive inertia which allows the disease-producing organisms to develop without opposition. Susceptibility is accordingly the absence of immunity.

Immunity is either natural or acquired.

Natural Immunity.—By this term is meant the natural and constant resistance which certain healthy animals exhibit toward certain diseases.

The white rat is peculiar in resisting anthrax. It is almost impossible to develop anthrax in a healthy white rat, but Roger found that such an animal would easily succumb to the disease if compelled to turn a revolving wheel until exhausted. Susceptibility which follows such an exhaustion of the vital powers cannot be regarded as other than accidental, and makes no exception to the statement that the white rat is immune to anthrax. Animals such as man, sheep, cows, rabbits, and white mice are susceptible to anthrax, while birds and reptiles are generally immune. The great difference in the morphology between mammals and birds and reptiles, together with the fact that their temperature, blood, and tissues all differ, makes this immunity reasonably intelligible. Morphological differences, however, will not suffice to explain all cases, for the Caucasian nearly always succumbs to yellow fever, while the negro is rarely affected ; and scarlatina, which is one of our commonest and most dangerous diseases of childhood, is said to be unknown among the Japanese. Nor is this all, for, close as is their resemblance in all respects except color, the house-mouse, field-mouse, and white mouse differ very much in their susceptibility to various diseases.

Acquired immunity is resistance which is the result of accidental circumstances. It may result—

A. By recovery from a mild attack of the disease. Most adults have suffered from rubeola, scarlatina, and varicella in childhood, and in consequence of the attacks are now immune to these diseases—*i. e.* will not become affected again. One attack of yellow fever is always a complete guard against another. Typhoid fever is rarely followed by a second attack.

B. By recovery from an attack of a slightly different disease. Sometimes the immunity is experimentally produced, as when by vaccination we produce the vaccine disease and afterward resist variola. Acquired immunity is a little less complete and not so permanent as natural immunity, for in the latter it is only when the functions of the individual are disturbed or his vitality depressed that the resistance is lost, while in the former time seems to lessen the power of resistance, so that rubeola and scarlatina may return in a few months or years, and for complete protection vaccination may need to be done as often as every seven years.

C. By the injection of antitoxic substances. At present there is much agitation over the newly-discovered antitoxin of diphtheria, the injection of about 500 units of which will give complete protection against the disease for a period lasting from a month to six weeks.

Immunity may be destroyed in numerous ways:

(*a*) *By variation from the normal temperature* of the animal under observation. Pasteur observed that chickens would not take anthrax, and suspected that this immunity might be due to their high body-temperature. After inoculation he plunged the birds into a cold bath, reduced their temperature, and succeeded in destroying their immunity. The experiment was a success, but the reasoning seems to have been faulty, as the sparrow, with a temperature equally high, readily falls a victim to anthrax without a cold bath.

(*b*) *By altering the chemistry of the blood* by changing the diet or by hypodermic injection. Leo found that when white rats were injected with or fed upon phloridzin an artificial glycosuria resulted which destroyed their natural resistance to anthrax. Hankin found that rats, which possess considerable immunity to anthrax, could be made susceptible by a diet of bread. Platania succeeded in producing anthrax in dogs, frogs, and pigeons, naturally immune, by subjecting them to the influence of curare, chloral, and alcohol.

(*c*) *By diminishing the strength of the animal.* Roger by compelling white rats to turn a revolving wheel until exhausted destroyed their immunity to anthrax.

(*d*) *By removing the spleen* (?). A large number of experiments have been performed by various investigators to show that the removal of the spleen does or does not affect immunity. From their work it seems proper to conclude that the spleen has little, if any, influence upon the vital resistance to disease.

I. Bardach,[1] Righi,[2] and Montuori[3] seem to have shown that the removal of the spleen lessens the ability of the organism to combat the infections.

II. Blumenreich and Jacoby,[4] on the contrary, found that the removal of the spleen was followed by a hyperleucocytosis, an increase in the bactericidal power of the blood, and consequent increase of immunity.

III. Milkinow-Raswedenow[5] found that the removal of the spleen was a weakening factor in the immunization of animals. The spleen itself, however, was of little importance in combating the micro-organismal infections.

Kurlow[6] concluded from his experiments that the in-

[1] *Ann. de l' Inst. Pasteur*, 1889, No. 2, p. 577, and 1891, No. 1, p. 40.
[2] *La Riforma Medica*, 1893, pp. 170, 171.
[3] *Ibid.*, Feb., 1893, 17, 18.   [4] *Berlin. klin. Wochenschrift*, May 24, 1897.
[5] *Zeitschrift für Hygiene*, 1896, xxi., 3.
[6] *Archiv für Hyg.*, 1889, Bd. ix., p. 450.

fluence of the spleen was not greater than that of any other organ in overcoming bacterial infections.

Kanthack[1] found that the removal of the spleen had practically no influence upon the natural immunity of animals to pyocyaneus infection.

(*c*) *By combining two different species of bacteria*, either of which, when injected alone, would be harmless or of slight effect. Roger found that when animals immune to malignant edema were simultaneously injected with 1 to 2 c.cm. of a culture of Bacillus prodigiosus and the bacillus of malignant edema, they would contract the disease. Pawlowski found that when rabbits, which are very susceptible to anthrax, were simultaneously injected with anthrax and prodigiosus, they recovered from the anthrax, as if the harmless microbe possessed the power of neutralizing the products of the pathogenic form.

Giarre found that if an adult guinea-pig, which is refractory to infection by pneumococci, were simultaneously inoculated with diphtheria, it readily died of septicemia.

Sometimes an apparent immunity depends upon the attenuation of the culture used for inoculation, and the erroneous results to which such a mistake may lead are obvious. Should a culture become attenuated, its virulence may sometimes be increased by inoculating it into the most susceptible animal, then from this to a less susceptible, and then to an immune animal. The virulence of anthrax is increased by inoculation into pigeons, and also by cultivation in an infusion of the tissues of an animal similar to the one to be inoculated.

It must be understood that the term "immunity" is a relative one, and that while "a white rat is immune against anthrax in amounts sufficiently large to kill a rabbit, it is perhaps not immune against a quantity sufficiently large to kill an elephant."

It is not to be expected that such intricate phenomena as these which have been mentioned could be observed

[1] *Centralbl. f. Bakt. u. Parasitenk.*, 1892, xii., p. 227.

and suffered to go unexplained. We have explanations, but, unfortunately, they are as intricate as the phenomena, and, though each may possess its grain of truth, not one will satisfy the demands of the thoughtful student. In brief review, the theories of immunity are the following :

1. THE EXHAUSTION THEORY.—This hypothesis was advanced by Pasteur in 1880, and suggests that by its growth in the body the micro-organism uses up some substance essential to its life, and that when this substance is exhausted the microbe can no longer thrive. The removal of the necessary material, if complete, will cause permanent immunity.

As Sternberg points out, were this theory true we must have within us a material of small-pox, a material of measles, a material of scarlet fever, etc., to be exhausted by its appropriate organism. It would necessitate an almost inconceivably complex body-chemistry and a rather stable condition of the same.

2. THE RETENTION THEORY. — In the same year Chauveau suggested that the growth of the bacteria in the body might originate some substance prejudicial to their further and future development. There seems to be a large kernel of truth in this, but were it always the case we would have added to our blood a material of small-pox, a material of measles, a material of scarlet fever, etc., so that we would become saturated with the excrementitious products of the bacteria, instead of having so many substances subtracted from our chemistry.

3. THE THEORY OF PHAGOCYTOSIS.—In 1881, Carl Roser suggested a relation between immunity and the already familiar phenomenon of phagocytosis. Sternberg in the United States and Koch in Germany observed the same thing, but little real attention was paid to the subject until 1884, when Metschnikoff appeared, with his careful observations upon the daphnia, as the great champion of the theory which is now known as "Metschnikoff's theory of phagocytosis."

Phagocytosis is the swallowing or incorporating of

particles by certain of the body-cells which are called
phagocytes. This activity of the cells toward inert
particles had been observed by Virchow as early as 1840,
and toward living bacteria by Koch in 1878, but was not
carefully studied until 1884. Metschnikoff divides the
phagocytes into *fixed phagocytes,* comprising the fixed
connective-tissue cells, endothelium, etc., and the *free
phagocytes,* which are the leucocytes. The terms "phag-
ocyte" and "leucocyte" are not to be regarded as synon-
ymous in this connection ; all leucocytes are not phagocytic, the *lymphocyte* having never been observed to
take up bacteria.

It is obvious that only those cells can be phagocytic
which are without a resisting cell-wall and possess
ameboid movement. When an ameba, in a liquid containing numerous diatoms and bacteria, is watched
through the microscope, an interesting phenomenon is
observed. The ameba will approach one of the vegetable cells, even though it may be at a distance, will
apprehend and surround it, and within the animal cell
the vegetable cell will be digested and assimilated. The
ameba has no eyes, no nose, no volition, and, so far as
we can determine, no nervous apparatus which gives
it tactile sense, yet it will approach the particle fitted
for its use and swallow it. The attraction which draws
the cells together has been called by Peffer *chemotaxis,
chemiotaxis,* or *chemotropism.*

Chemotaxis is the exhibition of an attractive force
between cells and their nutriment, ameboid cells and
food-particles, and sometimes between ameboid cells and
inert particles. This attractive force, when operating so
as to draw the ameba to the particle it will devour, is
further named *positive chemotaxis* in order to distinguish
it from a repulsive force sometimes exerted causing the
ameboid cells to fly from an enemy, as it were, and which
is called *negative chemotaxis.*

The force that operates and guides the ameba in its
movements is exactly the same as that which governs the

movement of the phagocytic cells of the human body, and observation of these phenomena is not difficult. If a small capillary tube be filled with sweet oil and placed beneath the skin, only a short time need pass before it will be found full of leucocytes—positive chemotaxis. If, instead of sweet oil, oil of turpentine be used, not a leucocyte will be found—negative chemotaxis.

Phagocytosis is almost universal in the micro-organismal diseases at some stage or another. If the blood of a patient suffering from relapsing fever be studied beneath the microscope, it will be found to contain numerous active mobile spirilla, all free in the liquid portion of the blood. As soon as the apyretic stage comes on not a single free spirillum can be found. Every one is seen to be enclosed in the leucocytes.

At the edge of an erysipelatous patch a most active warfare is waged between the streptococci and the cells. Near the centre of the patch there are many free streptococci and a few cells. At the margin there are free streptococci, and also a great many streptococci enclosed in cells (leucocytes) which are, for the most part, dead. In the newly-invaded tissue we find hosts of active living cells engaged in eating up the enemies as fast as they can. The phagocytologists tell us that at the centre the bacteria are fortified, actively growing, and virulent ; in the next zone the leucocytes which have feasted upon the bacteria are poisoned by them ; outside, the cells, which are more powerful and which are constantly being reinforced, are waging successful warfare against the streptococci. In this manner the battle continues, the cells now being obliged to yield to the bacteria and the patch spreading, while the cells subsequently reinforce and destroy the bacteria, so that the disease comes to a termination.

Metschnikoff introduced fragments of tissue from animals dead of anthrax under the skin of the back of a frog, and found it surrounded and penetrated by leucocytes containing many of the bacilli.

It need scarcely be pointed out that a loophole of doubt exists in all these illustrations: the bacteria may have been dead before the cells ingested them, and the phenomena of digestion and destruction which have gone on in their interiors may have been exerted upon dead bacteria. To the relapsing-fever illustration we may take exceptions, and state that the apyrexia may have marked the death of the spirilla, which were taken up by the leucocytes only when dead. In the erysipelas illustration the streptococci remote from the centre of the lesion may have met from the body-juices or some other cause a more speedy death than that from the digestive juices of the leucocyte.

Metschnikoff, however, is prepared to show us that the leucocytes do take up living pathogenic organisms. He succeeded in isolating two leucocytes, each containing an anthrax spore, and conveying them to artificial culture-media, where he watched them. The new environment being better adapted to the growth of the spore than for the nourishment of the leucocyte, the latter died, and the spore developed under his eyes into a healthy bacillus. Seeing that the animal cells take up bacteria, and seeing that the ameba can ingest and digest "threads of leptothrix ten times as long as itself," we need only put two and two together to see that Metschnikoff's theory rests upon a very substantial foundation. The more virulent the bacteria, the less ready the leucocytes are to seize them. The more immune the animal, the greater is the affinity of the leucocyte for the bacteria.

The organisms which are seized upon by the leucocytes do not remain in the blood, but are collected in the spleen and the lymphatic glands; and not the least important fact in favor of phagocytosis is that observed by Bardach, that excision of the spleen diminishes the resistance to infectious disease.

Quinin also furnishes a therapeutic support to the theory. It is known that quinin increases the destruction of leucocytes. Woodhead inoculated a number of rabbits with anthrax, giving quinin to some of them.

Those which had received the drug died earliest—a result probably dependent upon the destruction of part of the phagocytic army.

Ruffer found that the "phagocytes evince a distinct selective tendency between various kinds of organisms. They will leave the bacillus of tetanus in order to seize upon the Bacillus prodigiosus if simultaneously introduced; also the streptococci in diphtheria for the Klebs-Löffler bacilli. This is illustrated in the diphtheritic membrane, where at the surface one can see leucocytes taking in numbers of the bacilli, but leaving the streptococci almost untouched, with the immediate result that streptococci are often found in the deeper parts of the membrane, and with the remote result that secondary abscesses occurring in the course of diphtheria are never due to the bacillus of diphtheria, but to some other organism."

Hankin and Hardy found that the three varieties of leucocytes in the frog's blood play important parts in the destruction of anthrax bacilli, this destructive process being accomplished thus:

1. The eosinophile cells are first to approach and swallow the bacteria. As this takes place the eosinophile granules are seen to dissolve and act upon the bacteria.

2. The hyaline cells take up the remains of the bacteria destroyed by the eosinophile leucocytes.

3. The basophile cells come to the field loaded with basophilic granules, supposed to be antidotal to the poisons of the bacteria, surround the combatants, neutralize the bacterial poisons, and liberate the contesting cells.

Wyssokowitsch found that saprophytic micro-organisms are quickly eliminated from the blood when injected into the circulation. This elimination is not by excretion through organs nor by destruction in the streaming blood, but by collection in the small capillaries, where the blood-stream is slow and where the micro-organisms are taken up by the endothelial cells.

Wyssokowitsch found them most numerous in the liver, spleen, and bone-marrow, and found that in these situations they were destroyed in a short time—saprophytic in a few hours, pathogenic in from twenty-four to forty-eight hours. Spores of Bacillus subtilis remained as living entities in the spleen for three months.

An interesting communication upon phagocytosis is that of Bordet, whose experiments seem to show that the lack of disposition to take up bacteria on the part of the leucocytes may depend upon *negative chemotaxis.* He found that when a guinea-pig became very ill after the intraperitoneal introduction of a streptococcus of mild virulence, if an injection of a culture of Proteus vulgaris was given, the leucocytes, which had steadily refused to take up the streptococci, seized upon the bacilli with avidity. This seems to show that a chemical, or other negative, or inhibitory influence felt by the leucocyte, prevents it from taking up all the bacteria that come within reach.

4. THE HUMORAL THEORY.—It was observed that if anthrax bacilli were introduced into a few drops of rabbit's blood, they were instantly killed. This observation was one of immense importance, and from it and similar observations Buchner deduced the principles of his theory, which teaches that the destruction of pathogenic bacteria in the body is due to the *bactericidal action of the blood-plasma*, not to phagocytosis, which phenomenon amounts to nothing more than the burial of the dead bacteria in "cellular charnel-houses." The experiments of Buchner and his followers, conspicuous among whom is Nuttall, have shown that freshly drawn blood, blood-plasma, defibrinated blood, aqueous humor, tears, milk, urine, and saliva possess marked destructive influence upon the organisms brought in contact with them—an influence easily destroyed by heat.

The apparent paradox of rapid multiplication of anthrax bacilli in the rabbit's blood enclosed in the rabbit's body, and the reversed action in the test-tube, caused immediate and prolonged opposition to the theory. Each side of the question seemed well supported. The phagocytologists, however, showed that bacteria were often injured and their vegetative powers destroyed by sudden changes from one culture-medium to another, this being proved by Haffkine, who in experimenting with aqueous humor has shown that its germicidal actions are largely imaginary, and due to the dispersion of the organisms in a large amount of watery liquid. When the micro-organisms are introduced into it in such a manner as to remain together, they grow well. If the tube be shaken, so as to distribute them, they die. Again, Adami has shown that when blood is shed there is almost always a pronounced destruction of corpuscles, and suggests that the antibiotic property of the shed blood may be due to solution of the nucleins formerly in the substance of the leucocytes. Jetter endeavored to prove the germicidal action of the serum to be due to certain salts which it contained. His experiments, which consisted in observing the action of solutions of various salts in mixtures of water, glycerin, and gelatin, were justly condemned by Buchner on the ground that such mixtures, though they might contain constituents of blood-serum, were far from approximating the normal serum in composition.

Wyssokowitsch, however, surely argued against humoral germicide when he showed that the spores of Bacillus subtilis could reside in the spleen for three months uninjured.

In supporting their theory the humoralists experimented by placing beneath the skin micro-organisms enclosed in little bags of pith, collodium, etc. These bags allowed the fluids of the body free access to the bacteria, but would shut out the phagocytes. By these means Hüppe and Lubarsch have repeatedly seen the bacteria grow well, while the attempts of Baumgarten have failed. Such experiments are by no means conclusive, for we should remember that the operation necessary and the presence of the foreign body in which the bacteria are encased produce an inflammatory transudate which may have properties very different from those of the normal juices.

How much of the immunity which animals enjoy depends upon the antibactericidal action of their body-juices must remain an open question. In some cases the germicidal action of the blood seems to be unquestionable. Buchner has shown that the blood-serum of animals only possesses this germicidal power when freshly drawn, and that exposure of the serum to sunlight, its mixture with the serum from another species of animal, its mixture with distilled water or with dissolved corpuscles, and heating it to 55° C., check the bactericidal power. Buchner also points out that the bactericidal and globulicidal actions of the blood are simultaneously extinguished. Meltzer and Norris[1] found that lymph taken from the thoracic duct of the dog possessed marked bactericidal powers upon the typhoid bacillus.

The experiments of Pfeiffer seem to add additional support to the humoral theory of immunity. He found that when guinea-pigs were given experimental choleraic peritonitis, they could be saved from death from the affection by intraperitoneal injection of serum from an immunized animal. He also showed that when the culture of cholera, or a culture of typhoid bacilli, was injected into the peritoneum of a guinea-pig, the multiplication of the bacteria was rapid. If, however, a few drops of the immunized serum were introduced, a marked effect was observed, for the serum seemed to exert a germicidal effect upon the bacteria, and transform them from living entities into inanimate little granular masses.

Hankin is of the opinion that the germicidal substances of the blood-serum are derived from the eosinophile cells, and resides in the matter forming the eosin-granules.

Löwit,[2] in investigating the bactericidal power of the

---

[1] *Journal of Experimental Medicine*, vol. ii., No. 6, p. 701, Nov., 1897.

[2] Beiträge zur Pathol. Anatomie und zur Allgem. Pathologie, Bd. xxii., H. 1, p. 173.

blood in relation to its leucocytes, found that when a marked experimental hypoleucocytosis was produced, the bactericidal power of the blood was markedly diminished. The most interesting feature of his work was the discovery that bactericidal matter could be extracted from crushed leucocytes, and that it could be subjected to a temperature of 60° C. without change, thus differing markedly from the alexins.

Much discussion has arisen as to exactly what the protective substances are. Buchner has applied the term *alexin* to the protective proteid substances found in the blood of naturally immune animals. Hankin has given us, together with an extension of Buchner's idea, a considerable nomenclature of somewhat questionable utility. He divides the protective substances (alexins) into *sozins*, which occur in the blood of animals with natural immunity, and *phylaxins*, which occur in the blood of animals with acquired immunity. Both sozins and phylaxins are divisible into two groups—thus: a sozin which acts destructively upon bacteria is called a *myco-sozin;* one which neutralizes bacterial poisons, a *toxo-sozin.* A phylaxin which acts destructively upon bacteria is called a *myco-phylaxin;* one which neutralizes bacterial toxins, a *toxo-phylaxin.*

The *anti-microbic* serums obtained by Pfeiffer, Kolle, Löffler, and Abel from dogs and other animals immunized to typhoid fever belong in the group of myco-phylaxins. The toxo-phylaxins are the antitoxins.

5. The Theory of Antitoxins.—It is a well-known fact that individuals can accustom themselves to the use of certain poisons, as tobacco, opium, and arsenic, so as to experience no inconvenience from what would be poisonous doses for other individuals. This is purely a matter of tolerance, but is of interest in connection with the observations which are to follow.

Ehrlich has shown that animals can tolerate gradually increasing doses of ricin and abrin, provided that up to a certain point the increase of dosage is very small.

When this point is, however, safely passed, the increase in dosage can be very rapid, yet without signs of poisoning, seemingly because the drug is no longer simply tolerated, but tolerated and simultaneously neutralized. By experimentation Ehrlich has shown that during the period of simple tolerance the blood of the animal is unaltered, but that as soon as the tolerance becomes unlimited the blood contains a new substance, capable not only of protecting the animal by which it is produced, but also other animals into whose blood it is introduced. In the ricin experiments this substance was described as *antiricin;* in the experiments with abrin, as *antiabrin.*

These investigations of Ehrlich with the poisons of higher plants succeeded, but threw much light upon, the extraordinary work of Behring, Wernicke, and Kitasato, who experimented with the toxins of diphtheria and tetanus, and showed that in the blood of animals accustomed to these poisons, new substances—*antitoxins,* found by Brieger to be proteid in nature—were produced.

The antitoxic theory of immunity, being, in the cases cited at least, a fact capable of demonstration, has established itself at present as the most important hypothesis. According to it, acquired immunity, at least, depends upon the development in the blood of a neutralizing substance probably related to the nucleins.

It is of prime importance to remember that the antitoxin is an entirely new substance which does not occur in the blood of normal animals, even when they possess a high degree of natural immunity, except in rare instances, and then only in minute amounts not proportional to the degree of immunity. Calmette has called special attention to this fact, and points out that while fowls and tortoises resist abrin, their blood contains no anti-abrin; Vaillard has shown that, although the fowl resists tetanus, its blood contains no protective substance destructive to tetanus-toxin. Calmette finds that the blood of the ichneumon and hedgehog, which are immune to serpent's venom, contains some normal antitoxin, but only in small amount.[1] Fischl and v. Wunschheim found a small amount of a protecting substance in the blood of newborn infants, which prevented the operation of a fatal dose of diphtheria toxin upon guinea-pigs.[2]

Bolton[3] and the author have found some antitoxicity to diphtheria present in the blood of normal (not experimentally immunized) horses.

The origin of the antitoxin is a very important and interesting question. Is it in the blood, or in all the body juices? Does it come from the leucocytes? Dzerjgowsky[4] has estimated the quantity of antitoxin contained in the blood and organs of horses immunized against diphtheria. Of the constituents of the blood he found (1) the fibrin has no antitoxic power; (2) serum obtained normally and that got by expression from the clot, from the plasma of the same blood, have an equal antitoxic power; (3) the clot from the plasma, therefore, does not retain the active principle; (4) the plasma and the serum have an equal antitoxic power; (5) the red corpuscles, compared with the plasma, contain traces only of antitoxin; (6) serum containing the juice of the leucocytes is less rich in antitoxin than the plasma; (7) the extract of the leucocytes contains relatively little antitoxin, and the leucocytes themselves traces or none at all. Hence the white blood-corpuscles cannot be the place where the antitoxin is formed. The serous liquids contained in organs, such as the Graafian follicles, etc., contain as much antitoxin as the blood-serum—none of the organs contain as much of the antitoxin as the blood itself.

Dzerjgowsky is of the opinion, held probably by a

[1] *Ann. de l'Inst. Pasteur*, x., 12.
[2] *Zeitschrift für Heilkunde*, 1895, xvi., 429–482.
[3] *Jour. of Experimental Medicine*, vol. i., No. 3, July, 1896.
[4] *Archives des Sci. Biolog. de l'Institut. Imper. de Méd. Expér. à St. Peters-burg*, tome V., Nos. 2 and 3, 1897.

minority of scientists, that the antitoxin is the toxin in a modified (oxidized?) form, and supports his view by the fact that the antitoxins are specific for their respective toxins only, and by quoting the experiments of Kondrevitsky, who, killing animals two hours after an injection of toxin, found in the blood toxin alone; killing later, found some antitoxin, and still later much antitoxin.

The difference between this theory of neutralization by antitoxins and Chaveau's retention-hypothesis is quite marked. The retention-theory teaches that a bacterium leaves behind it a substance prejudicial to its future growth in the economy—a distinct metabolic product. The antitoxic theory shows the protective substance to be a product not of bacterial growth, but of tissue-energy, not depending upon the presence of the bacteria, but upon the presence of a poison.

The antitoxins do not usually act harmfully upon the bacteria, or preclude their growth in the animal body, but prevent their pathogenesis by annulling their toxicity— *i. e.*, enabling the body-cells to endure the injury—and placing them in a position exactly parallel with non-pathogenic bacteria.

Closely related to the antitoxins, if not-identical with them, are certain substances of an *anti-infectious* nature that can be generated in the blood of animals to which, in the process of immunization, the bacteria, instead of their poisons, have been administered. The anti-infectious serums are protective against the bacterial infections, but powerless against the toxins. They are the only results of immunization against cholera and typhoid fever. When antitoxic serums can be secured they are of far greater importance, and should always be selected for purposes of therapeutics.

The diseases which are at present controllable by anti-toxins are *toxic* diseases, caused by the entrance of toxin-producing bacteria into the body. The growth of these toxin-producers probably depends upon the inability of

6

the body-cells or bactericidal body-juices to properly cope with them, so that they develop and engender the poisonous substances which are the essential factors of disease-production. The more the body and its component elements are injured, the more successful the inroads of the bacteria, the more prolific the toxin-production, and the more severe the affection.

The presence of the antitoxin annuls the poison, maintains the vitality of the organism as a whole, sustains the integrity of its tissues, and so places the pathogenic bacterium on a very different footing in relation to the organism.

An antitoxin is a neutralizing or annulling agent, not a regenerating one, and therefore in therapeutics finds its proper sphere only in the beginning of the disease combated. Up to a certain point the symptoms of diphtheria and tetanus are due to the circulation of toxins in the blood, and can be successfully combated by antitoxic neutralization. Later in both diseases we have symptoms resulting from disorganization of the nervous system, degeneration of the heart-muscle, destruction of the kidneys, etc., and the neutralization of the poison can be of no avail because the lesions are irreparable, and the patient must succumb.

I have used the term "neutralization," in speaking of the antitoxins, in a rather free and scarcely warranted manner, and must call attention to the fact that their operation is probably not exactly analogous to chemical neutralization. From mixtures of toxin and antitoxin the unchanged poison has been recovered. The effect of an antitoxin may be a biologic one, by which the tissues are so stimulated as to endue the action of a substance ordinarily disorganizing in effect.

Buchner and Roux have both pointed out that when the toxins and antitoxins are mixed and introduced into animals of greater susceptibility than are ordinarily used, the presence of an unaltered toxin can easily be demonstrated. This proof is, however, of very little value, for let the amount of toxin endurance of a resistent animal be represented by x, and any addition to this as y. Then xy would certainly be fatal. If the least quantity of antitoxin that will protect the animal be expressed by z, then $xy + z$ is harmless. It is evident, however, that z does not necessarily have any influence upon x, but only need neutralize y in order to save the animal, and therefore it is obvious that the remaining x in such a mixture could readily destroy another more susceptible animal into which it might be injected.

I am of the opinion that the effect of the antitoxin really partakes of the nature of chemic neutralization from the following experiment: let x represent the least certainly fatal dose of diphtheria toxin for a guinea-pig, and y the least quantity of antitoxin that will protect against it; then

$x + y$ is harmless. That

$10\,x + 10\,y$ is also harmless is known to every one accustomed to test antitoxins. I have continued this and have found that

$50\,x + 50\,y$

$100\,x + 100\,y$ are also harmless.

According to Buchner, the antitoxins differ from the alexins in being new substances in the blood, in being without germicidal or chemical neutralizing power against the toxins, and in being stable compounds which can resist heat to 75° C., can resist a reasonable amount of exposure to light, and which are not altered by decomposition of the substances containing them.

The antitoxins are specific for one poison only. Ehrlich found that antiricin was powerless against abrin, and *vice versâ*. Diphtheria antitoxin is of no avail against tetanus, and *vice versâ*.

The immunity which the antitoxins produce is fugacious, varying considerably according to the particular substance employed. As a rule, it is limited to a few months—at least in the case of such antitoxins as we can produce experimentally.

A new principle discovered by Pfeiffer, and bearing directly upon the theories of immunity, is that the serum of animals immunized to certain diseases (cholera and typhoid) contains a germicidal substance. Metchnikoff has tried to show that the action of this body depends upon solution of the leucocytes, but Pfeiffer has disproved this by showing that the liquor puris from abscesses occurring in the experiment-animals did not contain the active substance.[1]

The work of van de Velde[2] is very interesting. An animal immunized by progressively increasing doses of strong filtered toxin produced a serum possessed of powerful anti-infectious and antitoxic powers; one immunized by the introduction into its body of the washed, precipitated bodies of diphtheria bacilli collected by filtration furnished a serum of appreciable anti-infectious, but no antitoxic properties; one immunized by the use of bacillus cultures developed antitoxic and anti-infectious serum identical with the first described; one immunized to weak toxin furnished serum of considerable anti-infectious, but slight antitoxic power, and still another animal that received toxin that had been heated developed neither anti-infectious nor antitoxic serum.

Seeing that the serums commercially manufactured are made by the use of strong filtered toxin, van de Velde examined a number of samples purchased in the market, and found that they were all possessed of both antitoxic and anti-infectious properties. It is important to remember the presence of both of these properties in the serum, as the successful use of the agent for immunizing depends upon the presence of the one, and the use in treatment upon the presence of the other.

Immunity and antitoxins stand in unknown relationship to one another. That an animal has considerable antitoxin in its blood is no guarantee that it is immune. I have seen a horse in each c.cm. of whose blood there

---

[1] *Centralbl. f. Bakt. u. Parasitenk.*, Bd. xix., Nos. 14 and 15.
[2] *Ibid.*, Nov. 24, 1897, Bd. xxii., Nos. 18 and 19.

were 300 immunizing units of diphtheria antitoxin, die of typical symptoms of diphtheria-poisoning after the administration of a comparatively small dose of the toxin.

From all that has gone before it must be clear to the reader that no single theory thus far advanced can explain immunity. Acquired immunity may depend in the great majority of cases upon antitoxins, but as yet we have no satisfactory explanation of natural immunity. The humoral theory may be applicable in some cases; in others one cannot deny the importance of the rôle played by the phagocytes.

CHAPTER IV.

METHODS OF OBSERVING BACTERIA.

WHOEVER would study bacteria must be equipped with a good microscope. The instruments generally provided for the use of medical students in college laboratories, as well as those seldom-employed "show microscopes" seen in physicians' offices, are ill adapted for the purpose. The essential features of a bacteriological instrument are lenses giving a *clear* magnification extending as high as one thousand diameters, and a good condenser for intensifying the lights thrown upon the objects. It naturally follows that the best work requires the best lenses. The cheapest good microscope which is at present offered to the public is the BB. Continental stand, made by Bausch and Lomb. This stand is provided with everything necessary, is fitted with very creditable objectives, including an excellent $\frac{1}{12}''$ oil-immersion lens, and seems capable of doing very good work. I do not recommend this as the best instrument obtainable, but as one that is both good and cheap. For those who desire the very best the rather costly outfits made by Carl Zeiss of Jena are unexcelled.

For those who may begin the use of the Abbe condenser and oil-immersion lenses without the advantage of personal instruction a few hints will not be out of place :

Always employ good slides without bubbles, and thin cover-glasses; No. 1 are best.

Place a drop of oil of cedar upon the cover-glass of the specimen to be examined ; rack the body of the instrument down until the oil-immersion lens touches the oil ;

keep on until it *almost* touches the glass, then look into the microscope and find the object by slowly and firmly racking *up*. As soon as the object comes into view leave the rack and pinion and focus with the fine adjustment.

Always select the light from a white cloud if possible ; if there are no white clouds, choose the clearest whitest light possible. *Never under any circumstances employ sunlight*, which is ruinous to the eyes and useful only for photomicrography.

In using low-power lenses the Abbe condenser must be moved away from the object and the light modified by the iris-diaphragm. The distance between condenser and object should correspond more or less closely with the distance between objective and object.

In using high powers the Abbe condenser must be brought near the object and the light modified by the iris-diaphragm.

If the oil-immersion lens is used, it is perhaps best to employ the plane side of the mirror. When with this lens a section of tissue is examined for details, the light must be modified by the iris-diaphragm, opening and closing it alternately until the best effect of illumination is achieved. If tissue be searched for stained bacteria, and no cellular detail is required, the diaphragm should be wide open to admit a great flood of light and extinguish everything except the deeply-colored bacteria.

When unstained bacteria are to be examined with the oil-immersion lens, the diaphragm should be closed so as to leave only a small opening through which the light can pass.

Bacteria may be examined either stained or unstained. The former condition would always be preferable if the process of coloring the organisms did not injure them. Unfortunately, it is generally the case that the drying, heating, boiling, macerating, and acidulating to which we expose the organisms in the process of staining alter

their shape, make their outlines less distinct, break up
their arrangement, and disturb them in a variety of other
ways. Because of the possible errors of appearance resulting from these causes, as well as because it must be
determined whether or not the individual is motile, in
making a careful study of a bacterium it must always be
examined in the living, unstained condition.

The simplest method of making such an examination
would be to take a drop of the liquid, place it upon a
slide, put on a cover, and examine.

While this method is simple, it cannot be recommended,
for if the specimen should need to be kept for a time
much evaporation takes place at the edges of the cover-
glass, and in the course of an hour or two has changed it
too much for further use. The immediate occurrence of
evaporation at the edges also causes currents of liquid to
flow to and fro beneath the cover, carrying the bacteria
with them and making it almost impossible to determine
whether the organisms under examination are motile or
not.

The best way to examine living micro-organisms is in
what is called the *hanging drop* (Fig. 6). A hollow-

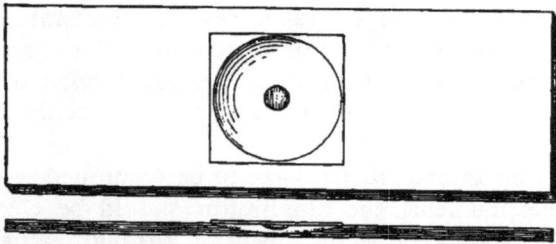

FIG. 6.—The "hanging drop" seen from above and in profile.

ground slide is used, and with the aid of a small camel's-
hair pencil a ring of vaselin is drawn on the slide about,
not in, the concavity at its centre. A drop of the material to be examined is placed in the centre of a large
clean cover-glass, and then placed upon the slide so

that the drop hangs in, but does not touch, the concavity.
The micro-organisms are now hermetically sealed in an
air-chamber, and appear under almost the same conditions as in the cul-
ture. Such a speci-
men may be kept
from day to day and
examined, the bac-
teria continuing to
live until the oxygen
or nutriment is ex-
hausted. By means
of a special appara-
tus (Fig. 7), in which
the microscope is
stood, the growing
bacteria may be
watched at any tem-
perature, and very
exact observations
made.

The hanging drop
should always be ex-
amined at the edge,
as the centre is too
thick.

In such a specimen
it is possible to de-
termine the shape,
size, grouping, divis-

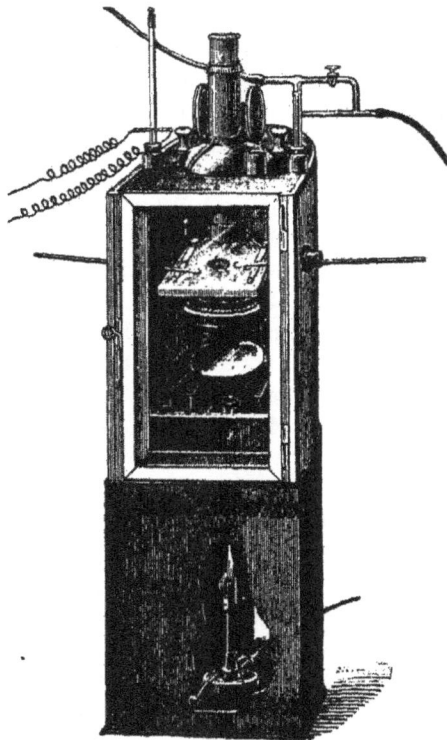

Fig. 7.—Apparatus for keeping objects under
microscopic examination at constant temperatures.

ion, sporulation, and motility of the organism under
observation.

Care should be exercised to use a rather small drop,
especially for the detection of motility, as a large one
vibrates very readily and masks the motility of the
sluggish forms.

When the bacteria to be observed are in solid or semi-
solid culture, a small quantity of the culture should be

mixed up in a drop of sterile bouillon or water and examined.

In the early days of study efforts were made to facilitate the observation of bacteria by the use of carmin and hematoxylon. Both of these reagents tinge the protoplasm of the organisms a little, but so unsatisfactorily that since Weigert introduced the anilin dyes for the purpose both of these tissue-stains have been rejected. The affinity between the bacteria and the anilin dyes is peculiar, and many times is so certain a reaction as to become an essential factor in the differentiation of species.

For the study of bacteria in the stained condition we now employ the anilin dyes only. These wonderful colors, as numerous as the rainbow hues, are coal-tar products. Hüppe classifies them as follows :

A. Dyes prepared from anilin oil.
1. Oxidation-products of pure anilin :
    Methylene blue,
    Chlorhydrin blue (basic indulin).
2. Oxidation-products of pure toluol :
    Safranin.
3. Oxidation-products of mixed anilin and toluol :
    (*a*) Rosanilin. When pure this is triamido-diphenyl-toluyl-karbinol.
    Fuchsin—rosanilin hydrochlorate. It is often mixed with the acetate and the pararosanilin acetate and hydrochlorate. The pure rosanilin hydrochlorate should always be chosen for purposes of staining.
    Azalein is rosanilin nitrite.
    Methylized and ethylized rosanilin :
        Iodin violet,
        Dahlia,
        Iodin green.
    (*b*) Pararosanilin. The colorless pure pararosanilin is triamido-triphenyl-karbinol.

Rubin-pararosanilin hydrochlorate.
Methylized, ethylized, and benzylized
    pararosanilid :
    Crystal violet,
    Gentian violet,
    Victoria blue,
    Methyl green,
    Auramin.
The rosanilins are more difficult to prepare
    than the pararosanilins, and are generally
    mixed with them. The pararosanilins
    color more sharply than the rosanilins.
4. Amido-azo combinations :
    Bismarck brown,
    Phenylene brown,
    Vesuvin.
5. Chinolin derivatives :
    Cyanin.
B. Naphthalin group.—Magdala red.

The best anilin dyes made at the present time, and
those which have become the standard for all bacteriological work, are made in Germany by Dr. Grübler. In
ordering the stain the name of this manufacturer should
always be specified.

A whole volume could easily be devoted to scientific
staining. Indeed, the technical difficulties encountered
are so great that no explanations can be too thorough to
be useful. The special methods essential for such bacteria as have peculiar staining reactions will be given
with the description of the organism. General methods
only will be discussed in this chapter.

Cover-glass Preparations for General Examination.
—The material to be examined must be spread in the
thinnest possible layer upon the surface of a perfectly
clean cover-glass, and dried. Here it may be remarked
that for bacteriological purposes thin covers (No. 1) are
generally required, because thick glasses interfere with
the focussing of the oil-immersion lenses, and that cover-

glasses can never be too clean. It is best to immerse them first in a strong mineral acid, then to wash them in water, then in alcohol, then in ether, and keep them in ether until they are to be used. Except that it sometimes cracks, bends, or fuses the edges of the glasses, a better and more convenient method of cleaning them is to wipe them as clean as possible, seize them in fine-pointed forceps, pass them repeatedly through a small Bunsen flame until it becomes greenish yellow, then slowly elevate the glasses above the flame, so as to allow them to anneal. This maneuvre removes the organic matter by combustion. It is not expedient to use covers twice for bacteriological work, though if well cleaned they may subsequently be employed for ordinary microscopic objects.

To return : After the material spread upon the cover has dried, it must be fixed to the glass by immersion for twenty-four hours in equal parts of absolute alcohol and ether, or, as is much easier and more rapid, be passed *three times through a flame.* Three is not a magic number, but experience has shown that when drawn through the flame three times the desired effect seems best accomplished. The Germans recommend that a Bunsen burner or a large alcohol lamp be used, that the arm holding the forceps containing the cover-glass inscribe a circle a foot in diameter, and that, as each revolution occupies a second of time, the glass be made to pass through the flame from apex to base three times. This is supposed to be exactly the requisite amount of heating. The rule is a good one for the inexperienced.

After fixing, the material is ready for the stain. Every laboratory should be provided with several *stock-solutions* of the more ordinary dyes. These stock-solutions are *saturated alcoholic* solutions made by adding 25 grams of the dye to 100 c.cm. of alcohol. Of these it is well to have fuchsin, gentian violet, and methylene blue always made up. The stock-solutions will not stain, but are the standards for the manufacture of the working stains.

For ordinary staining an aqueous solution made in a simple manner is employed. A small bottle is nearly filled with distilled water, and the stock-solution is added, drop by drop, until the color becomes just sufficiently intense to prevent the ready recognition of objects through it. Such a watery solution possesses the power of readily penetrating the dried protoplasm of the bacterium, taking the stain with it. Alcohol does not have this power.

As in the process of staining the cover is apt to slip from the fingers and spill the stain, it is well to be provided with cover-glass forceps (Fig. 8), which hold the

FIG. 8.—Stewart's cover-glass forceps.

glass in a firm grip and allow of all manipulations without danger to the fingers or clothes. The ordinary instruments are entirely unfitted for the purpose, as capillary attraction draws the stain between the blades and makes certain the soiling of the fingers. Sufficient stain is allowed to run from a pipette upon the smeared side of the cover-glass to flood it, but not overflow, and is · allowed to remain for a moment or two, after which it is thoroughly washed off with water. If the specimen is one prepared for temporary use, it can be examined at once, mounted in a drop of water, but under these conditions will not appear as advantageously as if dried and then mounted in Canada balsam.

Sometimes the material to be examined is too solid to spread upon the glass conveniently. Under such circumstances a drop of distilled water can be added and a minute portion of the material be mixed in it upon the glass.

The entire process is, in brief :

1. Spread the material upon the cover ; 2. Dry—do not heat ; 3. Pass three times through the flame ; 4. Stain

two to three minutes; 5. Wash thoroughly in water; 6. Dry; 7. Mount in Canada balsam.

This simple process suffices to stain most bacteria.

Ohlmacher[1] deserves credit for his observation that when the "fixed" preparation is immersed for a moment or two in a 2–4 per cent. solution of formalin, the brilliancy of the stain is considerably increased.

Staining Bacteria in Sections of Tissue.—It not infrequently happens that the bacteria to be examined are scattered among or enclosed in the cells of tissues. Their demonstration is then a matter of some difficulty, and the method employed is one which must be modified according to the kind of organism to be stained. Very much, too, depends upon the preservation of the tissue to be studied. As bacteria disintegrate rapidly in dead tissue, the specimen for examination should be secured as fresh as possible, cut into small fragments, and immersed in absolute alcohol from six to twenty-four hours to kill the cells and bacteria. Afterward they are removed from the absolute alcohol and kept in 80–90 per cent., which does not shrink the tissue. Bichlorid of mercury may also be used, but absolute alcohol seems to answer every purpose.

The ordinary *methods of imbedding* suffice. The simpler of these are probably as follows:

I. *Celloidin.*—From the hardening reagent (if other than absolute alcohol)—

12–24 hours in 95 per cent. alcohol,
6–12 "   " absolute alcohol,
12–24 "   " thin celloidin (consistence of oil),
6–12 "   " thick celloidin (consistence of molasses).

The solutions of celloidin are made in equal parts of absolute alcohol and ether.

Place upon a block of dry wood, allow to evaporate until the block can be overturned without dislodging the specimen; then place in 70–80 per cent. alcohol until

[1] *Medical News*, Feb. 16, 1896.

ready to cut.   The knife must be kept flooded with alcohol while cutting.

II. *Paraffin—*

12–24 hours in 95 per cent. alcohol,
6–12   "   " absolute alcohol,
4   "   " chloroform, benzole, or xylol,
4–8   "   " a saturated solution of paraffin in one of the above reagents.

Place in melted paraffin in an oven or paraffin waterbath, at 40°–45° C., until the volatile reagent is all evaporated, and the tissue impregnated with paraffin. Imbed in freshly melted paraffin in any convenient mould. In cutting, the knife must be perfectly dry.

When it is necessary, subsequently, to remove the imbedding material, dissolve the paraffin in chloroform, benzole, xylol, oil of turpentine, etc., which in turn can be removed with 95 per cent. alcohol.

Celloidin is soluble in absolute alcohol, ether, and oil of cloves. It is very convenient to fasten the cut sections upon the slide—paraffin sections by oil of cloves and collodion or gum arabic solution, celloidin sections by firmly pressing filter paper upon them and rubbing hard, then allowing a little vapor of ether to run upon them.

III. *Glycerin-Gelatin.*—As the penetration of the tissue by celloidin is attended with lessened staining-qualities of the tubercle bacillus, it has been recommended by Kolle[1] that the tissue be saturated with a mixture of glycerin, 1 part; gelatin, 2 parts; and water, 3 parts; cemented to a cork or block of wood, hardened in absolute alcohol and cut as usual for celloidin with a knife wet with alcohol.

For staining bacteria (other than the tubercle and lepra bacilli) in tissue, two universal methods can be recommended:

Loffler's Method.—The cut sections of tissue are stained for a few minutes in Loffler's alkaline methylene-blue solution (*q. v.*), and then differentiated in a 1 per

[1] Flügge's *Mikroörganismen.*

cent. solution of hydrochloric acid for a few seconds. The section is subsequently dehydrated in alcohol, cleared up in xylol, and mounted in balsam. ⌐—

Pfeiffer's Method.—The sections are stained for one-half hour in diluted Ziehl's carbol-fuchsin (*q. v.*), then transferred to absolute alcohol made feebly acid with acetic acid. The sections must be carefully watched, and as soon as the original, almost black-red color gives place to a red violet color the section is removed to xylol, where it is cleared preparatory to mounting in balsam.

For ordinary work the following simple method is recommended: After the sections are cut the paraffin must be, and the celloidin had better be, removed. From water the sections are placed in the same watery stain used for cover-glasses and allowed to remain five to eight minutes. They are next washed in water for several minutes, then decolorized in 0.5–1 per cent. acetic-acid solution. The acid removes the stain from the tissues, and ultimately from the bacteria as well, so that one must watch carefully, and as soon as the color almost disappears from the sections remove them to absolute alcohol. At this point the process may be interrupted to allow the tissue-elements to be counter-stained with alum carmin or any stain not requiring acid for differentiation, after which the sections are dehydrated in absolute alcohol, cleared in xylol, and mounted in Canada balsam.

As will be mentioned hereafter, certain of the bacteria which occur in tissue do not allow of the ready penetration of the color. For such forms a more intense stain must be employed. One of the best of these stains, which can be employed by the given method both for cover-glasses and tissues, is Löffler's alkaline methylene blue :

Saturated alcoholic solution of methylene blue, 30 ;
1 : 10,000 aqueous solution of caustic potash,   100.

Some bacteria, as the typhoid-fever bacillus, decolorize so rapidly as to contraindicate the use of acid for the differentiation, washing in water or alcohol being sufficient.

Gram's Method of Staining Bacteria in Tissue.— Gram was the fortunate discoverer of a method of staining bacteria in such a manner as to saturate them with an insoluble color. It will be seen at a glance what a marked improvement this is on the method given above, for now the stained tissue can be washed thoroughly in either water or alcohol until its cells are colorless, without fear that the bacteria will be decolorized. Its prosecution is as follows : The section is stained from five to ten minutes in a solution of a basic anilin dye—pure anilin (anilin oil) and water. This solution, first devised by Ehrlich, is known as Ehrlich's solution. The ordinary method of preparing it is to mix the following :

Pure anilin,   4 ;
Saturated alcoholic solution of gentian violet,   11 ;
Water,   100.

Instead of gentian violet, methyl violet, fuchsin, or any basic anilin color may be used. The mixture does not keep well—in fact, seldom longer than six to eight weeks, sometimes not more than two or three ; therefore it is best to prepare it in very small quantity by pouring about 1 c.cm. of pure anilin into a test-tube, filling the tube about one-half with distilled water, shaking the mixture well, then filtering as much as is desired into a small dish. To this the saturated alcoholic solution of the basic dye is added until the surface becomes distinctly metallic in appearance.

Friedländer recommends that the section remain from fifteen to thirty minutes in warm stain, and in many cases the prolonged process gives better results.

From the stain the section is given a rather hasty washing in water, and then immersed from two to three minutes in Gram's solution (a dilute Lugol's solution) :

7

Iodin crystals,	1 ;
Potassium iodid,	2 ;
Water,	300.

While the specimen is in the Gram's solution it appears to turn a dark blackish-brown color. When removed from the solution it is carefully washed in 95 per cent. alcohol until no more color is given off and the tissue assumes a grayish color. If it is simply desired to find the bacteria, the section is dehydrated in absolute alcohol for a moment, cleared up in xylol, and mounted in Canada balsam. If it is necessary to study the relation between the bacteria and the tissue-elements, a nuclear stain, such as alum carmin or Bismarck brown, may be subsequently used. Should a nuclear stain requiring acid for its differentiation be desirable, the process of staining must precede the Gram method altogether, so that the acid shall not act upon the stained bacteria.

The success of Gram's method rests upon the fact that *the combination of mycoprotein, basic anilin, and the iodids forms a compound insoluble in alcohol.*

The process described may be summed up as follows :

Stain in Ehrlich's anilin-water gentian violet five to thirty minutes ;
Wash momentarily in water ;
Immerse two to three minutes in Gram's solution ;
Wash in 95 per cent. alcohol until no more color comes out ;
Dehydrate in absolute alcohol ;
Clear up in xylol ;
Mount in Canada balsam.

This method stains a large variety of bacteria very beautifully, but, unfortunately, does not stain them all, and as some of those which do not stain are important, it seems well to mention the—

Spirillum of cholera and of chicken-cholera;
Bacillus mallei (of glanders);
Bacillus of malignant edema;
Bacillus pneumoniæ of Friedländer;
Micrococcus gonorrhœæ of Neisser;
Spirochæte Obermeieri of relapsing fever;
Bacillus of typhoid fever;
Bacillus of rabbit-septicemia.

. Gram's method is a method of staining bacteria in tissues, but the fact that the method colors some but not all bacteria is one of considerable importance from a differential point of view; and as the difficulty of separating the species of bacteria is so great that every such point must be eagerly seized for assistance, this method becomes one much employed for cover-glass preparations, where it is more easily performed than for sections.

Gram's Method for Cover-glass Preparations.—A thin layer of the bacteria to be examined is spread upon the cover-glass, dried, and fixed. The cover, held in the grip of a cover-glass forceps, is flooded with Ehrlich's solution. By holding the cover flooded with stain over a small flame for a moment or two the solution is kept warm, and the process of staining is continued from two to five minutes. If the heating causes the stain to evaporate, more of it must be dropped upon the glass, so that it does not dry up and incrust.

The stain is poured off, and the cover placed in a small dish of Gram's solution and allowed to remain one-half to two minutes, the solution being agitated. It is possible to apply the Gram solution in the same manner in which the stain is used, but as a relatively larger quantity should be employed, the dish seems preferable.

The cover is next washed in 95 per cent. alcohol until the blue color is wholly or almost lost, after which it can be counter-stained with eosin, Bismarck brown, vesuvin, etc., washed, dried, and mounted in Canada balsam. Given briefly, the method is:

Stain with Ehrlich's solution two to five minutes ;
Gram's solution for one-half to two minutes ;
Wash in 95 per cent. alcohol until decolorized ;
Counter-stain if desired ; wash the counter-stain
  off with water ;
Dry ;
Mount in Canada balsam.

Method of Staining Spores.—It has already been remarked that the peculiar quality of the spore-capsules protects them from the influence of stains and disinfectants to a certain extent. On this account they are much more difficult to color than the adult bacteria. Several methods are recommended, the one generally employed being as follows : Spread the thinnest possible layer of material upon a cover-glass, dry, and fix. Have ready a watch-crystalful of Ehrlich's solution, preferably made of fuchsin, and drop the cover-glass, prepared side down, upon the surface, where it should float. Heat the stain until it begins to steam, and allow the specimen to remain in the hot stain for five to fifteen minutes. The cover is now transferred to a 3 per cent. solution of hydrochloric acid in absolute alcohol for about one minute. Abbott recommends that the cover-glass be submerged, prepared side up, in a dish of this solution and gently agitated for exactly one minute, then removed, washed in water, and counter-stained with an aqueous solution of methyl or methylene blue.

In such a specimen the spores should appear red, the bacilli blue.

I have not generally found that spores color so easily, and for many species the best method seems to be to place the prepared cover-glass in a test-tube half full of carbol-fuchsin :

Fuchsin,   1 ;
Alcohol,   10 ;
5 per cent. aqueous solution of phenol crystals, 100,

and boil it for at least fifteen minutes, after which it is decolorized, either with 3 per cent. hydrochloric or 2–5 per cent. acetic acid, washed in water, and counter-stained blue.

Fiocca suggests the following rapid method: "About 20 c.cm. of a 10 per cent. solution of ammonium are poured into a watch-glass, and 10–20 drops of a saturated solution of gentian violet, fuchsin, methyl blue, or safranin added. The solution is warmed until vapor begins to rise, then is ready for use. A very thinly-spread cover-glass, carefully dried and fixed, is immersed for three to five minutes (sometimes ten to twenty minutes), washed in water, washed momentarily in a 20 per cent. solution of nitric or sulphuric acid, washed again in water, then counter-stained with a watery solution of vesuvin, chrysoidin, methyl blue, malachite green, or safranin, according to the color of the preceding stain. This whole process is said to take only from eight to ten minutes, and to give remarkably clear and beautiful pictures."

Method of Staining Flagella.—This is much more difficult than the staining of either the bacteria or their spores, because each species seems to behave differently in its relation to the stain, so that the chemistry of the micro-organismal products must be taken into consideration.

The best method introduced is that of Löffler. In it three solutions are used :

A. A 20 per cent. solution of tannic acid, 10 ;
   Cold saturated aqueous solution of ferrous sulphate, 5 ;
   Alcoholic solution of fuchsin or methyl violet, 1.

B. A 1 per cent. solution of caustic soda.

C. An aqueous solution of sulphuric acid of such strength that 1 c.cm. will exactly neutralize an equal quantity of Solution B.

Some of the bacteria to be stained are mixed upon a cover-glass with a drop of distilled water. This is the first dilution, but is too rich in bacteria to allow the flagella to show well, so that it is recommended to prepare a second dilution by placing a small drop of distilled water upon a cover and taking a small portion from the first cover to the second, spreading it over the entire surface. The material is allowed to dry, and is then fixed by passing it three times through the flame. When this is done with forceps there is some danger of the preparation becoming too hot, so Löffler recommends that the glass be held in the fingers while the passes through the flame are made.

The cover-glass is now held in forceps, and the mordant, Solution A, is dropped upon it until it is well covered. The cover is warmed until it begins to steam, and the stain replaced as it evaporates. It must not be heated too strongly ; above all things, must not boil. This solution is allowed to act from one-half to one minute, is then washed in distilled water, then in absolute alcohol until all traces of the solution have been removed. The real stain —Löffler recommends an anilin-water fuchsin (Ehrlich's solution)—which should have a neutral reaction, is now dropped on so as to cover the specimen, and heated for a minute until vapor begins to arise; it is then washed off carefully, dried, and mounted in Canada balsam. To obtain this neutral reaction enough of the 1 per cent. sodium-hydrate solution is added to an amount of the anilin-water-fuchsin solution having a thickness of several centimeters to begin to change the transparent into an opaque solution. Such a specimen may or may not show the flagella. If not, before proceeding farther it is necessary to study the products of the bacterium in culture-media. If by its growth the organism elaborates alkalies, Solution C, in proportion from 1 drop to 1 c.cm. in 16 c.cm. of the mordant A, must be added, and the process repeated again and again until the proper amount is determined. On the other hand, if the organism by its growth produces acid, Solution B must be added, drop by drop, until 1 in 16 cm. have been attained, and numerous experiments made to see when the flagella will appear. Löffler has fortunately worked out the amounts required for some of the species, and of the more important ones the following amounts of Solutions B and C must be added to 16 c.cm. of Solution A to attain the desired effect:

Cholera spirillum, ½–1 drop of Solution C;
Typhoid fever, 1 c.cm. of Solution B;
Bacillus subtilis, 28–30 drops of Solution B;
Bacillus of malignant edema, 36–37 drops of Solution B.

Part of the success of the staining depends upon having the bacteria thinly spread upon the glass, and as free from albuminous and gelatinous materials as possible. The cover-glass must be cleaned most painstakingly : too much heating in fixing must be avoided. After using and washing off the mordant, the preparation should be dried before the application of the anilin-water-fuchsin solution.

Pitfield[1] has devised a simple and good method of staining flagella. A single solution at once mordant and stain is employed. It is made in two parts, which are filtered and mixed.

A. Saturated aqueous solution of alum, 10 c.cm.;
   Saturated alcoholic solution of gentian-violet, 1 c.cm.

B. Tannic acid, 1 gr.;
   Distilled water, 10 c.cm.

The solutions should be made with cold water, and immediately after mixing the stain is ready for use. The cover-slip is carefully cleaned, the grease being burned off in a flame. After it has cooled the bacteria are spread upon it, well diluted with water. After drying thoroughly in the air, the stain is gradually poured on and by gentle heating brought almost to a boil; the slip

[1] *Med. News*, Sept. 7, 1895.

covered with the hot stain is laid aside for a minute, then washed in water and mounted. In such preparations I have always been able to see the flagella well, but usually find that while the flagella are very distinct, the bodies of the bacteria are scarcely visible.

Bunge suggests a mordant consisting of a concentrated aqueous tannin solution and a 1 : 20 solution of liq. ferri sesquichloridi in water. The best mixture seems to be 3 parts of the tannin solution to 1 part of the diluted iron solution. To 10 c.cm. of this mixture 1 c.cm. of a concentrated aqueous fuchsin solution is added. It is not necessary to prepare this mordant fresh for each staining, as it seems to improve with age. The use of acid and alkaline solutions added to the mordant is dispensed with.

The bacteria are carefully fixed to the glass, stained with the mordant for five minutes, warming a little toward the end, washed, dried, and subsequently colored with carbol-fuchsin warmed a little.

Bacteria can best be measured by an eye-piece micrometer. As these instruments vary somewhat in construction, the unit of measurement for each objective magnification or the method of manipulating the adjustable instruments must be learned from dealers' catalogues.

Photographing bacteria requires special apparatus and methods, which are fully described in text-books upon the subject.

CHAPTER V.

STERILIZATION AND DISINFECTION.

BEFORE considering the cultivation of bacteria and the preparation of media for that purpose it is necessary to discuss methods of destroying bacteria whose accidental presence might ruin our experiments.

The dust of the atmosphere, as has already been shown, is almost constant in its micro-organismal contamination, so that the spores of moulds and bacilli, together with yeasts and micrococci, constantly settle from it upon our glassware, enter our pots, kettles, funnels, etc., and would ruin every culture-medium with which we operate did we not take measures for their destruction.

Micro-organisms may be killed by heat or by the action of chemicals, the processes being generically termed sterilization. The term sterilization is usually employed to denote the destruction of bacteria by heat, in contradistinction to disinfection, which usually means the destruction of the bacteria by the use of chemical agents. A chemical agent causing the death of bacteria is called a *germicide*. An object which is entirely free from bacteria and their spores is described as *sterile*. Certain substances whose action is detrimental to the vitality of bacteria and prevents their growth amid otherwise suitable surroundings are termed *antiseptics*.

The study of sterilization, disinfection, and antisepsis will naturally lead us through the following subdivisions:

I. The sterilization and protection of instruments and glassware used in experimentation.

II. The sterilization and protection of culture-media.

III. The disinfection of the instruments, ligatures, etc. and the hands of the surgeon, and the use of antiseptics.

IV. The disinfection of sick-chambers and their contents, as well as the dejecta and discharges of patients suffering from contagious and infectious diseases.

The Sterilization and Protection of Instruments and Glassware Used in Experimentation.—Sterilization may be accomplished by either moist or dry heat. For the perfect sterilization of objects capable of withstanding it dry heat is preferable, because more certain in its action. If we knew just what organisms we had to deal with, we might be able in many cases to save time and gas, but while some simple non-spore-producing forms are killed at a temperature of 60° C., others can withstand boiling for an hour ; it is therefore best to employ a temperature high enough to kill all with certainty. Platinum wires used for inoculation are held in the direct flame until they become incandescent. In sterilizing such wires attention must be bestowed upon the glass handle, which should be held in the flame for at least half its length for a few moments when used for the first time each day. Carelessness in this respect may cause the loss of much time by contaminating cultures.

Knives, scissors, and forceps may be exposed for a very brief time to the direct flame, but this affects the temper of the steel when continued too long. They may also be boiled, steamed, or carbolized.

All glassware is sterilized by exposure to a sufficiently high temperature, 150° C. or 302° F., for one hour in the well-known hot-air closet (Fig. 9). A temperature of 150° C. is sufficient to kill all known bacteria and their spores if continued for an hour.

Rubber stoppers, corks, wooden apparatus, and other objects which are warped, cracked, charred, or melted by so high a temperature must be sterilized by moist heat in the steam apparatus for at least an hour before they can be pronounced sterile.

It must always be borne in mind that after sterilization has been accomplished the same sources of contamination that originally existed are still present, and begin to operate as soon as the objects are removed from the sterilizing apparatus.

To Schröder and Van Dusch belong the credit of

having first shown that when the mouths of flasks and tubes are closed with plugs of sterile cotton no germs can filter through. This observation has been of inestimable value to every bacteriologist. Before sterilizing

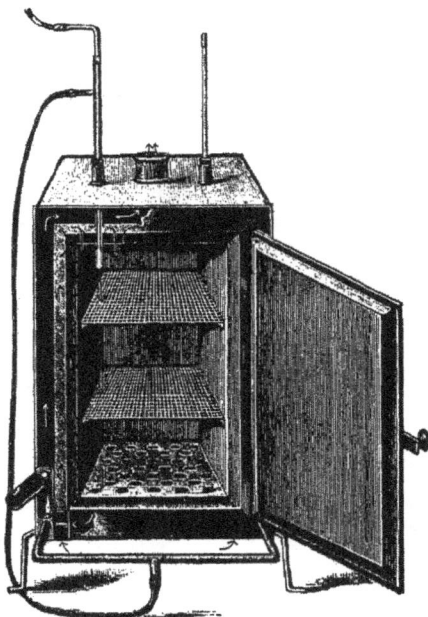

FIG. 9.—Hot-air sterilizer.

flasks and tubes we plug them with ordinary raw cotton, and are sure that afterward their interiors will remain free from the access of germs until opened. Instruments may be sterilized wrapped in cotton, to be opened only when ready for use; or instruments and rubber goods sterilized by steam can subsequently be wrapped in sterile cotton and kept for use. It is of the utmost importance to carefully protect every sterilized object, and to allow as little dust to collect upon it as possible, in order that the object of the sterilization be not defeated. As the spores of moulds falling upon cotton sometimes grow and allow their mycelia to work their way through and drop into a culture-medium, Roux has introduced little paper caps with which the cotton stoppers are protected from the dust. These are easily made by curling a small square of paper into a "cornucopia," fastening by turning up the edge or putting in a pin. The paper is placed over the stopper before the sterilization, after which no contamination of the cotton can occur.

Sterilization of Culture-media.—As almost all of the culture-media contain about 80 per cent. of water, which would be evaporated in the hot-air closet, so that the material would be destroyed, hot-air sterilization is not appropriate for them. Sterilization by streaming steam is the best and surest method. The prepared media are placed in flasks or tubes carefully plugged with cotton and previously sterilized with dry heat, and then sterilized in what is known as Koch's steam apparatus (Fig. 10) or in Arnold's

FIG. 10.—Koch's steam sterilizer.

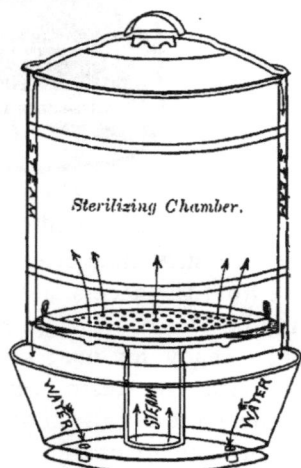

FIG. 11.—Arnold's steam sterilizer.

steam sterilizer (Fig. 11), which is more convenient and more generally useful.

The temperature of boiling water, 100° C., does not

kill many spores, so that the exposure of culture-media to streaming steam is of little use unless applied in a systematic manner—*intermittent sterilization*—based upon a knowledge of sporulation.

In carrying out the intermittent sterilization the culture-medium is exposed for fifteen minutes to the passage of streaming steam in the apparatus or to some temperature judged to be sufficiently high, so that the bacteria contained in it are killed. As the spores remain uninjured, the medium is stood aside in a cool place for twenty-four hours, and the spores allowed to develop into perfect bacteria.

When the twenty-four hours have passed the culture-medium is again placed in the apparatus and exposed to

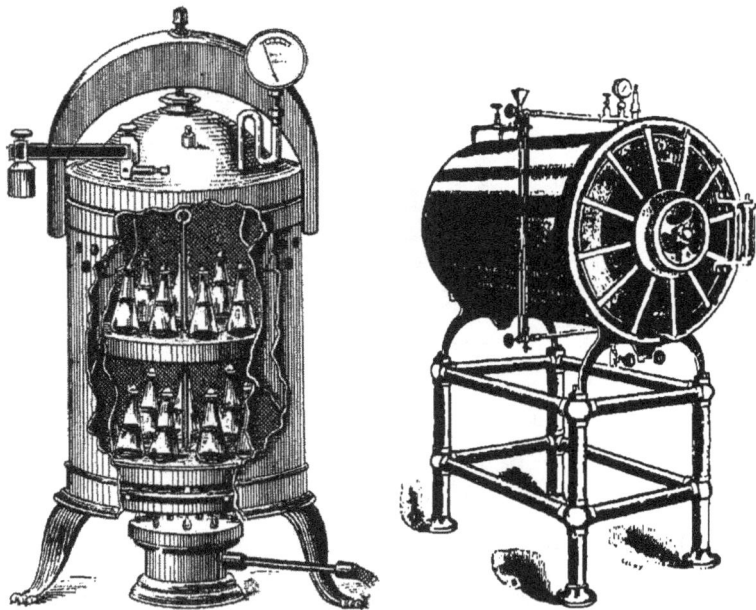

FIG. 12.—Autoclave for rapid sterilization FIG. 13.—Kny-Sprague steam sterilizer. by superheated steam under pressure.

the same temperature, until these newly-developed bacteria are also killed. Eventually, the process is repeated a third time, lest a few spores remain alive and capable of spoiling the material. When properly sterilized in this way, culture-media will remain free from contamination until time and evaporation cause them to dry up.

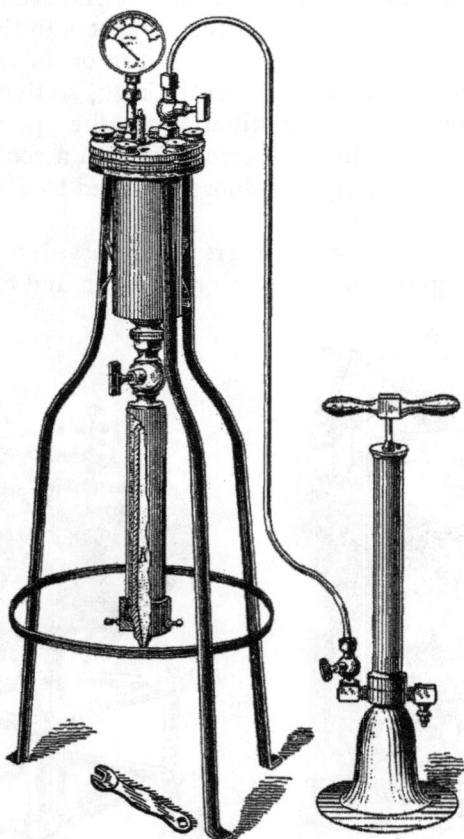

FIG. 14.—Pasteur-Chamberland filter arranged to filter under pressure.

If hermetically sealed, a sterile medium will keep indefinitely.

If it should be necessary to sterilize culture-media at once, not waiting the three days required by the intermittent method, it may be done by superheated steam in an autoclave (Fig. 12). Here under a pressure of two or three atmospheres sufficient heat is generated to destroy the spores. The objections to this method are that it sometimes turns the agar-agar dark, and that it is likely to destroy the gelatinizing power of the gelatin, which after sterilization remains liquid.

Liquids may also be sterilized by filtration—*i. e.* by passing them through unglazed porcelain or some other material whose interstices are sufficiently fine to resist the passage of the bacteria. This method is largely employed

FIG. 15.—Kitasato's filter: *a*, porcelain bougie; *b*, attachment for suction-pump; *c*, reservoir; *d*, sterile receiver.

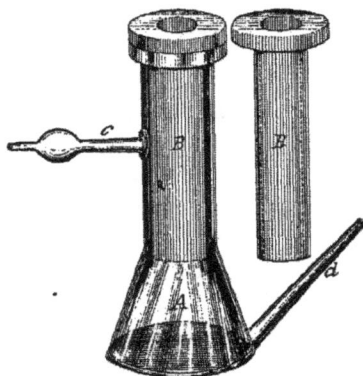

FIG. 16.—Reichel's bacteriologic filter of unglazed porcelain: *A*, sterile receiver; *B*, porcelain filter; *c, d*, attachments for pump.

for the sterilization of the unstable toxins and antitoxins, which are destroyed by heat. Various substances have been used for filtration, as stone, sand, powdered glass, etc., but experimentation has shown porcelain to be the only reliable filter against bacteria. Even this material, whose interstices are so small as to allow the liquid to pass through with great slowness, is only certain in its action for a time, for after it has been used considerably the bacteria seem able to work their way through. To be certain of the efficacy of such a filter the fluid first passed through must be tested by cultivation methods. The complicated Pasteur-Chamberland and the simple Kitasato and Reichel filters are shown in Figures 14, 15, and 16.

After having been used a porcelain filter must be disinfected, scrubbed, *dried thoroughly*, and then heated in a Bunsen burner or blowpipe flame until all the organic matter is consumed. In this firing process the filter first turns black as the organic matter chars, then becomes white as it is consumed. The greatest care must be exercised in cleansing, and especially must care be taken that the porcelain is dry before entering the fire, as it will certainly crack if moist.

Before using a new filter it should be sterilized by dry heat, then connected with receivers and tubes, also carefully sterilized. It should not be forgotten that the filtered material is still a good culture-medium and must be handled with the greatest care.

While the filtration of water, peptone solution, and bouillon is comparatively easy, gelatin and blood-serum pass through with great difficulty, and speedily gum the filter, so that it is useless until fired.

A convenient apparatus used by the author for the rapid filtration of large quantities is shown in the accompanying illustration (Fig. 17).

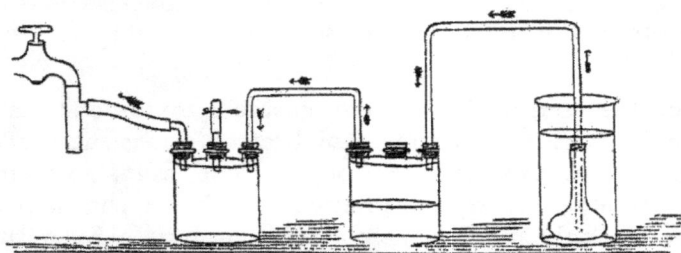

Fig. 17.—Apparatus for the rapid filtration of toxins, etc.

The Disinfection of Instruments, Ligatures, Sutures, the Hands, etc.—There are certain objects used by the surgeon which cannot well be rendered incandescent, exposed to dry heat at 150° C., steamed, or intermittently heated without injury. For these objects disinfection must be practised. Ever since Sir Joseph Lister introduced antisepsis, or disinfection, into surgery there has been a struggle for the supremacy of this or that highly-recommended germicidal substance, with two results— viz. that a great number of feeble germicides have been discovered, and that belief in the efficacy of all germicides has been somewhat shaken; hence the origin of the successful *aseptic* surgery of the present day, which strives to prevent the entrance of germs into, rather than their destruction after admission to, the wound.

For a complete discussion of the subject of antiseptics in relation to surgery the reader must be referred to the large text-books of surgery, where much space is thus occupied. A short list of useful germicides of which the respective values are given, and a brief discussion of some of the more important measures, can alone find space in these pages. The antiseptic value of some of the principal substances used may be expressed as follows, the figures indicating the strength of the solution necessary to prevent the development of bacteria :

Pyoktanin (methyl violet) . 1 : 2,000,000—1 : 5000.
Formalin . . . . 1 : 25,000—1 : 5000.
Bichlorid of mercury . . . 1 : 14,300.
Hydrogen peroxid . . . . . 1 : 20,000.
Formalin . . . . 1 : 20,000.
Nitrate of silver . . . . . . 1 : 12,500.
Creolin . . . . 1 : 5000 to 1 : 200 (does not kill
                                       anthrax).
Chromic acid . . . . 1 : 5000.
Thymol . . . . 1 : 1340.
Salicylic acid . . . . 1 : 1000.
Potassium bichromate . . . 1 : 909.
Trikresol . . . . 1 : 1000—1 : 500.
Zinc chlorid . . . . 1 : 526.
Potassium permanganate . 1 : 285 ; not prompt in action.
Nitrate of lead . . . . . . 1 : 277.
Izal . . . . 1 : 200.
8

Boracic acid . . . . 1 : 143.
Chloral hydrate . . . . . . 1 : 107.
Ferrous sulphate . . . . . 1 : 90—1 : 200, Sternberg.
Calcium chlorid . .   . . . 1 : 25.
Creosote . . . . 1 : 20.
Carbolic acid . . . . .   . 1 : 20 :: 1 : 50.
Alcohol . . . . 1 : 10.
Ether.   Pure ether will not kill anthrax spores immersed
    in it for eight days.

The value of antiseptics, like that of disinfectants, is
always relative, the destructive as well as the inhibitory
power of the solution varying with the micro-organism
upon which it acts.   The following table, from Boer,
will illustrate this :

Methyl Violet (*Pyoktanin*).

	Restrains.	Kills.
Anthrax bacillus	1 : 70,000	1 : 5000
Diphtheria	1 : 10,000	1 : 2000
Glanders	1 : 2500	1 : 150
Typhoid	1 : 2500	1 : 150
Cholera spirillum	1 : 30,000	1 : 1000

Large numbers of both strongly and feebly antiseptic
substances have purposely been omitted from the above
lists, compiled from Sternberg and Micquel, as either inappropriate for ordinary use or as having been replaced
by better agents.

The newest, and one of the best germicides for all purposes is formaldehyde.   Its use as a vapor for the sterilization of infected rooms was first suggested by Trillat in
1895, but it did not make much stir in the medical world
until a year or more had passed and a 40 per cent. solution of the gas, under the name of "Formalin," had
been placed upon the market.   The original method consisted of the evolution of the gas from methyl alcohol by
volatilizing it in a steam apparatus, and passing the vapor
over a heated metal plate.   At present the original autoclave has been replaced by the apparatus shown in Fig.
19, in which a solution of formochloral is volatilized by
heating under a pressure of three atmospheres.

The gas is very penetrating, easily diffusing itself, and is said to have enormous bactericidal powers. In experiments conducted by Prof. Robinson, of Bowdoin College, the gas penetrated mattresses and killed cultures in tubes wrapped up in them.

There seems to be little doubt of the ability of the

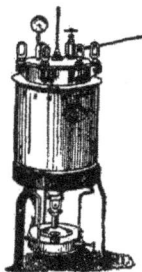

FIG. 18.—The Trillat autoclave.

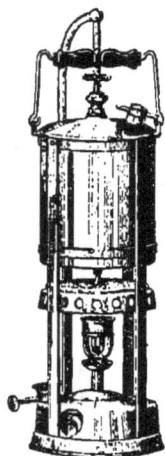

FIG. 19.—Sanitary formaldehyde regenerator.

formaldehyde gas to disinfect, but there are few apparatus upon the market at present that seem capable of discharging a sufficient volume of the gas with sufficient rapidity to do the work. The physician, therefore, who desires to disinfect with confidence should choose an apparatus that has been shown by competent experiments to fill the requirements.

The "formalin," or 40 per cent. solution of the gas, when fresh and tightly corked, is fatal to most bacteria in dilutions of from 1: 5000 to 1: 25,000. It can be employed with great advantage to spray the walls and floors of rooms. It cannot be employed upon the skin or mucous membranes, because of its marked irritating effect.

The disinfection of the skin, both the hands of the surgeon and the part about to be incised, is a matter of importance. It is almost impossible to secure absolute sterility of the hands, so deeply do the skin-cocci penetrate between the layers of the scarf-skin. The method at present generally employed, and recommended by Welch and Hunter Robb, is as follows: The nails must be trimmed short and perfectly cleansed. The hands are washed thoroughly for ten minutes in water of as high a temperature as can comfortably be borne, soap and a brush previously sterilized being freely used, and afterward the excess of soap washed off in clean hot water. The hands are then immersed for from one to two minutes in a warm saturated solution of permanganate of potassium, then in a warm saturated solution of oxalic acid, until complete decolorization of the permanganate occurs, after which they are washed free from the acid in clean warm water or salt-solution. Finally, they are soaked for two minutes in a 1 : 500 solution of bichlorid of mercury, after which they are ready for use.

Lockwood,[1] of St. Bartholomew's Hospital, recommends after the use of the scissors and penknife, scrubbing the hands and arms for three minutes in hot water and soap to remove all grease and dirt. The scrubbing brush ought to be steamed or boiled before use, and kept in 1 : 1000 biniodid of mercury solution. When the soap-suds have been thoroughly washed away with plenty of clean water, the hands and arms are thoroughly washed and soaked for not less than two minutes in a solution of biniodid of mercury in methylated spirit; 1 part of the biniodid in 500 of the spirit. Hands that cannot bear 1 : 1000 bichlorid and 5 per cent. carbolic solutions, bear frequent treatment with the biniodid. After the spirit and biniodid have been used for not less than two minutes, the solution is washed off in 1 : 2000 or 1 : 4000 biniodid of mercury solution.

Catgut cannot be sterilized by boiling without deterioration. The present method of preparing it is to dry it in a hot-air chamber and then boil it in cumol, which is afterward evaporated and the skeins preserved in sterile test-tubes or special receptacles plugged with sterile cotton. Cumol was first introduced for this purpose by

[1] *Brit. Med. Jour.*, July 11, 1896.

Krönig, as its boiling-point is 168°–178° C., and thus sufficiently high to kill spores. The use of cumol for the sterilization of catgut has been carefully investigated by Clarke and Miller.[1]

Ligatures of silk and silkworm-gut are boiled in water immediately before using, or are steamed with the dressings, or placed in test-tubes plugged with cotton and steamed in the steam sterilizer.

At present, in most hospitals, instruments are boiled before using in a 1–2 per cent. soda solution. Plain water has the disadvantage of rusting the instruments, and during the operation they are either kept in the boiled water or in carbolic solution. Andrews makes special mention of the fact that the instruments must be completely immersed to prevent rusting.

During the operation the wound is frequently washed with normal salt solution, applied by sterile marine or gauze sponges.

The water and the salt solution used for surgical purposes are to be sterilized before using, either by steaming for a prolonged period, or by the intermittent method. Large hospitals are generally furnished with special apparatus for supplying sterile distilled water in large quantity.

To La Place belongs the credit of observing that the efficacy of bichlorid of mercury is greatly increased by the addition of a small amount of acid, by which the penetration is increased and the formation of insoluble albuminates lessened.

The knowledge that the action of germicides is chemical, and that the destruction of the bacteria is due to the combination of the germicide with the mycoprotein, is apt to lessen our confidence in the permanence of their action. Geppert has shown of bichlorid of mercury that in the reaction between it and anthrax spores the vitality of the latter seems lost, but that the precipitation of the bichlorid from this combination by the action of ammonium sulphid restores the vitality of the spore.

[1] *Bull. of the Johns Hopkins Hospital,* Feb. and March, 1896.

Again, the fact that some of the antiseptics, as nitrate of silver and bichlorid of mercury, are at once precipitated by albumins, and thus lose their germicidal and antiseptic powers, limits the scope of their employment. I think it may. be safely said that carbolic acid is the most reliable and most generally useful of all the germicides and antiseptics.

The Disinfection of Sick-chambers, Dejecta, etc.— What has just been remarked concerning the unreliability of many of the germicidal substances is eminently *a propos* of the disinfection of dejecta. It is useless to mix bichlorid of mercury with typhoid stools or tubercular sputum rich in albumin, and imagine these substances rendered harmless in consequence. It should not be forgotten that the sick patient is less the means of conveying the contagium than the objects with which he is in contact, which can be carried to other rooms or houses during or after the progress of the disease. A careful consideration of the condition of the sick-room will lead us to a clear understanding of its bacteriological condition.

The Air of the Sick-room.—It is impossible to sterilize or disinfect the atmosphere of a room during its occupancy by the patient. The disinfecting capacity of the solutions given above must make obvious the concentration of their useful solutions, and show the foolishness of placing beneath the bed or in the corners of a room small receptacles filled with carbolic acid or chlorinated lime. These can serve no purpose for good, and may be potent for harm by obscuring the disagreeable odors emanating from materials which should be removed from the room by the still more disagreeable odors of the disinfectants. The practice of such a custom is only comparable to the old faith in the virtue of asafetida tied in a corner of the handkerchief as a preventive of cholera and smallpox.

During the period of illness a chamber in which the patient is confined should be freely ventilated, so that its atmosphere is constantly changing and replacing the closeness so universally an accompaniment of fever by fresh, pure air—a comfort to the patient and a protection to the doctors and nurses.

After recovery or death one should rely less upon fumigation than upon the disinfection of the walls and floor, the similar disinfection of the wooden part of the furniture, and the sterilization of all else. The fumes of sulphur may do some good—when combined with steam, much good—but are greatly overestimated, and the sulphurous vapors are rapidly giving way to the more penetrating and germicidal formaldehyde vapor. To apply this, the room to be sterilized is carefully closed, the carefully selected apparatus set in action, and the discharged vapor allowed to act undisturbed for some hours, after which the windows and doors are all thrown open to fresh air and sunlight.

Instead of the gas, a 40 per cent. solution, which can be sprayed upon the ceiling, walls, floor, and contents of the room from a large atomizer, is sometimes used. Experience has not shown, however, that this possesses any distinct advantages.

So far as is at present known, the disinfection by formaldehyde is complete and leaves nothing to be desired. Only one point is to be considered, already often mentioned—that is, the apparatus. Of those experimented with by the author, few have given satisfaction.

The Dejecta.—A little thought will direct attention to those of the dejections which are dangerous to the community and promote efforts for their complete sterilization. In cases of diphtheria the vomit, expectorations, and nasal discharges are most important. They should be received in old rags or in Japanese paper napkins—not handkerchiefs or towels—and should be burned. The sputum of tuberculous patients should either be collected in a glazed earthen vessel which can be subjected to boiling and disinfection, or, as is an excellent plan, should be received in Japanese rice-paper napkins, which can at

once be burned.   These napkins are not quite as good
as the small pasteboard boxes (Fig. 20) recommended by

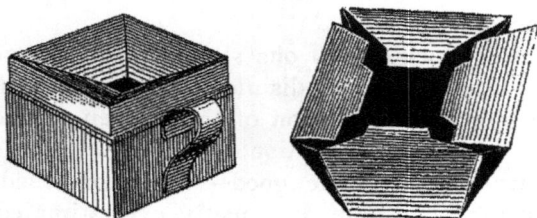

FIG. 20.—Pasteboard cup for receiving infectious sputum.   When used the
pasteboard can be removed from the iron frame and burned.

some city boards of health, because, being highly absorbent, the sputum is apt to soak through and soil the fingers, etc.   Tuberculous patients should be provided with
rice-paper instead of handkerchiefs, and should have their
towels, knives, forks, spoons, plates, etc. kept strictly
apart from the others of the household (though the patients, whose mental acuity makes their sensibilities very
pronounced, need never be told of their isolation), and
frequently boiled for considerable lengths of time.

The excreta from typhoid-fever and cholera cases require particular attention.   These, and indeed all alvine
matter possibly the source of infection or contagion,
should be received in glazed earthen vessels and immediately intimately mixed with a 5 per cent. solution
of chlorinated lime (containing 25 per cent. of chlorin)
if semi-solid, or with the powder if liquid, and allowed
to stand for an hour before being thrown into the
drain.

The Clothing, etc.—All bed-clothing which has been
used in the sick-room, all towels, napkins, handkerchiefs,
night-robes, underclothes, etc. which have been used by
the sick, and all towels, napkins, handkerchiefs, caps,
aprons, and outside dresses worn by the nurse, should be
regarded as infected and subjected to sterilization.   The
only satisfactory method of doing this is by prolonged
subjection to steam in a special apparatus; but, as this

is only possible in hospitals, the next best thing is boiling for some time in the ordinary wash-boiler. When possible, the clothes should be soaked in 1 : 2000 bichlorid solution before or after boiling, and in drying should hang in the sun and wind. Woollen underwear can be treated exactly as if of cotton. The woollen clothing of the patient, if infected, requires special treatment. Fortunately, the infection of the outer woollen garments is unusual. The only reliable method for their purification is prolonged exposure to hot air at 110° C. In private practice it becomes a grave question what shall be done with these articles. Prolonged exposure to fresh air and sunlight will aid in rendering them harmless ; when it is certain that articles of wool are infected, they may be sent to the city hospital or to certain of the moth-destroying and fumigating establishments which can be found in all large cities, and be baked.

The Furniture, etc.—The wholesale destruction of furniture practised in earlier times has at present become unnecessary. The doctor, if he properly performs his functions, will save much trouble and money for his patient by ordering the immediate isolation of his charge in an uncarpeted, scantily- and cheaply-furnished room the moment an infectious disease is *suspected*, before much infection can have occurred. However, if before his removal the patient has occupied another bed, its clothing should be promptly handled in the above-described manner.

After the illness the walls of the rooms, including the ceiling should be sprayed with formalin, or, where it cannot be obtained, may be rubbed with fresh bread, which Löffler has shown to be efficacious, though scarcely practicable, in collecting the bacteria, or, if possible, should be whitewashed. If the walls are hung with paper, they may be dampened with 1 : 1000 bichlorid-of-mercury solution before new paper is hung.

Aronson[1] says: "For the disinfection of living-rooms

[1] *Verein für Offentliche Gesundheitspflege*, Berlin, April 26, 1897.

there is no method that can compare in the remotest degree, as regards certainty and simplicity, with that by means of formaldehyde gas. For example, any one who has seen the process of cleansing walls by rubbing them down with bread, as carried out by the disinfecting corps, will agree with me that, however effective it may be from a theoretical point of view, it is absolutely inefficient in practice. The possibility of disinfecting rooms and all their contents with certainty, by means of a simple, cheap, harmless, and easily managed method must be hailed as a great advance.''

The floor should be scoured with 5 per cent. carbolic-acid solution or 1 : 1000 bichlorid of mercury, and all the wooden articles wiped off two or three times with the same solution employed for the floor. In this scouring no soap can be used, as it destroys the virtue of the germicide. If a straw mattress was used, it should be burned and the cover boiled. If a hair mattress was used, it can be steamed or baked by the manufacturers, who generally have ovens for the purpose. Curtains, shades, etc., should receive proper attention; but, of course, the greater the precautions exercised in the beginning, the fewer the articles which will need attention in the end. They should be removed before the case has developed.

Strehl has succeeded in demonstrating that when 10 per cent. formalin solution is sponged upon artificially infected curtains, etc., the bacteria are killed by the action of the disinfectant. This knowledge will be an important adjunct to our means for disinfecting the furniture of the sick-chamber.

The patient, whether he lives or dies, may also be a means of spreading the disease unless specially cared for. After convalescence the body should be bathed with a weak bichlorid-of-mercury solution or with a 2 per cent. carbolic-acid solution, or with 25–50 per cent. alcohol, before the patient is allowed to mingle with society, and the hair should either be cut off or carefully washed with the above solution. In desquamative diseases it seems best to have the entire body anointed with cosmolin once daily, the unguent being well rubbed in, in order to prevent the particles of epidermis being distributed through the atmosphere. Carbolated cosmolin may be better than the plain, not because of the very slight antiseptic value it possesses, but because it helps to allay the itching which may be part of the desquamative process.

After the patient is about the room again, common sense will prevent the admission of strangers until all the disinfective measures have been adopted, and after this, touching, and especially kissing him, should be omitted for some time.

The dead who die of infectious diseases should be washed in a strong disinfectant solution, and should be given a private funeral in which the body, if exposed, should not be touched. In my judgment, the body is best disposed of by cremation.

It seems, however, to be an error to suppose that a dead body can remain for an indefinite period a source of infection. Esmarch [1] has made a series of laboratory experiments to determine what the fate of pathogenic bacteria in the dead body really is. From his results it seems clear that in septicemia, cholera, anthrax, malignant edema, tuberculosis, tetanus, and typhoid the pathogenic bacteria all die sooner or later, generally more rapidly in conditions of decomposition than in good preservation of the tissues. Lack of oxygen may be a cause of their disappearance.

[1] *Zeitschrift für Hygiene*, 1893.

CHAPTER VI.

CULTIVATION OF BACTERIA; CULTURE-MEDIA.

ACCURACY of observation requires that the bacteria be separated from their natural surroundings and artificially grown upon certain prepared media of standard composition, in such a manner that only organisms of the same kind are together.

One after another various organic and inorganic mixtures have been suggested, but, although almost any compound containing organic matter, even in small amounts, will suffice for the nourishment of bacteria, a certain few have met with particular favor as being most valuable.

Rather than give a complete review of the work which has already been done, in the following pages the most useful preparations only will be considered.

Our knowledge of the biology of the bacteria has shown that they grow best in a mixture containing at least 80 per cent. of water, of a neutral or feebly alkaline reaction, and of a composition which, for the pathogenic forms at least, should approximate the juices of the animal body. It might be added that transparency is a very desirable quality, and that the most generally useful culture-media are those that can be readily liquefied and solidified.

Bouillon is one of the most useful and most simple of the media. Its preparation is as follows: To 500 grams of finely-chopped lean, boneless beef, 1000 c. cm. of clean water are added and allowed to stand for about twelve hours on ice. At the end of this time the liquor is decanted, that remaining on the meat expressed through a cloth, and then, as the entire quantity is seldom regained,

124

enough water added to bring the total amount up to 1000 c.cm. This liquid is called the *meat-infusion*. To it 10 grams of Witte's or Fairchild's dried beef-peptone and 5 grams of sodium chlorid are added, and the whole boiled until the albumins coagulate. The reaction is then carefully tested, in order that whatever sarcolactic acid may have been present in the meat may be neutralized by the addition of a few drops of a saturated aqueous solution of sodium carbonate. The solution is added drop by drop, and the reaction frequently tested with litmus-paper. When a neutral reaction, or, better, a faint alkaline reaction, is attained, the mixture is well stirred, boiled again for about half an hour to precipitate the alkaline albumins formed, and filtered. The use of phenolphthalein to determine the reaction of the culture-media is much more reliable than litmus, and in many laboratories has replaced it. The method of using it suggested by Timpe is to continue the addition of the carbonate of sodium solution until a drop of it produces a red spot upon phenolphthalein-paper. Such a paper can easily be made by using a solution of 5 grams of phenolphthalein to 1 liter of 50 per cent. alcohol. The bibulous paper is cut into strips, moistened with the solution, and then hung up to dry. It keeps quite well. Acids do not change the appearance of the paper, but small traces of alkali turn it red.

If it is necessary to be extremely accurate concerning the acidity or alkalinity of the culture-medium, the method of titration with phenolphthalein can be employed. For this purpose a small quantity of the culture-fluid—say 10 c.cm.—receives an addition of a drop or two of a weak alcoholic solution of phenolphthalein (1 : 300), and then drop by drop from a burette a dilute soda solution is added until a faint rose color occurs, when a simple calculation will show that if so much is required to bring about the required color in the 10 c.cm., so much more will be required for the total amount. The occurrence of the rose color marks the change from a neutral

to a faintly alkaline reaction.   Definite varying degrees
of alkalinity can be secured by adding measured quantities of the soda solution beyond that necessary to bring
about the beginning alkalinity.   In using phenolphthalein one should remember that the first sign of rose color
marks a change to alkalinity, and that this is a higher
degree of alkalinity than that required to turn red litmus
blue.

The bouillon thus prepared is a clear fluid of a straw
color, much resembling normal urine in appearance.   It

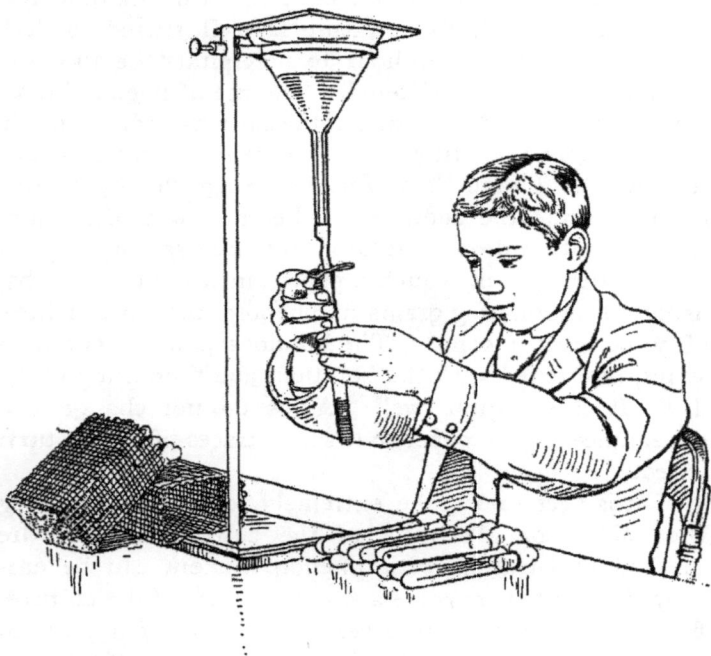

FIG. 21.—Funnel for filling tubes with culture-media (Warren).

is dispensed in previously sterilized tubes with cotton
plugs—about 10 c.cm. to each—and is then sterilized by
steam three successive days for fifteen to twenty minutes
each, according to the directions already given for fractional sterilization.   (See p. 109.)

For the preparation of bouillon, as well as gelatin, agar-agar, and glycerin agar still to be described, beef-extract (Liebig's) may be employed, but for the most delicate work this is rather undesirable, because of its unstable composition and because of the precipitation of meat-salts, which can scarcely be filtered out of the agar-agar, owing to the fact that they only crystallize when the solution cools. When it is desirable to prepare the bouillon from beef-extract, the method is very simple. To 1000 c.cm. of clean water 10 grams of Whitte's dried beef-peptone, 5 grams of sodium chlorid, and about 2 grams of beef-extract are added. The solution is boiled until the constituents are dissolved, neutralized, if necessary, and filtered *when cold*. If it is filtered while hot, there is always a subsequent precipitation of meat-salts, which clouds it.

Bouillon and other liquid culture-media are best dispensed and kept in small receptacles—test-tubes or flasks —in order that a single contaminating organism, should it enter, may not spoil the entire bulk. A very convenient simple apparatus used by bacteriologists for filling tubes with liquid media is shown in Figure 21. It need not be sterilized before using, as the culture-medium will be sterilized by the intermittent method after the tubes are filled. The test-tubes and flasks into which the culture-medium is filled must, however, be previously sterilized by dry heat. The dry-heat sterilization is done, of course, after the cotton plugs are in place.

Bouillon is the basis of most of the culture-media. The addition of 10 per cent. of gelatin makes it "gelatin;" that of 1 per cent. of agar-agar makes it "agar-agar." The preparation of these media, however, varies somewhat from that of the plain bouillon.

Gelatin.—The culture-medium known as gelatin has decided advantages over the bouillon, not only because it is an excellent food for bacteria, and, like the bouillon, transparent, but because it is also *solid*. Nor is this all : it is a transparent solid which can be made liquid or solid at will. It is prepared as follows : To 1000 c.cm. of meat-infusion or to 1000 c.cm. of water containing 2 grams of beef-extract in solution, 10 grams of peptone, 5 grams of salt, and 100 grams of gelatin ("Gold label" is the best commercial article) are added, and boiled for about an hour over a moderately hot flame. Double boilers are very slow, and if proper care is exercised there is little danger of the gelatin burning. It must be stirred occasionally, and the flame should be so distributed by wire gauze as not to act upon a single point of the bottom of the kettle. At the end of the hour the albumins of the meat-infusion will be coagulated and the gelatin thoroughly dissolved. Günther has shown that the gelatin congeals better if allowed to dissolve slowly in warm water before boiling. The liquid is now cooled to 60° C. and neutralized—*i. e.* alkalinized. As the gelatin is itself acid, a relatively larger amount of the sodium-carbonate solution will be needed than was required for the bouillon. When the proper reaction is attained, as much water as has been lost by vaporization during the process of boiling, intimately mixed with the white of an egg, is added, well stirred in, and the whole boiled for half an hour, then filtered.

If the filter-paper be of good quality and properly folded (pharmaceutical filter), and if the gelatin be properly dissolved, the whole quantity should pass through before cooling too much. Should only half go through before cooling, the remainder must be returned to the pot, heated to boiling once more, and then passed through a new filter-paper. As a matter of fact, gelatin generally filters readily. A wise precaution is to catch the first few centimeters in a test-tube and boil them, so that if a cloudiness shows the presence of uncoagulated albumin, the whole mass can be boiled again. The finished gelatin is at once distributed into sterilized tubes, and then sterilized like the bouillon by the fractional method.

Of course, the gelatin or any other culture-medium can be kept *en masse* indefinitely, but should a contaminating micro-organism accidentally enter, the whole quantity will be spoiled; if, on the other hand, it is kept in tubes, several of them may be lost without much inconvenience. Under proper precautions of sterilization and protection it should all keep well.

Agar-agar.—Agar-agar is the commercial name of a Japanese sea-weed which dissolves in boiling water with resulting thick jelly when cold. The jelly, which solidifies between 30° and 40° C., cannot again be melted except by the elevation of its temperature to the boiling-point, so that this culture-medium, which is nearly transparent, is almost as useful as gelatin. In addition to its readiness to liquefy and solidify, it is sufficiently firm to allow of the incubation-temperature—*i. e.* 37° C.—at which gelatin is always liquid, and no better than bouillon.

The preparation of this medium is generally described in the text-books as one "requiring considerable patience and much waste of filter-paper." In reality, it is not difficult if a good heavy filter-paper be obtained and no attempt be made to filter the solution until the agar-agar is perfectly dissolved. It is prepared as follows : To 1000 c.cm. of bouillon made as described above, preferably of meat instead of beef-extract, 10 grams of agar-agar are added. The mixture is boiled for an hour, or, if possible, two. At the end of the first hour it is cooled to about 60° C., and after neutralization, which may not be necessary if the bouillon was neutral, an egg beaten up in water is added, and the liquid is boiled again until the egg is entirely coagulated. The reaction of the agar-agar should be *neutral* rather than alkaline, as, for an unknown reason, alkalinity seems to interfere slightly with filtration.

After the boiling, which should be brisk, has caused the thorough solution of the agar-agar, it is filtered, just as the gelatin was, through a carefully-folded pharmaceutical filter wet with boiling water. It may expedite matters to pour in about one-half of the solution, keep the remainder hot, and subsequently add it when necessary.

9

Experience shows that 1000 c. cm. of agar-agar rarely go through one paper, and I always expect when beginning the filtration to be compelled to boil the material which remains on the paper again, and pour it through a new filter.

The formerly much-employed hot-water and gas-jet filters seem unnecessary. If properly prepared, the whole quantity will filter in from fifteen to thirty minutes.

If made from beef-extract, the agar-agar almost always precipitates a considerable amount of meat-salts as it cools. This should be anticipated, but, so far as I can determine, cannot always be prevented. The amount is certainly lessened by making the bouillon first, filtering it *cold*, then adding the agar-agar, and dissolving and filtering it.

The difficulty of filtering the agar-agar has led Flügge and others to adopt a method of sedimentation. An ingenious apparatus for this purpose has lately been devised by Bleisch. The methods can be simplified by using a small pharmaceutical percolator, the bottom of which is closed by a rubber cork containing a tube which extends nearly to the top of the percolator and is attached to a rubber tube with a pinchcock below. The melted agar-agar is poured into this, and kept in the steam apparatus until the sedimentation is sufficient to allow clear fluid to be drawn from the top. As the clear agar-agar is drawn off the tube is pulled down through the rubber cork, and more drawn off until only the sediment is left.

Agar-agar is dispensed in tubes like the gelatin and bouillon, sterilized by steam by the intermittent process, and after the last sterilization, before cooling, each tube is inclined against a slight elevation, so as to offer an extensive flat surface for the culture.

After the agar-agar jelly solidifies its contraction causes some water to collect at the lower part of the tube. This should not be removed, as it keeps the material moist, and also because it has a distinct influence upon the character of the growth of the bacteria.

Glycerin Agar-agar.—For an unknown reason certain of the bacteria which will not grow upon the agar-agar as prepared above will do so if 3–7 per cent. of glycerin be added. Among these is the tubercle bacillus, which, not growing at all upon plain agar-agar, will grow well when glycerin is added—a fact discovered by Roux and Nocard. The glycerin may also be added to gelatin or any other medium.

Blood Agar-agar was recommended by R. Pfeiffer for the cultivation of the influenza bacillus. It is ordinary agar-agar whose surface is coated with a little blood secured under antiseptic precautions from the finger-tip, ear-lobule, etc., of man, or the veins of one of the lower animals. Some bacteriologists prepare a hemoglobin agar-agar by spreading a little powdered hemoglobin upon the surface of the agar-agar. This has the disadvantage that powdered hemoglobin is not sterile, and the medium must be sterilized after its addition.

The blood agar-agar should be kept in the incubator a day or two before use so as to insure perfect sterility.

Blood-serum.—The great advantage possessed by this medium is that it is itself a constituent of the body, and hence offers opportunities for the development of the parasitic forms of bacteria under the most natural conditions possible. It is the most difficult of all the media to prepare. The blood must be obtained from a slaughter-house in an appropriate receptacle, the best things for the purpose being tall narrow jars of about 1 liter capacity, with a tightly-fitting lid. The jars are sterilized by heat or by washing with alcohol and ether, are carefully dried, closed, and carried to the slaughter-house where the blood is to be obtained. As the blood flows from the severed vessels of the animal the jars are filled one by one. It seems advisable to allow the first blood to escape, as it is likely to become contaminated from the hair. By waiting until a coagulum forms upon the hair the danger of contamination is obviated. The jars when full are allowed to stand undisturbed until quite firm coagula form within them. If these have any tendency to cling to the glass, each one should be given a few violent twists, so as to break away the fibrinous attachments. After this the jars are carried to the laboratory and stood upon ice for forty-eight hours, by which time the clots will have retracted considerably, and a moderate amount of clear serum can be removed by sterile pipettes and placed in

FIG. 22.—Koch's apparatus for coagulating and sterilizing blood-serum.

sterile tubes. If the serum obtained is red and clouded from the presence of corpuscles, it may be pipetted into sterile cylinders and allowed to sediment for twelve hours, then repipetted into tubes. It is evident that such complicated maneuvring will offer many possible chances of infection; hence the sterilization of the serum is of the greatest importance.

If it is desirable to use the serum as a liquid medium, it is exposed to a temperature of 60° to 65° C. for one hour upon each of five consecutive days. If it is thought best to coagulate the serum and make a solid culture-medium, it may be exposed twice, for an hour each time—or three times if there is distinct reason to think it contaminated—to a temperature just short of the boiling-point. During the process of coagulation the tubes should be inclined, so as to offer a large surface for the growth of the culture. The serum thus prepared may be white, or have a reddish-gray color if many corpuscles are present, and is opaque. It cannot be melted, but once solid remains so.

Koch devised a very good apparatus (Fig. 22) for coagulating blood-serum. The bottom should be covered with cotton, a single layer of tubes placed upon it, and the temperature elevated until coagulation occurs. The repeated sterilizations may be conducted in this apparatus, or may be done equally well in the steam apparatus, the cover of which is not completely closed, for if the temperature of the serum is raised too high it is certain to bubble.

Löffler's blood-serum mixture, which seems rather better for the cultivation of some species than the blood-serum itself, consists of 1 part of a beef-infusion bouillon containing 1 per cent. of glucose and 3 parts of liquid blood-serum. After being well mixed this is distributed in tubes, and sterilized and coagulated like the blood-serum itself. Most organisms grow more luxuriantly upon it than upon either plain blood-serum or other culture-media. Its special usefulness is for the Bacillus diphtheriæ, which grows upon it with rapidity and with quite a characteristic appearance.

Alkaline Blood-serum.—According to Lorrain Smith, a very useful culture-medium can be prepared as follows: To each 100 c.cm. of blood-serum add 1–1.5 c.cm. of a 10 per cent. solution of sodium hydrate and shake it gently. Put sufficient of the mixture into each of a series of test-tubes, and, laying them upon their sides, sterilize like blood-serum, taking care that their contents are not heated too quickly, as then bubbles are apt to form. The result should be a clear, solid medium consisting chiefly of alkali-albumins. It is especially useful for the bacillus diphtheriæ.

Deycke's Alkali-albuminate.—1000 grams of meat are macerated twenty-four hours with 1200 c.cm. of a 3 per cent. solution of potassium hydrate. The clear brown fluid is filtered off and pure hydrochloric acid carefully added while a precipitate forms. The precipitated albuminate is collected upon a cloth filter, mixed with a small quantity of liquid, and made distinctly alkaline. To make solutions of it of definite strength it can be dried, pulverized, and redissolved.

The most useful formula used by Deycke was a 2½ per cent. solution of the alkali-albuminate with 1 per cent. of peptone, 1 per cent. of NaCl, and gelatin or agar-agar enough to make it solid.

Potatoes.—Without taking time to review the old method of boiling potatoes, opening them with sterile knives, and protecting them in the moist chamber, or the much more easily conducted method of Esmarch in which the slices of potato are sterilized in the small dishes in which they are afterward kept and used, we will at once pass to what seems the most simple and satisfactory method of using this valuable medium—that of Bolton and Globig :[1]

With the aid of a cork-borer a little smaller in diameter than the test-tube ordinarily used a number of cylinders are cut from potatoes. Rather large potatoes should be used, the cylinders being cut transversely, so that a number, each about an inch and a half in length, can be cut from one potato. The skin is removed from the cylinders by cutting off the ends, after which each cylinder is cut in two by an oblique incision, so as to leave a broad, flat surface. The half-cylinders are placed each in a test-tube previously sterilized, and then are exposed three times, for half an hour each, to the passing steam of the sterilizer. This steaming cooks the potato and also sterilizes it. Such cultures are apt to deteriorate rapidly, first by turning very dark ; second, by drying so as to be useless. Abbott has shown that if the cut cylinders be allowed to stand for twelve hours in running water before being dispensed in the tubes, they do not turn dark. Drying may be prevented by

[1] *The Medical News*, vol. l., 1887, p. 138.

adding a few drops of clean water to each tube before sterilizing. It is not necessary to have a special small chamber blown in the tube to contain this water; only a small quantity need be added, and this will not touch the potato, which does not reach the bottom of the rounded tube.

A potato-juice has also been suggested, and is of some value. It is made thus: To 300 c. cm. of water 100 grams of grated potato are added, and allowed to stand on ice over night. Of the pulp 300 c. cm. are expressed through a cloth and cooked for an hour on a water-bath. After cooking, the liquid is filtered and receives 4 per cent. of glycerin. It may or may not need neutralization. Upon this medium the tubercle bacillus grows well, especially when the reaction of the medium is acid, but loses its virulence.

Milk.—Milk is useful as a culture-medium. As when the milk stands the cream which rises to the top is a source of inconvenience, it is best to secure from a dairy fresh milk from which the cream has been removed by a centrifugal machine. It is placed in sterile tubes and sterilized by steam by the intermittent method. The opaque nature of this culture-medium often permits the undetected development of contaminating organisms. A careful watch should therefore be kept upon it lest it spoil.

Litmus Milk.—This is milk to which just enough of a saturated watery solution of pulverized litmus is added to give a distinct blue color. Cow's milk is inclined to be acid in reaction, and a small amount of sodium carbonate may be necessary to give it a distinct blue. The use of litmus is probably the best method of determining whether bacteria by their growth produce acids or alkalies.

The watery solution of litmus, being a vegetable infusion, is likely to spoil; hence it should always be treated like the culture-media and sterilized by steam every time the receptacle in which it is kept is opened.

Petruschky's Whey.—In order to differentiate between acid and alkaline producers among the bacteria, Petruschky has recommended a neutral whey colored with litmus. It is made as follows:

To a liter of fresh skimmed milk 1 liter of water is added. The mixture is violently shaken. About 10 c.cm. are now taken out as a sample to determine how much hydrochloric acid must be added to produce coagulation of the milk, and, having determined the least quantity required for the whole bulk, it is added. After coagulation the whey is filtered off, exactly neutralized and boiled. After boiling it is generally found clouded and acid in reaction. It is therefore filtered again, and again neutralized. Litmus is finally added to the neutral liquid, so that it has a violet color, which can readily be changed to blue or red by alkalies or acids.

The medium is a very useful aid in differentiating the typhoid and colon bacilli, showing well the alkaline formation of the typhoid bacillus.

Peptone Solution, or Dunham's solution, is very useful for the detection of certain faint colors. It is a perfectly clear, colorless solution, made as follows:

Sodium chlorid,	0.5	Boil until the ingredients
Witte's dried peptone,	1.	dissolve; then filter, fill
Water,	100.	into tubes, and sterilize.

It is one of the best media for the detection of indol. In it the bacillus pyocyaneus produces its blue color. A very important fact in regard to peptone has been pointed out by Garini,[1] who found that many of the peptones upon the market were impure, and on this account failed to show the indol reaction for bacteria known to produce indol. He recommends the use of the biuret reaction for testing the peptone to be employed. The reagent used is Fehling's copper solution, with which pure peptone strikes a violet color not destroyed upon boiling,

[1] *Centralbl. f. Bakt. u. Parasitenk.*, 1893, xiii., p. 790.

while impure peptone gives a red or reddish-yellow precipitate. Both the peptone and copper solution should be in a dilute form to make successful tests. The addition of 4 c.cm. of the following solution—

Rosalic acid,	0.5,
80 per cent. alcohol,	100.

makes it become an excellent reagent for the detection of acids and alkalies. The solution is pale rose in color. If the bacterium produces acids, the color fades; if alkalies, it intensifies. As the color of rosalic acid is destroyed by glucose, it cannot be used in culture-media containing it.

Theobald Smith calls attention to the fact that Dunham's solution is unsuited to the growth of many bacteria, some failing altogether to grow in it, and recommends that, instead, bouillon free of dextrose shall be used. All bacteria grow well in it, and the indol-reaction is pronounced in sixteen-hour-old cultures. His method of preparation[1] is as follows: beef-infusion, prepared either by extracting in the cold or at 60° C., is inoculated in the evening with a rich fluid culture of some acid-producing bacterium (Bacillus coli), and placed in the thermostat. Early next morning the infusion, covered with a thin layer of froth, is boiled, filtered, peptone and salt added and the neutralization and sterilization carried on as usual.

To test for the presence of indol, the bacterium is planted in the culture-medium, allowed to grow for upward of twelve hours, and then subjected to the combined action of a nitrite and chemically pure sulphuric acid. In making the test, Smith adds to each tube 1 c.cm. of a 0.01 per cent. solution of KNO_2, freshly prepared, and 10 drops of chemically pure H_2SO_4. The presence of indol is characterized by the production of a red color.

[1] *Journal of Exp. Medicine,* vi., Sept. 5, 1897, p. 546.

It is not intended that the student shall infer that there are no culture-media other than these, which have been selected because of their usefulness and popularity. Many other compounds and as many simple substances are employed; for example, eggs, white of egg, urine, bread, sputum, sugar solutions, hydrocele fluid, and aqueous humor.

CHAPTER VII.

CULTURES, AND THEIR STUDY.

THE objects which we have had before us in the preparation of the culture-media were numerous. We have prepared them so as to allow us to separate—or, rather, to *isolate*—bacteria, to keep them in healthy growth for considerable lengths of time, to enable us to observe their biologic peculiarities, and to introduce them without difficulty into the bodies of animals.

The isolation of bacteria was impossible until the fluid culture-media of the early observers were replaced by the solid media, and was exceedingly crude until Koch gave us the solid, transparent media and the well-known "plate-cultures."

A growth of artificially-planted micro-organisms in which an immense number are massed together is called a *culture*. If such a growth contains but one kind of organism, it is known as a *pure culture*.

It has become the habit at present to use the term "culture" rather loosely, so that it does not always signify a growth of micro-organisms artificially planted, but may signify a growth taking place under natural conditions; thus, typhoid bacilli are said to exist in the spleens of patients dead of that disease "in pure culture," because no other bacteria are there; and sometimes, when in expectorated fragments of cheesy matter from tuberculosis pulmonalis the tubercle bacilli are very numerous and unmixed with other bacteria, the term "pure culture" is again used to describe the condition.

Three principal methods are at present employed to enable us to secure pure cultures of bacteria, but before beginning a description of them it is well to observe that the peculiarities of certain pathogenic forms enable us to use special means, taking advantage of their eccentricities, for their isolation, and that the general methods are in reality more useful for the non-pathogenic than for the pathogenic forms.

All three methods depend upon the observation of Koch, that when germs are equally distributed throughout some liquefied nutrient medium which can be solidified in a thin layer, the growth of the germs takes place in little scattered groups or families, called *colonies*, distinctly separated from each other and capable of transplantation to tubes of culture-media.

Plate-cultures.—The plate-cultures, originally made by Koch, require considerable apparatus, and of late years have given place to the more ready methods of Petri and Von Esmarch. So great, however, is the historic interest attached to the plates that it would be a great omission not to describe Koch's method in full.

Apparatus.—Half a dozen glass plates, about 6 by 4 inches in size, free from bubbles and scratches and ground at the edges, are carefully cleaned, placed in a sheet-iron box made to receive them, and then put in the hot-air closet, where they are sterilized. The box, which is tightly closed, allows the sterilized plates to be kept on hand indefinitely before using.

FIG. 23.—Complete levelling apparatus for pouring plate-cultures, as taught by Koch.

A moist chamber, or double dish, about 10 inches in diameter and 3 inches deep, the upper half being just enough larger than the lower to allow it to close over it, is carefully washed. A sheet of bibulous paper is placed in the bottom, so that some moisture can be retained, and a 1 : 1000 bichlorid solution is poured in and brought in contact with the sides, top, and bottom by turning the dish in all directions. The solution is emptied out, and the dish, which is always kept closed, is ready for use.

A levelling apparatus is required (Fig. 19). This consists of a wooden tripod with adjustable screws, and a glass dish covered by a flat plate of glass upon which a low bell-jar stands. The glass dish is filled with broken ice and water, covered with the glass plate, and then exactly levelled by adjusting the screws under the legs of the tripod. When level the cover is placed upon it, and it is ready for use.

Method (Fig. 24).—A sterile platinum loop is dipped into the material to be examined, a small quantity se-

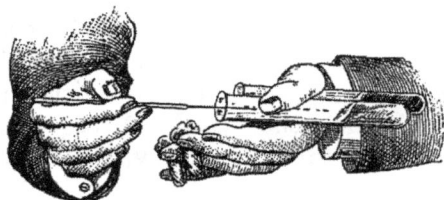

FIG. 24.—Method of holding tubes during inoculation.

cured, and stirred about so as to distribute it evenly through a tube of the melted gelatin. If the material under examination is very rich in bacteria, one loopful may contain a million individuals, which, if spread out in a thin layer, would develop so many colonies that it would be impossible to see any one clearly; hence the necessity for a dilution. From the first tube a loopful of gelatin is carried to a second tube of melted gelatin and stirred well, so as to distribute the organisms evenly through it. In this tube we may have no more than ten thousand organisms, and if the same method of dilution be used again, the third tube may have only a few hundreds, and a fourth only a few dozen colonies.

After the tubes are prepared, one of the sterile glass plates is caught by its edges, removed from the iron box, and placed beneath the bell-glass upon the cold plate covering the ice-water of the levelling apparatus. The plug of cotton closing the mouth of tube No. 1 is removed, and to prevent contamination during the outflow of the gelatin the mouth of the tube is held in the flame of a Bunsen burner for a moment or two. The gelatin is then cautiously poured out upon the plate, the mouth of the tube, as well as the plate, being covered by the bell-glass to prevent contamination by germs in the air. The apparatus being level, the gelatin spreads out in an even, thin layer, and, the plate being cold from the ice

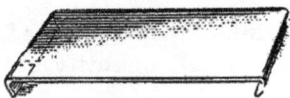

FIG. 25.—Glass bench.

beneath, it immediately solidifies, and in a few moments can be removed to the moist chamber prepared to receive it. As soon as plate No. 1 is prepared, the contents of tube No. 2 are poured upon plate No. 2, allowed to spread out and solidify, and then superimposed on plate No. 1 in the moist chamber, being separated from the plate already in the chamber by small glass benches (Fig. 25) made for the purpose and sterilized. After the contents of all the tubes are thus distributed, the moist chamber and its contents are allowed to stand for some hours, to permit the bacteria to grow. Where each organism falls a colony develops, and the success of the whole method depends upon the isolation of a colony and its transfer to a tube of culture-medium where it can grow unmixed and undisturbed.

The description must have made evident the fact that only such culture-media can be used for plate-cultures as can be melted and solidified at will—viz. gelatin, agar-agar, and glycerin agar-agar. Blood-serum and Löffler's mixture are entirely inappropriate.

The great drawback to this excellent method is the cumbersome apparatus required and the comparative impossibility of making plate-cultures, as is often desirable, in the clinic, at the bedside, or elsewhere than in the laboratory. The method therefore soon underwent modifications, the most important being

Petri's Dishes.—These small dishes (Fig. 26), about 4 inches in diameter and ½ inch deep, with accurately fitting lids, are about as convenient as anything that has been devised in bacteriological technique. They dis-

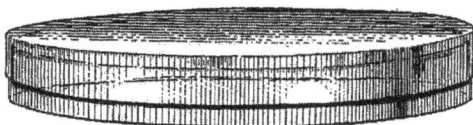

FIG. 26.—Petri dish for making plate-cultures.

pense with plates and plate-boxes, with moist chambers and benches, and usually with the levelling apparatus, though this is still employed in connection with the Petri dishes in some laboratories.

The method of the employment of Petri dishes is very simple. The dishes are carefully cleaned, polished, and sterilized by hot air, care being taken that they are placed in the hot-air closet right side up, and after sterilization are kept covered and in that position. The dilution of the material under examination is made with gelatin or agar-agar tubes in the manner described above, the plugs are removed, the mouth of the tube is cautiously held for a moment in the flame, then the contents of each tube are poured into one of the sterile dishes, whose top is elevated just sufficiently to allow the mouth of the tube to enter. The gelatin is spread over the bottom of the dish in an even layer, is allowed to solidify, labelled, and then stood away for the colonies to develop.

Esmarch Tubes.—This method, devised by Esmarch, converts the walls of the test-tube into the plate and dispenses with all other apparatus. The tubes, which are inoculated and in which the dilutions are made, should contain less than half the usual amount of gelatin or agar-agar. After inoculation the cotton plugs are pushed into the tubes until even with their mouths, and then covered with a rubber cap, which protects them from wetting. A groove is next cut in a block of ice, and the tube, held almost horizontally, is rolled in this until the entire surface of the glass is covered with a thin layer of the solid medium (Fig. 27). Thus the tube becomes the plate upon which the colonies develop.

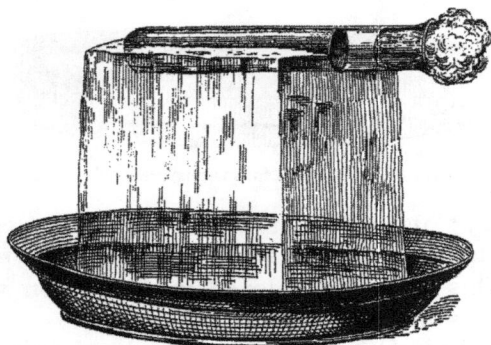

FIG. 27.—Esmarch tube on block of ice (redrawn after Abbott).

Several little points need to be observed in carrying out Esmarch's method. The tube must not contain too much culture-medium, or it cannot be rolled into an even layer. In rolling the contents should not touch the cotton plug, lest it be glued to the glass and its subsequent usefulness be injured. No water must be admitted from the melted ice.

The offspring of each bacterium growing upon the film of gelatin constituting a plate-culture form a mass which has already been pointed out as a *colony*. These small bacterial families may be seen through a microscope when still much too small for detection by the naked eye, and because of their minuteness should always be studied with the microscope.

The original plates of Koch are very inconvenient for such examination, because it is impossible to remove them from the moist chamber and lay them upon the stage of the microscope without exposing them to the danger of contamination by the atmosphere, so that the advantages of Petri dishes and Esmarch tubes, where the examination may be made through the glass tube or through the bottom of the inverted dish, will be more than ever apparent.

The colonies should be viewed from time to time in their growth, drawings being made of the appearances, so as to form a series showing the developmental cycle. Most colonies will be found to originate as spherical, circumscribed, slightly granular, yellowish, greenish, or brownish dots, and later to send out offshoots or filaments or to develop concentric rings or characteristic liquefactions. A few appear from the very first as woolly clumps of entangled threads.

Some of the most diverse forms of colonies are represented in the accompanying illustrations (Figs. 28–32).

FIGS. 28, 29, 30.—The various appearances of colonies of bacteria under the microscope: *a*, colony of Bacillus liquefaciens parvus (Lüderitz); *b*, colony of Bacillus polypiformis (Liborius); *c*, colony of Bacillus radiatus (Lüderitz).

A pure culture, when obtained from colonies growing upon a plate, must always be made from a *single colony*, the transplantation being accomplished under a low power of the microscope. The naked eye can rarely be depended upon to recognize the purity of a colony or its isolation.

Selecting as isolated, large, and characteristic a colony as possible, it is brought to the centre of the field. A platinum wire, securely fused into a glass handle about 8 inches long, is sterilized by being made incandescent in a Bunsen flame, cooled, and then cautiously manipulated until, while it is watched through the microscope,

10

it is seen to touch the colony and take part of its contents away. *In this maneuvre the wire must not touch the objective, the glass, or anything except the colony.* Having secured the adhesion of a few bacteria to the sterile wire, the pure culture is made by introducing them into a sterile culture-medium.

If the pure culture is to be made in bouillon, the tube is held obliquely, so that when the cotton plug is cautiously removed no germs can fall in from the air. The plug is removed by a twisting movement. The wire, without being allowed to touch the mouth or sides of the tube, is plunged into its contents and stirred about until the bacteria are detached, and is then re-

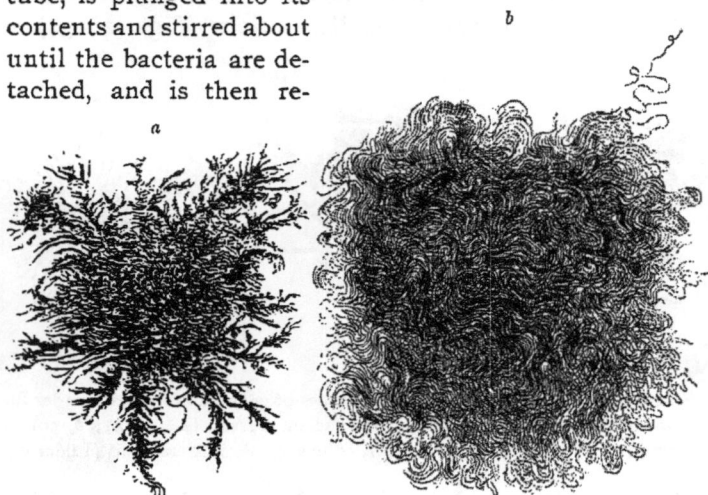

FIGS. 31, 32.—The various appearances of colonies of bacteria under the microscope : *a*, colony of Bacillus muscoides (Liborius) ; *b*, colony of Bacillus anthracis (Flügge).

moved and the plug replaced. The wire should be immediately sterilized by heating to incandescence, lest the bacteria be pathogenic and capable of doing subsequent harm.

If the culture is to be made in gelatin, a different method is employed. The tube is either held horizontally, or, as is perhaps better, inverted ; the cotton plug is removed cautiously ; the wire bearing the bacteria from the colony is introduced until its point enters the centre of the gelatin, and is then carefully pushed on until a vertical puncture from the surface to the bottom of the gelatin is made. This is the *puncture-culture*— "stichcultur" of the Germans.

If the bacteria are only to be planted upon the surface of the culture-medium, the wire is drawn over the surface of a tube of obliquely solidified gelatin, agar-agar, blood-serum, etc. with a steady, slow movement, so as to scatter the germs along its path and cause the development of the bacteria in an enormous colony or mass of colonies in a line following the longest diameter of the exposed surface from end to end. This is the *stroke-culture*— "strichcultur."

The method of holding the tubes, cotton plugs, and platinum wire during the process of inoculation is shown in Figure 20.

Sometimes it is desirable to preserve an entire colored colony as a microscopic specimen. To do this a perfectly clean cover-glass, not too large in size, is momentarily warmed, then carefully laid upon the surface of the gelatin or agar-agar containing the colonies. Sufficient pressure is applied to the surface of the glass to exclude bubbles underneath, but the pressure must not be too great, as it may destroy the integrity of the colony. The cover is gently raised by one edge, and if successful the whole colony or a number of colonies, as the case may be, will be found adhering to it. It is treated exactly as any other cover-glass preparation, is dried, fixed, stained, and mounted, and kept as a permanent specimen. It is called an *adhesion preparation*—"klatsch präparat."

Very often, when one is in a hurry, pure cultures from single colonies may be secured by a very simple manipulation suggested by Banti.[1] The inoculation is made into the water of condensation at the bottom of an agar-

[1] *Centralbl. f. Bakt. und Parasitenk.*, 1895, xvii., No. 16.

agar tube, without touching the surface. The tube is then inclined so that the water flows over the agar, after which it is stood away in the vertical position. Colonies will grow where bacteria have been floated upon the agar-agar, and may be picked up later in the same manner as from a plate.

In other cases pure cultures may best be secured by animal inoculation. For example, when the tubercle bacillus is to be isolated from milk or urine which contains rapidly growing bacteria that would outgrow the slow-developing tubercle bacillus, it is better to inject some of the fluid into the abdominal cavity of a guinea-pig and await the development of tuberculosis, and then seek to secure the bacillus from the unmixed material in the softened lymphatic glands. Anthrax bacilli are also more easily secured in pure culture by inoculating a mouse and recovering the bacilli from a spleen or the heart's blood after death, than by going to the trouble of making plates and picking out the colonies.

In many cases when it is desired to isolate the micrococcus tetragenus, the pneumococcus, and others, it is easier to inoculate the most susceptible animal and recover the germ from the organs than to plate it out and search for the colony among many others which may be similar to it.

The development of bacteria in liquids is of less interest than that upon solid media. The growth generally manifests itself by a diffused turbidity. Sometimes flocculi float in the otherwise clear medium. Some forms grow most rapidly at the surface of the liquid, and produce a distinct membranous pellicle called a *mycoderma*. In such a growth multitudes of degenerated bacteria and large numbers of spores are to be observed. On the other hand, it occasionally happens that the growth occurs chiefly below the surface, and may produce gelatinous masses which are known as *zoöglea*.

In gelatin the bacteria exhibit a great variety of appearances, many of which are beautiful and interesting.

Certain bacteria, as the tubercle bacillus, will not grow at all upon gelatin. Some forms which are rigidly aërobic will only grow upon or near the surface ; others, anaërobic, only in the deeper parts. The majority, however, grow both upon the surface and in the puncture made by the wire. Sometimes the consistence of the gelatin is unaltered ; sometimes it is liquefied throughout, sometimes only at the surface. Sometimes offshoots extend from the colonies into the gelatin, giving the culture

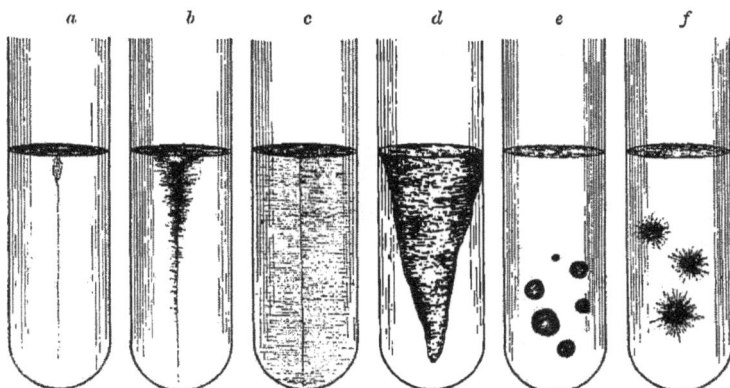

FIG. 33.—Various forms of gelatin puncture-cultures : *a*, Bacillus typhi abdominalis; *b*, B. anthracis; *c*, B. mycoides; *d*, B. mesentericus vulgatus; *e*, B. of malignant edema; *f*, B. radiatis.

a bristling appearance. Figure 33 will serve to illustrate different varieties of gelatin growth.

The growth in gelatin is generally so far removed from the walls of the tube (a central puncture nearly always being made in the culture-medium, in order that the growth be symmetrical) that it is next to impossible to make a microscopical examination of it with any power beyond that given by a hand-lens.

Much attention has been given of late to the preparation of microtome sections of the gelatin growth. To accomplish this the glass is warmed sufficiently to allow the gelatin to be removed and placed in Müller's fluid (bichromate of potassium 2.–2.5, sulphate of sodium 1, water 100), where it is hardened. When quite firm it is washed in water, passed through alcohols ascending in strength from 50 to 100 per cent., imbedded in celloidin, cut wet, and stained like a section of tissue.

A ready method of doing this has been suggested by Winkler,[1] who bores a hole in a block of paraffin with the smallest-size cork-borer, soaks the block in bichlorid solution for an hour, pours liquid gelatin into the cavity, allows it to solidify, inoculates it by the customary puncture of the platinum wire, allows it to develop sufficiently, and when ready cuts the sections under alcohol, subsequently staining them with much-diluted carbol-fuchsin.

Very pretty museum specimens of plate- and puncture-cultures in gelatin can be made by simultaneously killing the micro-organisms and permanently fixing the gelatin with formalin, which can either be sprayed upon the gelatin or applied in dilute solution. As gelatin fixed in formalin cannot subsequently be liquefied, such preparations will last indefinitely.

The growths which occur upon agar-agar are in many ways less characteristic than those in gelatin, but as this medium does not liquefy except at a high temperature (100° C.), it has that great advantage over gelatin. The colorless or almost colorless condition of the preparation also aids in the detection of such chromogenesis as may be the result of the micro-organismal growth.

Sometimes the growth is colored, sometimes not; sometimes the production of a soluble pigment colors the agar-agar as well as the growth; sometimes the growth is one color and the agar-agar another. Sometimes the growth is filamentous, sometimes a smooth, shining band. Occasionally the bacterium does not grow upon agar-agar unless glycerin be added (tubercle bacillus); sometimes it will not grow even then (gonococcus).

Still less characteristic are the growths upon potato. Most bacteria produce rather smooth, shining, irregu-

[1] *Fortschritte der Medicin*, Bd. xi., 1893, No. 22.

larly-extending growths, which often show very beautiful colors.

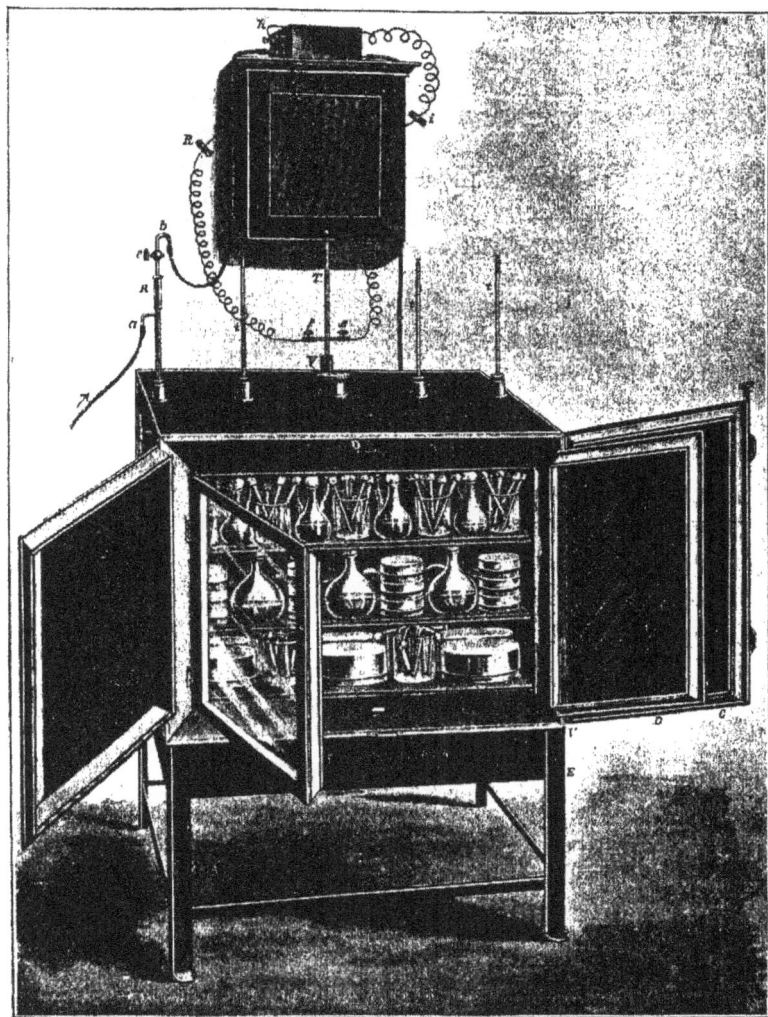

FIG. 34.—New model incubating-oven with electro-regulator.

In milk and litmus milk one must observe the presence or absence of acid-production, the coagulation which may

or may not accompany it, and the subsequent gelatinization or digestion of the coagulum.

Blood-serum is liquefied by some bacteria. The majority of organisms are not very characteristic in their
development upon it. Others, as the bacillus of diphtheria, are, however, characterized by their shape, color,
and rapidity of development at given temperatures.

While most of the saprophytic bacteria will grow well
at the ordinary temperature of a well-warmed room, the
important pathogenic forms require to be kept at the
temperature of the body. To do this accurately an incubating oven becomes a necessity. Various forms, of
wood and metal, are in the market, the one shown in the
illustration (Fig. 34) being one of the newest and best.

It scarcely need be pointed out that gelatin cultures
cannot be grown in the incubating oven, as the medium
will not remain solid at temperatures above 20–22° C.

CHAPTER VIII.

THE CULTIVATION OF ANAËROBIC BACTERIA.

THE cultivation of micro-organisms which will not grow where the least amount of oxygen is present is always attended with much difficulty, and can seldom be accomplished with certainty. Many methods have been suggested, but not one can be described as satisfactory.

Koch originally cultivated anaërobic bacteria upon plates by covering the surface of the soft gelatin with a thin film of mica previously sterilized by incandescence. Some anaërobic forms will grow quite well by such a simple exclusion of the air, but the strictly anaërobic forms will not develop at all.

Hesse originated the plan, still sometimes followed, of making a deep puncture in recently boiled and rapidly sterilized gelatin or agar-agar, then covering the surface with sterilized oil, through which no oxygen was supposed to penetrate (**Fig. 35**).

Liborius suggested the plan of having a tube nearly full of gelatin or agar-agar, boiling it just before inoculation, so as to expand and drive out whatever air it might contain, making the inoculation while the culture-medium was still fluid, cooling rapidly in ice-water, and sealing up the tube in a blowpipe as near the surface of the gelatin as possible.

Esmarch used a regular "Esmarch tube," into the central cavity of which melted sterile gelatin was poured to exclude the air.

Buchner invented a method by which, by the use of pyrogallic acid, the oxygen was absorbed from the atmosphere in which the culture was kept, and the growth allowed to continue in the nitrogen and carbonic acid

153

which remained (Fig. 36).   His method was to place the
tube which had been inoculated in a much larger outer
test-tube containing alkaline pyrogallic acid.   The large

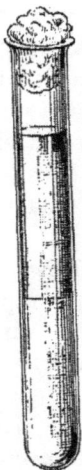

FIG. 35.— Hesse's
method of making
anaërobic cultures.

FIG. 36.—Buchner's
method of making an-
aërobic cultures.

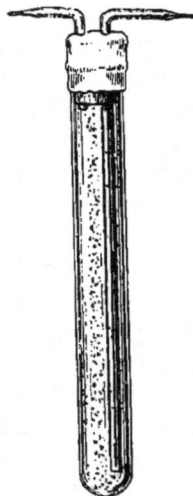

FIG. 37.—Fränkel's meth-
od of making anaërobic cultures.

tube was closed with a rubber cap, and the absorption of
the oxygen allowed to progress.

Gruber, instead of absorbing the oxygen as Buchner
does, prefers to use an air-pump and exhaust the contents
of the tube.   He uses a tube having a slender neck and
a perforated rubber stopper.   After the inoculation is
made the air is pumped out and the slender neck sealed
in the blowpipe.   After this the tube can be warmed and
the melted gelatin or agar-agar rolled on its sides, as suggested by Esmarch, if desired.

Better than any of the preceding is the method of
Fränkel, which removes the air and replaces it by hydrogen.   Fränkel prepares an ordinary Esmarch tube,
removes the cotton stopper, and replaces it by a carefully
sterilized rubber cork containing two tubes (Fig. 37).   The

tubes are connected with a hydrogen generator, and the gas is allowed to pass through until all the oxygen is forced out and replaced by the hydrogen, after which the ends of the tubes are sealed in the flame (Fig. 36).

Liborius has designed a special tube for accomplishing the same thing.

Kitasato and Weil found the addition of 0.3–0.5 per cent. of sodium formate to be of use in aiding the rapidity of the development of anaërobic cultures. Liborius found that 2 per cent. of glucose added to the culture-medium also increased the rapidity of the process.

The methods now generally employed by bacteriologists for the anaërobic cultivations embrace all the essentials of the foregoing methods. One of the best arrangements for the purpose is that devised by Dr. Ravenel. His inoculations are deeply made in culture-media as free from air as possible. The tubes are loosely plugged, and are placed in an air-tight chamber the bottom of which contains pyrogallic acid—pyrogallic acid 1, solution of caustic potash 1, water 10. The apparatus is connected by two tubes with an exhaust-pump on one side, and with a hydrogen apparatus on the other, by which means the atmosphere is exhausted, and replaced by hydrogen until only pure hydrogen remains, after which the chamber is permanently sealed and the germs allowed to grow. Such a chamber can be constructed to hold a number of tubes or Petri dishes, yet not be too large to be stood in an incubator. Whatever oxygen may have escaped the exhaustion or have entered by the process of leakage is at once absorbed by the pyrogallic acid in the lower chamber of the apparatus.

Apparatus for plating out strictly anaërobic bacteria that have met with great favor are those invented by Botkin (Fig. 38) and Novy (Fig. 39). The first mentioned combines the replacement of the air by hydrogen and the absorption of the oxygen possibly remaining by alkaline pyrogallic acid; the other simply replaces the oxygen by hydrogen. In using Botkin's apparatus the uncovered Petri dishes are placed one above the other in the rack C, and covered with the bell-glass A. Liquid paraffin is poured in the dish B until it is about half full. From a Kipp's apparatus hydrogen gas enters the little rubber tube *a*, subsequently escaping by the tube *b*. When only pure hydrogen escapes the rubber tubes *a* and *b* are withdrawn, and the apparatus remains filled with hydrogen. Lest a little oxygen should remain, it is best to have the dishes at the top and bottom of the rack filled with alkaline pyrogallic acid. Tetanus can be cultivated in this apparatus.

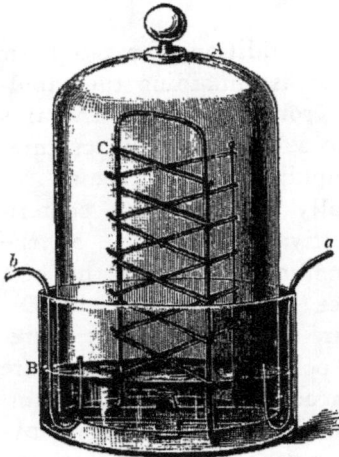

FIG. 38.—Botkin's apparatus for making anaërobic cultures.

The jars recently introduced by Novy are similar in principle, depending upon the replacement of the air by hydrogen. They are

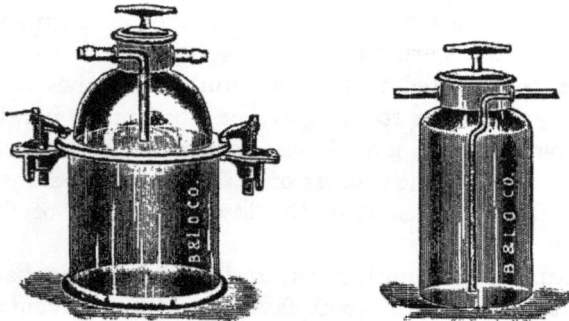

FIG. 39.—Novy's jars for anaërobic cultures.

so constructed that when the stopper occupies a certain relative position to the neck the gas can enter and exit, but when the stopper is turned a little the jar is hermetically sealed. Alkaline pyrogallic acid in a test-tube, or in the bottom of the jar, will serve to absorb any remaining oxygen. The larger jar (Fig. 39, *a*) is intended for Petri dishes, the smaller one (*b*) for test-tube cultures.

Roux has suggested the simplest method of cultivating anaërobic bacteria. The germs are distributed through freshly boiled, still liquid, gelatin or agar-agar, as in making the dilutions for plate-cultures, then drawn into a long, slender sterile piece of glass tubing of small calibre. When the tube is full the ends, which should have been narrowed, are closed in a flame, and the culture is hermetically sealed in an air-tight chamber. The chief difficulty is in transplanting the growing colony. To do this the tube must be opened with a file or a diamond at the point where the colony desired is observed.

CHAPTER IX.

EXPERIMENTATION UPON ANIMALS.

BACTERIOLOGY has to-day become a science whose principal objects are to discover the cause, explain the symptoms, and prepare the cure of diseases. We cannot hope to achieve these objects except by the introduction of bacteria into animals, where their effects and the effects of their products can be studied.

No one should more heartily condemn wanton cruelty to animals than the physician and the naturalist. Indeed, it is hard to imagine a class of men so much of whose lives is spent in relieving pain, and who know so much about pain, being guilty of the wholesale butchery and torture accredited to them by a few of the laity, whose eyes, but not whose brains, have looked over the pages of physiological text-books.

Experimentation upon animals has given us almost all our knowledge of physiology, most of our valuable therapeutics, and the only scientific methods of treating tetanus and diphtheria.

Experiments upon animals we must make, and, as animals differ in their susceptibility to diseases, large numbers and different kinds must be employed.

The bacteriological methods are not cruel. Two principal modes of introducing bacteria are employed: the subcutaneous injection and the intravenous injection.

Subcutaneous injections into animals are made exactly as hypodermic injections are given to man.

Any hypodermic syringe that can be conveniently cleaned and disinfected may be employed for the purpose. Those expressly designed for bacteriological work and most frequently employed are shown in Fig. 40.' Those

of Meyer and Roux resemble ordinary hypodermic
syringes; that of Koch is supposed to possess the decided
advantage of not having a piston to come into contact
with the fluid to be injected. This is, however, somewhat disadvantageous inasmuch as the cushion of compressed air that drives out the contents is elastic, and unless carefully watched will follow the injection into the
body of the animal. In making subcutaneous injections
there is no disadvantage or danger from the entrance of

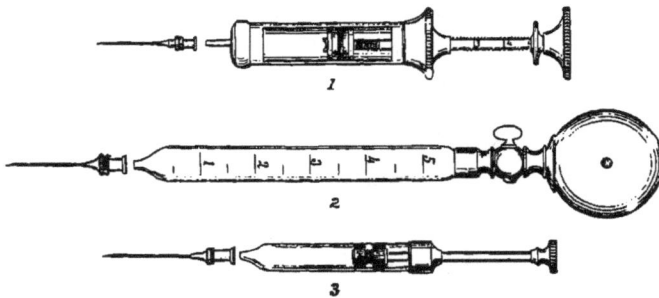

FIG. 40.—1, Roux's bacteriological syringe; 2, Koch's syringe; 3, Meyer's
bacteriological syringe.

air beneath the skin, but in intravenous injections it is
commonly supposed to be dangerous.

All syringes should be disinfected with carbolic acid
solutions *before* and *after* using, the carbolic acid being
allowed to act for some time and then washed out
with sterile water. Syringes should not be boiled, as
it ruins the packings, whether of asbestos, leather, or
rubber.

The intravenous injections differ only in that the needle
of the syringe is introduced into a vein. This is easy in a
large animal like a horse, but is very difficult in a small
animal, and wellnigh impossible in anything smaller than
a rabbit. Such injections when given to rabbits are generally made into the ear-veins, as those most conspicuous
and accessible (Fig. 41). A peculiar and important fact
to remember is, that the less conspicuous posterior vein

is much better adapted to the purpose than the anterior. The introduction of the needle should be made from the hairy surface of the ear.

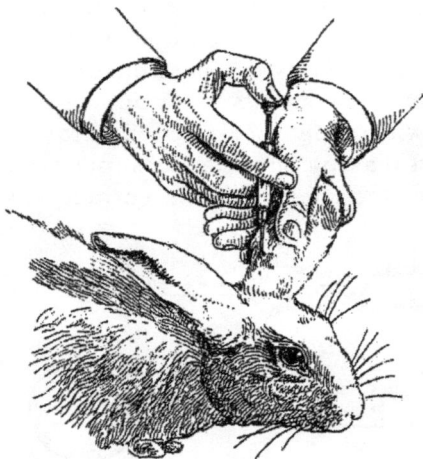

FIG. 41.—Method of making an intravenous injection into a rabbit. Observe that the needle enters the posterior vein from the hairy surface.

If the ear is manipulated for a moment or two before the injection is begun, vasomotor dilatation occurs and the blood-vessels all become larger and more conspicuous. The vein should be compressed at the root of the ear until the needle is introduced, and the injection made as near the root as possible.

The introduction of bacteria into the lymphatics is only possible by injecting liquid preparations of them into some organ with comparatively few blood-vessels and large numbers of lymphatics. The testicle is best adapted to this purpose, the needle being introduced deeply into the organ.

Sometimes the inoculation can be made by the platinum wire, a very small opening made in the skin by a snip of the scissors being sufficient.

Sometimes intra-abdominal and intra-pleural injections are made, and in cases where it becomes necessary to determine the presence or absence of tuberculosis or glanders in tissues it may be necessary to introduce small pieces of the suspected tissue under the skin or into the abdominal cavities. To do this is not difficult. The hair is carefully, closely cut over the point of election, which is generally on the abdomen near the groin, the skin picked up with forceps, a snip made through it, and the points of the scissors introduced for half an inch

or so and then separated. By this maneuver a subcutaneous pocket is formed, into which the tissue is easily
forced. The opening should not be large enough to require subsequent stitching.

Small animals, like rabbits and guinea-pigs, can be held
in the hand, as a rule. Rabbit-holders of various forms
can be obtained from dealers. Dogs, cats, sheep, and goats
can be tied and held in troughs. A convenient form of
mouse-holder, invented by Kitasato, is shown in Fig. 42.

In all these experiments one must remember that the
amount of material introduced into the animal must be
in proportion to its size, and that injection-experiments
upon mice generally are so crude and destructive as to
warrant the comparison drawn by Fränkel, that to inject
a few minims of liquid into the pleural cavity of a mouse
is "much the same as if one would inject through a fire-
hose three or four quarts of some liquid into the respiratory organs of a man."

The blood of animals, when it is necessary to experiment with it, is best secured from
a large vein, generally the jugular. From small animals, such as
guinea-pigs, it may be secured by
introducing a small cannula into
the carotid artery.

Our observations of animals by
no means cease with their death.
Indeed, he cannot be a bacteriologist who is not already a good
pathologist and expert in the recognition of diseased organs.

When an autopsy is to be made
upon a small animal, it is best to
wash it for a few moments in a
disinfecting solution, to kill the

FIG. 42. Mouse-holder.

germs present upon the hair and the skin, as well as to
moisten the hair and enable it to be kept out of the
incision.

11

The animal should be tacked to a board if small, or tied, by cords fastened to the legs, to the corners of a table if large, and should be dissected with sterile knives and scissors. When a culture is to be made from the interior of an organ—say the spleen—it should be incised deeply with a sterile knife and the culture made from its centre.

Fragments intended for subsequent microscopical examination should be cut very small (cubes of 1 c.cm.), placed in absolute alcohol for a few hours, then transferred to weaker alcohol, 80–90 per cent., for preservation. The technique of imbedding and staining the tissues can be found in almost any reliable text-book on pathology or on the special subject of microscopical technique.

CHAPTER X.

THE RECOGNITION OF BACTERIA.

THE most difficult thing in bacteriology is to be able to recognize the bacteria which come under observation.

A certain few micro-organisms are so characteristic in shape and grouping as to be separated by a microscopic examination. Some, as the tubercle bacillus, are characteristic in their reaction to the anilin dyes, and can be differentiated at once by this peculiarity. Some, as the Bacillus mycoides, are so characteristic in their agar-agar growth as to eliminate others. The red color of Bacillus prodigiosus and the blue of Bacillus janthinus will speak almost positively for them. The potato culture of the Bacillus mesentericus fuscus and its close relative the vulgatus is quite sufficient to enable us to pronounce upon them. Unfortunately, however, there are several hundreds of described species which lack any one distinct character that may be used for differential purposes, and require that for their diagnosis we shall wellnigh exhaust the bacteriological technique in an almost fruitless effort to recognize them.

A series of useful tables has been compiled by Eisenberg, and is now almost indispensable to the worker. Unfortunately, in tabulating bacteria we constantly meet species described so insufficiently as to make them worse than useless on account of the confusion caused.

The only way to recognize a species is to study it thoroughly and compare it, step by step, with the descriptions and tables of known species compiled by Eisenberg and others.

CHAPTER XI.

THE BACTERIOLOGIC EXAMINATION OF THE AIR.

IT has been repeatedly emphasized—and indeed at the present time almost every one knows—that micro-organisms float almost everywhere in the air, and that their presence there is a constant source of danger, not only of contamination in our bacteriologic researches, but also a menace to our health.

Such micro-organisms are neither ubiquitous nor equally disseminated, but are much more numerous where the air is dusty than where it is pure—much more so where men and animals are accustomed to live, than upon the ocean or upon high mountain-tops. The purity of the atmosphere bears a distinct relation to the purity of the soil over which its currents blow.

The micro-organisms that occur in the air are for the most part harmless saprophytes which have been separated from their nutrient birthplace and carried about by the wind. They are almost always taken up from dried materials, experiment having shown that they arise from the surfaces of liquids in which they grow with much difficulty. They are by no means all bacteria, and a plate of sterile gelatin exposed for a brief time to the air will generally grow moulds and yeasts as well as bacteria.

The bacteria present are occasionally pathogenic, especially in localities where the discharges of diseased animals have been allowed to collect and dry. For this reason the atmosphere of the wards of hospitals and of rooms in which infectious cases are being treated is much more apt to contain them than the air of the street. However, the dried expectoration of cases of tuberculosis, of in-

164

fluenza, and sometimes of pneumonia, causes the specific bacteria of these diseases to be far from uncommon in street-dust.

Günther points out that the majority of the bacteria which occur in the air are cocci, sarcina being very abundant. Most of them are chromogenic and do not liquefy gelatin. It is unusual to find a considerable variety of bacteria at a time ; generally not more than two or three species are found.

It is an easy matter to determine whether bacteria are present in the air or not, all that is necessary being to expose sterile plates or Petri dishes of gelatin to the air for a while, close them, and observe whether or not bacteria grow upon them.

To make a quantitative estimation is, however, much

FIG. 43.—Hesse's apparatus for collecting bacteria from the air.

more difficult. Several methods have been suggested, of which the most important may be considered.

The method suggested by Hesse is simple and good. It consists in making a measured quantity of the air to be examined pass through a horizontal sterile tube about 70 cm. long and 3.5 cm. wide (Fig. 43), the interior of which is coated with gelatin in the same manner as an Esmarch tube. The tube, having been prepared, is closed at both ends with sterile corks carrying smaller glass tubes closed with cotton. When ready for use the tube at one end is attached to a hand-pump, the cotton is removed from the other end, and the air passed through very slowly, the bacteria having time to precipitate upon the gelatin as they pass. When the required amount has passed the tubes are again plugged, the apparatus stood away for a time, and subsequently, when they have grown, the colonies are counted. The number of colonies in the tube will represent pretty accurately the number of bacteria in the amount of air which passed through the tube.

In such a cylindrical culture it will be noted that if the air is passed through with the proper slowness, the colonies will be much more numerous near the end of entrance than that of exit. The first to fall will probably be those of heaviest specific gravity—*i. e.* the moulds and yeasts.

A still more exact method is that of Petri, who uses small filters of sand held in place in a wide glass tube by small wire nets (Fig. 44). The sand used is made to pass through a sieve whose openings are of known size, is heated to incandescence, then arranged in the tube so that two of the little filters, held in place by their wire-gauze coverings, are

FIG. 44.—
Petri's sand
filter for air-
examination.

superimposed. One or both ends of the tube are closed with corks having a narrow glass tube. The apparatus is heated and sterilized in a hot-air sterilizer, and is then ready for use. The method of employment is very simple. By means of a hand-pump 100 liters of air are made to pass through in from ten to twenty minutes. The sand from the upper filter is then carefully mixed with sterile melted gelatin and poured into sterile Petri dishes, where the colonies develop and can be counted. Sternberg remarks that the chief objection to the method is the presence in the gelatin of the slightly opaque sand, which interferes with the recognition and counting of the colonies. This objection has, however, been removed by Sedgwick and Miquel, who use a soluble material—granulated or pulverized sugar—instead of the sand. The apparatus used for the sugar-experiments differs a little from the original of Petri, but the principle is the same, and can be modified to suit the experimenter. Petri points out in relation to his method that the filter catches a relatively greater number of bacteria in proportion to moulds than the Hesse apparatus, which depends upon sedimentation.

A particularly useful form of apparatus is a granulated sugar-filter suggested by Sedgwick and Tucker, which has an expansion above the filter, so that as soon as the sugar is dissolved in the melted gelatin it can be rolled out into a lining like that of an Esmarch tube. This cylindrical expansion is divided into squares which make the counting of the colonies very easy (Fig. 45).

The number of germs in the atmosphere will naturally be very variable. Roughly, the number may be estimated at from 100 to 1000 per cubic meter.

FIG. 45.—Sedgwick's expanded tube for air-examination.

In reality, the bacteriologic examination of air is of very little value, as so many possibilities of error may occur. Thus, when the air of a room is quiescent there may be very few bacteria in it; let some one walk across the floor and dust at once rises, and the number of bacteria is considerably increased : if the person be a woman with skirts, more bacteria will probably be raised from the floor than would be disturbed by a man ; if the room be swept, the increase is enormous. From these and similar contingencies it becomes very difficult to know just when and how the air is to be examined, and the value of the results is correspondingly lessened.

The most valuable examinations are those which aim at the discovery of some definite organism or organisms regardless of the number per cubic meter.

CHAPTER XII.

BACTERIOLOGIC EXAMINATION OF WATER.

UNLESS water has been specially sterilized or distilled and received and kept in sterile vessels, it always contains some bacteria. The number will bear a very distinct relation to the amount of organic matter in the water, though experiment has shown that certain pathogenic and non-pathogenic bacteria can remain vital in perfectly pure distilled water for a considerable length of time. Ultimately, owing to the lack of nutriment, they undergo a granular degeneration.

The majority of the water-bacteria are bacilli, and as a

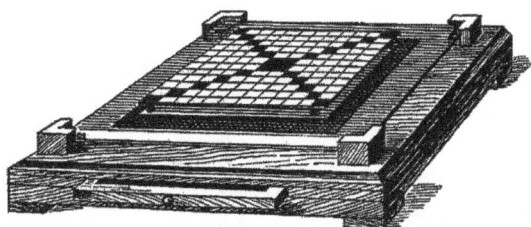

FIG. 46.—Wolfhügel's apparatus for counting colonies of bacteria upon plates.

rule they are non-pathogenic. Wright,[1] in his examination of the bacteria of the water from the Schuylkill River, found two species of micrococci, two species of cladothrices, and forty-six species and two varieties of bacilli. Of course, at times the most virulent forms of pathogenic bacteria—those of cholera and typhoid fever —occur in polluted water, but this is the exception, not the rule.

The method of determining quantitatively the number

[1] Memoirs of the National Academy of Sciences, Third Memoir.

of bacteria in water is very simple, and can generally be prosecuted without much apparatus. The principle depends upon the equal distribution of a given quantity of the water to be examined through a sterile liquid medium, and the subsequent solidification of this medium in a

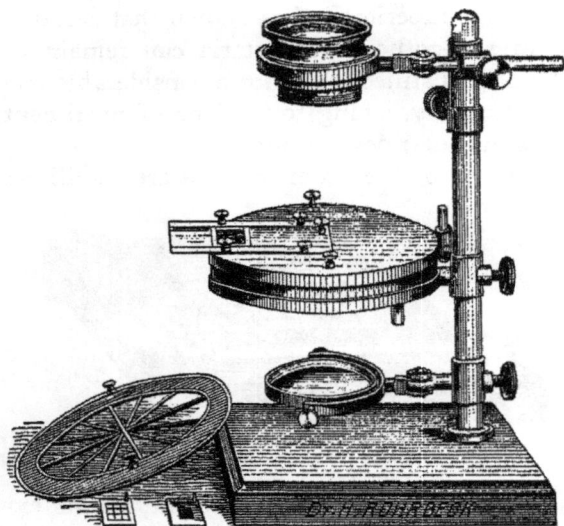

FIG. 47.—Heyroth's instrument for counting colonies of bacteria in Petri dishes.

thin layer, so that all the colonies which develop may be counted.

The method, which originated with Koch, may be performed with the Koch plates or with Petri dishes or with Esmarch rolls. It is always best to make a number of these plate-cultures with different amounts of the water to be examined, using, for example, 0.01, 0.1, 0.5, and 1.0 c.cm. added to a tube of gelatin, agar-agar, or glycerin agar-agar.

The exact method must depend somewhat upon the quality of the water to be examined. If the number of bacteria per cubic centimeter is small, large quantities may be used, but if there are millions of bacteria in every cubic centimeter, it may be necessary to dilute the water to be examined in the proportion of 1 : 10 or 1 : 100 with sterile water, mixing well, and making the plate-cultures from the dilutions.

It is best to count all the colonies if possible, but when there are hundreds or thousands scattered over the plate, an average estimation of a number of squares ruled upon a glass background (Fig. 46), as suggested by Wolfhügel, is most convenient. In his apparatus a large plate of glass is divided into small square di-visions, the diagonals being spe-cially indicated by color. The plate or Petri dish is stood upon the glass, and the number of colonies in a number of small squares is easily counted, and the total number of colonies es-timated. In counting the colo-nies a lens is indispensable. Special apparatuses have been devised for counting the colo-nies in Petri dishes (Fig. 47) and in Esmarch tubes (Fig. 48).

FIG. 48.—Esmarch's instrument for counting colonies of bacteria in tubes.

The majority of the water-bacteria are rapid liquefiers of gelatin, for which reason it seems better to employ agar-agar than gelatin for making the cultures.

In ordinary hydrant-water the bacteria number from 2–50 per cubic centimeter; in good pump-water, 100–500; in filtered water from rivers, according to Günther, 50–200 are present; in unfiltered river-water, 6000–20,000. According to the pollution of the water the number may reach as many as 50,000,000.

The waters of wells and springs are dependent for their purity upon the character of the earth or rock through which they filter, and the waters of deep wells are much more pure than those of shallow wells, unless contamina-tion takes place from the surface of the ground.

Ice always contains bacteria if the water contained

them before it was frozen.   In Hudson-River ice Prudden found an average of 398 colonies in a cubic centimeter.

A sample of water when collected for examination should be placed in a clean sterile bottle or in a hermetically-sealed pre-sterilized glass bulb, and must be examined as soon as possible, as the bacteria multiply rapidly in water which is allowed to stand for a short time.   In determining the species of bacteria found in the water reference must be made to the numerous monographs upon the subject, and to tables such as those compiled by Eisenberg.

The discovery of certain important pathogenic bacteria, as those of cholera and typhoid, will be considered under the specific headings.

Unfortunately, the bacteriologic examination of waters does not throw satisfactory light upon their exact hygienic usefulness.   Of course, if cholera or typhoid-fever bacteria are present, the water is harmful, but the quality of the water cannot be gauged by the number of bacteria it contains.

The drinking-water furnished large cities is not infrequently contaminated with sewage, and contains intestinal bacteria—Bacillus coli communis.   For the ready determination of this organism, which is an important one as an indicator that the water is polluted, Smith[1] has made use of the fermentation-tube in addition to the plate.   His method is to add to each of the fermentation-tubes containing 1 per cent. dextrose-bouillon a certain quantity of water.   The evolution of 50–60 per cent. of gas by the third day is a strong indication that the colon bacillus is present.   Plates may be used to confirm the presence of the bacillus, but are hardly necessary, as there is scarcely another bacterium met with in water that is capable of producing so much gas.

Filtration with sand, etc. diminishes the number of bacteria for a time, but, as the organisms multiply in

[1] *American Journal of the Medical Sciences,* 1895, 110, p. 301.

the filter, the benefit is not permanent. The filters must frequently be renewed. Porcelain filters seem to be the only positive safeguard, and even these, the best of which seems to be the Pasteur-Chamberland, allow the bacteria to pass through if used too long without renewal or without firing.

CHAPTER XIII.

BACTERIOLOGIC EXAMINATION OF SOIL.

ALMOST all soil contains bacteria in its upper layers. Their number and character, however, depend somewhat upon the surrounding conditions. Near the habitations of men, where the soil is cultivated, the excrement of animals, largely made up of bacteria, is spread upon it to increase its fertility, this being a treatment which not only adds new bacteria to those already present, but also enables those present to grow very much more luxuriantly because of the increased amount of organic matter they receive.

The researches of Flügge, C. Fränkel, and others show that the bacteria of the soil do not penetrate very deeply—that they gradually decrease in number until the depth of a meter is reached, then rapidly diminish until at a meter and a quarter they rather abruptly cease to be found.

Many of the soil-bacteria are anaërobic, and for a careful consideration of them the reader must be referred to monographs upon the subject. The estimation of their number seems to be devoid of any distinct practical importance. C. Fränkel has, however, originated a very accurate method of determining it. By means of a special boring apparatus (Fig. 49)

FIG. 49.—Fränkel's instrument for obtaining earth from various depths for bacteriologic study.

earth can be secured from any depth without digging and without danger of mixing that secured with that of the superficial strata. With sterile liquefied gelatin a definite

174

amount of this soil is mixed thoroughly and the mixture solidified upon the walls of an Esmarch tube. The colonies are counted with the aid of a lens. Flügge found in virgin earth about 100,000 colonies in a cubic centimeter.

Samples of earth, like samples of water, should be examined as soon as possible after being secured, for, as Günther points out, the number of bacteria changes because of the unusual environment, exposure to increased amounts of oxygen, etc.

The most important bacteria of the soil are those of tetanus and malignant edema, in addition to which, however, there are a great variety which are pathogenic for rabbits, guinea-pigs, and mice.

In the "Bacteriological Examination of the Soil of Philadelphia," Ravenel[1] came to the conclusion that—

1. Made soils, as commonly found, are rich in organic matter and excessively damp through poor drainage.

2. They furnish conditions more suited to the multiplication of bacteria than do virgin soils, unless the latter are contaminated by sewage or offal.

3. Made soils contain large numbers of bacteria per gram of many different species, the deeper layers being as rich in the number and variety of organisms as the upper ones. After some years the number in the deeper layers probably becomes proportionally less. Made soils are more likely than others to contain pathogenic bacteria.

In 71 cultures that were isolated and carefully studied by Ravenel, there were two cocci, one sarcina, and five cladothrices; all the others were bacilli.

[1] Memoirs of the National Academy of Sciences, First Memoir, 1896.

CHAPTER XIV.

TO DETERMINE THE THERMAL DEATH-POINT.

SEVERAL methods may be employed for this purpose. Roughly, it may be done by keeping a bouillon-culture of the micro-organism to be studied in a water-bath whose temperature is gradually increased from that of the body to 75° C.

Into a fresh bouillon-culture thus exposed to heat, the experimenter cautiously, and at given intervals, introduces a platinum loop or a capillary pipette, and withdraws a drop of the culture which he inoculates into fresh bouillon and stands aside to grow. It is economy to make the transplantations rather infrequently at first and frequently later on in the experiment, when the temperature is ascending. In an ordinary determination it would be well to make a transfer at 40° C., one at 45° C., another at 50°, still another at 55°, and then beginning at 60° make one for every additional degree up to 75° C. or above. The day following the experiment it will be observed that all the cultures grow except those heated beyond a certain point, as 60° C. and upward, when it can properly be concluded that 60° C. is the thermal death-point. If all the transplantations grow, of course the maximum temperature that the bacteria can endure was not reached, and the experiment must be performed again with higher temperatures.

When more accurate information is desired, and one wishes to know how long the micro-organism can endure some such temperature as 60° C. without losing its vitality, a dozen or more bouillon-tubes may be inoculated with the germ to be studied, and stood in the water-bath at the temperature to be investigated. The first can be

176

removed as soon as it is certainly heated through, another
in five minutes, another in ten minutes, or at whatever
intervals the thought and experience of the experimenter
shall suggest.

In both of the described procedures one must be careful that the temperature in the test-tube is identical with
that of the water in the bath. There is no reason why a
sterile thermometer should not be placed in the heated
tube in the first case, and in the second experiment one
of the test-tubes exposed under conditions similar to the
others might contain a thermometer which would show
the temperature of the contents of the tube containing it,
and so be an index of the rest.

Another method of accomplishing the same end is to
use Sternberg's bulbs. These are small glass bulbs
blown on one end of a piece of glass tubing, which is
subsequently drawn out to capillarity at the opposite end.
If such a bulb be heated, and its capillary tube dipped
into inoculated bouillon, in cooling, the fluid is drawn in
so as to fill it one-third or one-half. A number of these
tubes are filled in this manner with freshly inoculated
culture-medium and then floated, tube upward, upon
a water-bath whose temperature is gradually elevated,
the bulbs being removed from time to time as the
required temperatures are reached. Of course, as the
bulbs are already inoculated, all that is necessary is to
stand them aside for a day or two, and observe whether
or not the bacteria grow, judging the death-point exactly
as in the other case.

**To Determine the Antiseptic and Germicidal Value
of Reagents.**—There are various methods whose modifications can be elaborated according to the extent and
thoroughness of the investigation to be made.

I. *The Antiseptic Value.*—Remembering that an antiseptic is a substance that inhibits bacterial growth, the
method that will at once suggest itself is that of adding
varying quantities of the antiseptic to be investigated to
culture-media in which the bacteria are subsequently

planted. It is always well to use a considerable number of tubes. Bouillon is generally employed. If the antiseptic is non-volatile, it may be added before sterilization, which is to be preferred; but if it is volatile, it must be added by means of a sterile pipette, with the greatest precaution as regards asepsis, immediately before the test is to be made. Control-experiments—*i. e.* without the addition of the antiseptic—should always be made.

The results of antiseptic action are two: retardation of growth and complete inhibition of growth. As the tubes used for the study of the antiseptic are watched in their development, it will usually be noticed that those containing very small quantities develop almost as rapidly as the control-tubes; those containing more, a little more slowly; those containing still more, very slowly, until at last there comes at time when the growth is not deferred, but prevented.

Sternberg points out that certain circumstances may modify the results obtained. They are:

1. The composition of the nutrient media, with which the antiseptic may be incompatible.

2. The nature of the test-organism, no two organisms being exactly alike in their susceptibility.

3. The temperature at which the experiment is conducted, a relatively greater amount of the antiseptic being necessary at temperatures favorable to the organism than at temperatures unfavorable.

4. The presence of spores which are always more resistant than the asporogenous forms.

II. *The Germicidal Value.*—Koch's original method of doing this was to dry the micro-organisms upon sterile shreds of linen or silk, and then soak them for varying lengths of time in the germicidal solution. After the bath in the reagent the threads were washed in clean, sterile water and then transferred to fresh culture-media, and their growth or failure to grow observed. It will be observed that this method is aimed at the determination of the *time* in which a certain solution will kill.

Sternberg suggested a method in which the time should remain constant (two hours' exposure), and the object be the determination of the exact dilution of the reagent required to destroy the bacteria. " Instead of subjecting a few of the test-organisms attached to a silk thread to the action of the disinfecting agent, a certain quantity of the recent culture—usually 5 c.cm.—has been mixed with an equal quantity of a standard solution of the germicidal agent, . . . and after two hours' contact one or two öse-fuls would be introduced into a suitable nutrient medium to test the question of disinfection."

A very simple and popular method of determining the germicidal value is to make a series of dilutions of the reagent to be tested; add to each a couple of loopfuls of a fresh liquid culture, and at varying intervals of time transfer a loopful to fresh culture-media. By a little ingenuity this method may be made to yield information as to both time and strength.

When it is desired to secure information concerning the progress of the germicidal action of reagents, body-fluids, etc., especially in the unusual and interesting cases in which the material subjected to the test may exert a restraining action for a time only, or bring about destruction of some or many, but not all of the germs, the use of the Petri dish can be introduced.

For example, it is desired to determine whether a blood-serum is germicidal or not. Into about 5 c.cm. of the serum contained in a test-tube, two or three loopfuls of any desired bacterium, in liquid culture, are added. The tube is well agitated and immediately one loopful is transferred to a tube of melted gelatin, distributed through it, and poured into a Petri dish. After one minute the operation is repeated, in five minutes again, and so on as often as is desired.

The dishes are stood away until the bacteria develop into colonies, which are then counted with a Wolfhügel apparatus. On the first dish there may be 100 colonies; on the second, 80; on the third, 50; on the fourth, 20; on the fifth, 30; on the sixth, 150; on the seventh, 1000, etc.; indicating that the serum exerted a destructive action upon some but not all of the bacteria, and that this power disappeared after the lapse of a certain time, allowing the bacteria to develop *ad libitum.*

When the germicide to be studied is a gas, as in the case of sulphurous acid or formaldehyd, a different method must, of course, be adopted.

It may be sufficient simply to place a few test-tube cultures of various bacteria, some with plugs in, some with plugs out, in a closed room in which the gas is afterward evolved. The germicidal action is shown by the failure of the cultures to grow upon transplantation to fresh culture-media. This crude method may be supplemented by an examination of the dust of the room. Pledgets of sterile cotton are rubbed upon the floor, washboard, or any dust-collecting surface present, and subsequently dropped into culture-media. Failure of growth under such circumstances is very certain evidence of good disinfection. These tests are, however, very severe, for in the cultures there are immense numbers of bacteria in the deeper portions of the bacterial mass upon which the gas has no opportunity to act, and in the dust there are many sporogenous organisms of extreme resisting power. Failure to kill all the germs exposed in such manner is no indication that the vapor cannot destroy all the ordinary pathogenic organisms.

More refined is the method of saturating sterile sand or fragments of blotting-paper or absorbent cotton with cultures and exposing them, moist or dry, to the action of the gas. Such materials are best made ready in Petri dishes, which are opened immediately before and closed immediately after the experiment. A piece of cotton or blotting-paper or a little sand transferred to fresh culture-media will not give any growth where the disinfection has been thorough. By transplanting from different depths the sand may be used incidently to show to what depth the gas is capable of penetrating.

Easier of execution, but rather more severe, is a method in which cover-glasses are employed. A number of them are spread with cultures of various bacteria, allowed to dry, and then exposed to the gas as long as required. The cover-glasses are afterward dropped into culture-media to permit the growth of the germs not destroyed.

Animal-experiments may also be employed to determine whether or not a germ that has survived exposure to the action of reagents has its pathogenic power destroyed. An excellent example of this is seen in the case of the anthrax bacillus, a virulent form of which will kill rabbits, but after being grown in media containing an insufficient amount of a germicide to kill it will often lose its rabbit-killing power, though still able to fatally infect guinea-pigs, or may lose its virulence for both rabbits and guinea-pigs, though still able to kill white mice.

PART II. SPECIFIC DISEASES AND THEIR BACTERIA.

A. THE PHLOGISTIC DISEASES.

I. THE ACUTE INFLAMMATORY DISEASES.

CHAPTER I.

SUPPURATION.

SUPPURATION was at one time supposed to be an inevitable outcome of the majority of wounds, and, although bacteria were observed in the discharges, the old habit of thought and insufficiency of information caused most surgeons to believe that they were spontaneously developed there.

Lord Lister, whose name we cannot sufficiently honor, conceived that Pasteur's observations upon the germs of life floating in the atmosphere, if they explained the contamination of his sterile infusions, might also explain the changes in wounds, and upon this idea based that most successful system of treatment known as "antiseptic surgery."

The further development of antiseptic surgery, and the extremes to which it was carried by specialists, almost attain to the ridiculous, for not only were the hands of the operator, his instruments, sponges, sutures, ligatures, and dressings kept constantly saturated with irritating germicidal solutions, but at one time the air over the wound was carefully saturated with pulverized antiseptic lotions during the whole operation by means of a steam atomizer. This rather monstrous outcome of the application of Lister's system to surgery was the very natural result of the erroneous idea that the germs which cause

182

the suppurative changes in wounds entered the exposed tissues principally from the atmosphere, and that the hands and instruments of the operator, while certainly means of infection, were secondary in importance to it.

The researches of more recent date, however, have shown not only that the atmosphere cannot be disinfected, but also that the air of ordinarily quiet rooms, while containing the spores of numerous saprophytic organisms, very rarely contains many pathogenic bacteria. We now also know that a direct stream of air, such as is generated by an atomizer, causes more bacteria to be conveyed into a wound than would ordinarily fall upon it, thereby increasing instead of lessening the danger of infection. It may therefore be stated, with a reasonable amount of certainty, that the atmosphere is rarely an important factor in the process of suppuration.

We have already called attention to the fact that various micro-organisms are so intimate in their relation to the skin that it is almost impossible to get rid of them, and have cited in this relation the experiments of Welch, Robb, and Ghriskey, whose method of disinfecting the hands has been recommended as the best. The investigations of these observers have shown that, no matter how rigid the disinfection of the patient's skin, the cleansing of the operator's hands, the sterilization of the instruments, and the precautions exercised, a certain number of wounds in which sutures are employed will always suppurate. The cause of the suppuration is a matter of vast importance in surgery and in surgical bacteriology, yet it is one which it is impossible to remove. We carry it constantly with us upon our skins.

STAPHYLOCOCCUS EPIDERMIDIS ALBUS.

Welch has described, under the name *Staphylococcus epidermidis albus,* a micrococcus which seems to be habitually present upon the skin, not only upon the surface, but also deep down in the Malpighian layer. He is of the opinion that it is the same organism which is familiar to us under the name of Staphylococcus pyogenes albus, but in an attenuated condition. If his opinion be correct, and we have seated deeply in our derm a coccus which can at times cause abscess-formation, the conclusions of Robb and Ghriskey, that sutures of catgut when tightly drawn may be a cause of skin-abscesses by predisposing to the development of this organism, are certainly justifiable.

Not only does the coccus occur in the attenuated form described, but we have very commonly present upon the skin, generally as a harmless saprophyte, the important *Staphylococcus pyogenes albus*, which is a common cause of suppuration.

STAPHYLOCOCCUS PYOGENES ALBUS.

Although, as stated, the Staphylococcus pyogenes albus is a common cause of suppuration, it rarely occurs alone, the studies of Passet showing that in but 4 out of 33 cases which he investigated was this coccus found by itself. When pure cultures of the coccus are injected subcutaneously into rabbits and guinea-pigs, abscesses sometimes result; sometimes there is no result. Injected into the circulation of these animals, the staphylococci sometimes cause septicemia, and after death can be found in the capillaries, especially of the kidneys. From these illustrations it will be seen that the organism is feebly pathogenic.

In its vegetative characteristics the Staphylococcus albus is almost identical with the species next to be described, but differs from it in that there is no golden color produced. Upon the culture-media it grows white.

STAPHYLOCOCCUS PYOGENES AUREUS.

Generally present upon the skin, though in smaller numbers, is the dangerous and highly virulent *Staphylococcus pyogenes aureus* (Fig. 50), or "golden staphylococcus" of Rosenbach. As the morphology of this organism, and indeed the generality of its characters, are identical with those of the preceding species, it seems convenient to describe them together, pointing out such

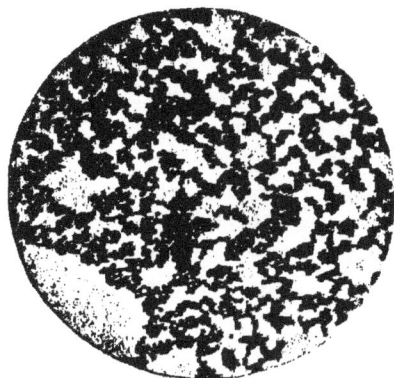

FIG. 50.—Staphylococcus pyogenes aureus, from an agar-agar culture; × 1000 (Günther).

differences as occur step by step. In doing this, however, it must not be forgotten that, although the Staphylococcus albus has been described first, the Staphylococcus aureus is the more common organism of the suppurative diseases.

Although they had been seen earlier by several observers, the staphylococci were not isolated and carefully described until 1884, when Rosenbach worked upon them. The results of his study, followed by Passet and a host of others, have now given us pretty accurate information about them.

The cocci are distributed rather sparingly in nature, seeming not to find a purely saprophytic existence a suitable one. They occur, however, wherever man and animals have been, and can be found in the dust of houses, hospitals, and especially surgical wards where proper precautions are not exercised. They are common upon the skin, they live in the nose, mouth, eyes, and ears of man, they are nearly always beneath the fingernails, and they sometimes occur in the feces, especially in children.

The cocci are rather small, measuring about 0.7 μ in diameter. When examined in a delicately-stained condition the organisms may be seen to consist of hemispheres separated from each other by a narrow interval. The contiguous surfaces are flat, thus differing from the gonococcus, whose contiguous surfaces are concave. The grouping is not very characteristic. In both liquid and solid culture-media the organisms either occur in solid masses or are evenly distributed. It is only in the organs or tissues of a diseased animal that it is possible to say that a true staphylococcus grouping is present.

The organism stains brilliantly with aqueous solutions of the anilin dyes. In tissues it can be beautifully stained by Gram's method.

The staphylococci grow well either in the presence or absence of oxygen at a temperature above 18° C., the most rapid development being at about 37° C. Upon the surface of gelatin plates small whitish points can be observed in forty-eight hours (Fig. 51). These rapidly

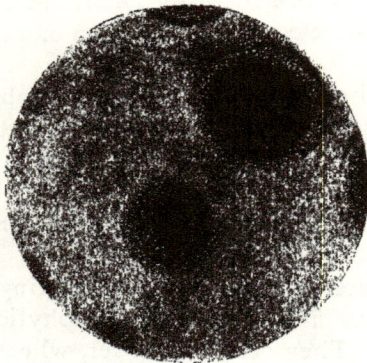

Fig. 51.—Staphylococcus pyogenes aureus: colony two days old, seen upon an agar-agar plate; ×40 (Heim).

extend to the surface and cause extensive liquefaction. Hand in hand with the liquefaction is the formation of an orange color, which is best observed at the centre of the colony. Under the microscope the colonies appear as round disks with circumscribed, smooth edges. They are distinctly granular and dark-brown. When the colonies are grown upon agar-agar plates the formation of the pigment is much more distinct.

In gelatin punctures the growth occurs along the whole length of the needle-track, and causes an extensive liquefaction in the form of a long, narrow, blunt-pointed, inverted cone (Fig. 52) full of clouded liquid, at the apex

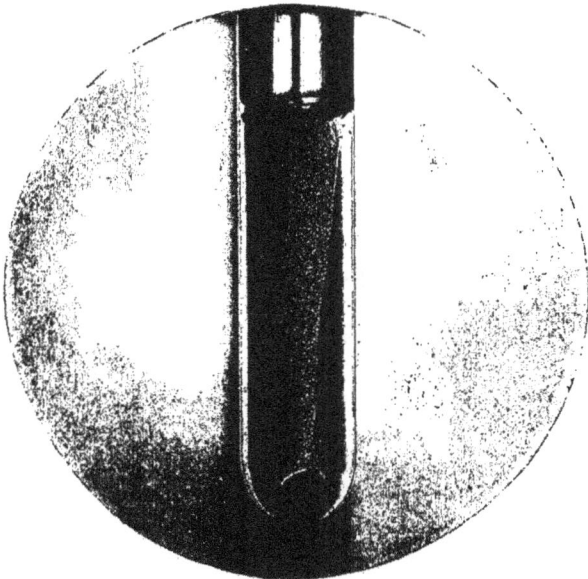

FIG. 52.—Staphylococcus pyogenes aureus: puncture-culture three days old in gelatin (Fränkel and Pfeiffer).

of which a collection of golden or orange-yellow precipitate is always present. It is this precipitate in particular that gives the organism its name, "golden staphylococcus."

The most characteristic growth is upon agar-agar. Along the whole line of inoculation an orange-yellow, moist, shining growth occurs. When the growth takes place rapidly, as in the incubator, it exceeds the rapidity of color-production, so that the centre of the growth is distinctly golden ; the edges may be white.

Upon potato the growth is luxuriant, producing an orange-yellow coating over a large part of the surface. The potato-cultures give off a sour odor.

When grown in bouillon the organism causes a diffuse cloudiness.

In milk coagulation takes place, and is followed by gradual digestion of the casein.

The Staphylococcus albus is exactly the same as the aureus, with the exception that in all media it is constantly colorless.

Experiments have shown that the Staphylococcus aureus, like its congener, the albus, exists in an attenuated form, and there is every reason to believe that in the majority of instances it inhabits the surface of the body in that condition.

When virulent the golden staphylococcus is a dangerous and often deadly organism. Its pathogeny among animals is decided. When introduced subcutaneously, abscesses almost invariably follow, except in a certain few comparatively immune species, and not infrequently lead to a fatal termination. In such cases the organisms may be cultivated from the blood of the large vessels, though by far the greater number collect in, and frequently obstruct, the capillaries. In the lungs and spleen, and still more frequently in the kidneys, infarcts are formed by the bacterial emboli. The Malpighian tufts of the kidneys sometimes are full of cocci, and become the centres of small abscesses.

The coccus is almost equally pathogenic for man, though the fatal outcome is much more rare. It enters the system through scratches, punctures, or abrasions, and when virulent generally causes an abscess, as various experimenters who inoculated themselves have discovered to their cost. Garré applied the organism in pure culture to the uninjured skin of his arm, and in four days developed a large carbuncle with a surrounding zone of furuncles. Bockhart suspended a small portion of an agar-agar culture in salt-solution, and scratched it gently into the deeper layers of the skin with his finger-nail; a furuncle developed. Bumm injected the coccus suspended in salt-solution beneath his skin and that of several other persons, and produced an abscess in every case.

The Staphylococcus aureus is not only found in the great majority of furuncles, carbuncles, abscesses, and other inflammatory diseases of the surface of the body, but also plays an important rôle in a number of deeply-seated diseases of the internal organs. Becker and others obtained it from the pus of osteomyelitis, demonstrating that if, after fracturing or crushing a bone, the staphylo-coccus was injected into the circulation, osteomyelitis would result. Numerous bacteriologists have demon-strated its presence in ulcerative endocarditis. Rodet has been able to produce osteomyelitis without previ-ous injury to the bones; Rosenbach was able to produce ulcerative endocarditis by injecting some of the staphy-lococci into the circulation in animals whose cardiac valves had been injured by a sound passed into the carotid artery; and Ribbert has shown that the injection of cultures of the organism may cause the valvular lesion without the preceding injury.

The Staphylococcus aureus is an easy organism to ob-tain, and can be secured by plating out a drop of pus in gelatin or in agar-agar. Such a preparation, however, generally does not contain the Staphylococcus aureus alone, but shows colonies of the Staphylococcus albus as well. In addition to these two principal forms, one sometimes discovers an organism identical with the pre-ceding, except that its growth on agar-agar and potato is of a brilliant lemon-yellow color, and its pathogeny for animals much less. This is the *Staphylococcus citreus* of Passet. It is not quite so common, and not so patho-genic as the others, and consequently much less im-portant.

Streptococcus Pyogenes.

Another organism whose colonies are frequently obtained from the pus containing the staphylococci is the *Streptococcus pyogenes* of Rosenbach (Fig. 53). It was found by him in 18 of 33 cases of suppurative lesions studied, fifteen times alone and five times with the Staphylococcus aureus. It is a spherical organism of variable size (0.4–1 μ in diameter), constantly

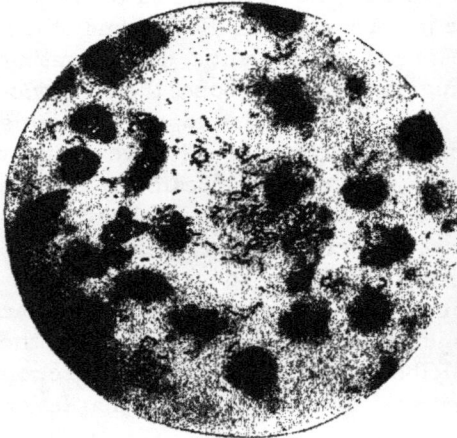

FIG. 53.—Streptococcus pyogenes, from the pus taken from an abscess; × 1000 (Fränkel and Pfeiffer).

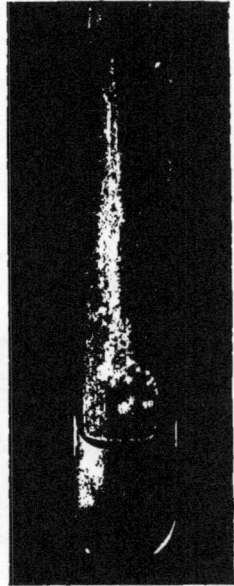

FIG. 54.—Streptococcus pyogenes: culture upon agar-agar two days old (Fränkel and Pfeiffer).

associated in pairs and chains of from four to twenty individuals. A special variety of it, known as Streptococcus longus, sometimes forms chains of more than one hundred members.

The organism stains well with ordinary aqueous solutions of the anilin dyes, and also by Gram's method. Like the coccus already described, it is not motile and does not seem to form spores, though sometimes a large individual —much larger than the others in its chain—may be observed, and may suggest the thought of arthro-sporulation.

Upon gelatin plates very small colonies of translucent appearance are observed. When superficial, they spread out to form flat disks about 0.5 mm. in diameter. The microscope shows them to be irregular and granular, to have a slightly yellowish color, and to have numerous irregularities around the edges, due to projecting chains of the cocci. No liquefaction occurs.

In gelatin puncture-cultures no liquefaction is observed. The minute spherical colonies grow along the whole needle-track and form a slightly opaque granular line.

Upon agar-agar an exceedingly delicate transparent growth develops slowly along the line of inoculation. It consists of almost transparent, colorless small colonies which do not become confluent.

The growth upon blood-serum much resembles that upon agar-agar. The streptococcus does not seem to grow upon potato.

In bouillon the cocci develop rather slowly, seeming to prefer a neutral or feebly acid reaction. The culture-medium remains clear, while numerous small flocculi are suspended in it. When the flocculi-formation is very distinct the name *Streptococcus conglomeratus* is used to describe the organism. These masses sometimes adhere to the sides of the tube; sometimes they form a sediment. Rarely, there is general clouding of the medium (*Streptococcus diffusus*).

In mixtures of bouillon and blood-serum or ascitic fluid the streptococcus grows much better, especially at incubation temperatures, and in such mixtures the luxuriant development causes the liquid to appear clouded.

The organism seems to grow well in milk, which is coagulated and digested.

The streptococcus is not very sensitive to acids, and can be grown quite well in media with a slightly acid reaction.

Sternberg found that the streptococci succumb to a temperature of 52°–54° C. continued for ten minutes.

Their vitality in culture is not great. Unless frequently transplanted they die. In bouillon they are said to die in five to ten days. On solid media they seem to retain their vegetative and pathogenic powers much longer. They resist drying well. Their growth in artificial media is accompanied by the production of an acid which probably acts destructively upon the bacteria themselves.

The Streptococcus pyogenes is generally not very pathogenic for animals. Subcutaneous injections into mice and rabbits are, as a rule, without either general or local manifestations of importance. If, however, an ear of a rabbit is carefully inoculated with a small amount of a pure culture, a small patch resembling erysipelas usually results. The disturbance passes away in a few days and the animal recovers.

If, however, the streptococcus is highly virulent, the rabbit dies in from twenty-four hours to six days from a general septicemia. The cocci may be found in large numbers in the heart's blood and in the organs. In less virulent cases minute disseminated abscesses are sometimes found.

According to Marmorek,[1] the virulence can be increased to a remarkable degree by rapid passage through rabbits, and maintained by the use of a culture-medium consisting of three parts of human blood-serum and one of bouillon. The blood of the ass, and ascitic and chest fluids may also be used. By these means Marmorek succeeded in intensifying the virulence of his culture to such a degree that one hundred millionth of a c.cm. injected into the ear vein was fatal to a rabbit.

Petruschky[2] found the virulence of the culture to be well retained if the culture was planted in gelatin, transplanted every five days, and when grown kept on ice.

Holst[3] succeeded in keeping an exceedingly virulent Streptococcus brevis on artificial culture-media for eight

---

[1] *Ann. de l' Inst. Pasteur*, Tome ix., No. 7, July 25, 1895, p. 593.
[2] *Centralbl. für Bakt. und Parasitenk.*, Bd. xviii., No. 16, May 4, 1895, p. 551.   [3] *Ibid.*, Bd. xix., No. 11, Mar. 21, 1896.

years without any particular precautions and found its virulence unchanged.

Probably the virulence and attenuation are peculiarities of the organism itself.

Dried streptococci are said by Frosch and Kolle to retain their energies longer than those growing on culture-media.[1]

Like the staphylococci, the streptococcus is frequently associated with internal diseases, and has been found in erysipelas, ulcerative endocarditis, periostitis, otitis, meningitis, emphysema, pneumonia, lymphangitis, phlegmons, sepsis, and in the uterus in cases of infective puerperal endometritis. In man the streptococci occur in the most active forms of suppuration. Its relation to diphtheria is of interest, for, while, in all probability, the great majority of cases of pseudomembranous angina are caused by the Klebs-Löffler bacillus, yet an undoubted number of cases are met with in which, as in Prudden's 24 cases, no diphtheria bacilli can be found, but which seem to be caused by a streptococcus exactly resembling that under consideration.

There is no clinical difference in the picture of the throat-lesion produced by the two organisms, and the only positive method of diagnosticating the one from the other is by means of a careful bacteriologic examination. Such an examination should always be made, as it has much weight in connection with the treatment. Of course, in streptococcus angina no benefit could be expected from the diphtheria antitoxic serum.

Hirsh[2] has shown that under pathological conditions streptococci are by no means rare organisms in the intestinal canal of infants, and may cause a streptococcic enteritis. In these cases the organisms are found in large numbers in the stomach and in the stools, and later in the course of the disease in the blood and urine of the living child and in the internal organs of the cadaver.

[1] Flügge's *Die Mikroörganismen.*
[2] *Centralbl. für Bakt. und Parasitenk.*, Bd. xxii., Nos. 14 and 15, p. 369.

13

Liebman[1] reports two cases of streptococcic enteritis that were cerefully studied bacteriologically.

Flexner,[2] in a series of autopsies upon cases of death from various diseases, found the bodies invaded by numerous micro-organisms, causing what he has called "terminal infections," and hastening the fatal issue. Of 793 autopsies at Johns Hopkins Hospital, 255 from chronic heart or kidney diseases, or both, were sufficiently well studied bacteriologically to meet the needs of a statistical inquiry. Tubercular infection was not included. Of the 255 cases, 213 gave positive bacteriological results. "The micro-organisms causing the infections, 38 in all, were the Streptococcus pyogenes, 16 cases; Staphylococcus pyogenes aureus, 4 cases; Micrococcus lanceolatus, 6 cases; gas bacillus (B. Aërogenes capsulatus), three times alone and twice combined with the Bacillus coli communis; the gonococcus, anthrax bacillus, Bacillus proteus, the last combined with the Bacillus coli, the Bacillus coli alone, a peculiar capsulated bacillus, and an unidentified coccus."

It is interesting to observe how many cases were accompanied by the streptococcus. All the streptococci may not have been streptococcus pyogenes, but for convenience in his statistics they were regarded by Flexner as identical.

The streptococcus of Rosenbach is thought by many to be identical with a streptococcus described by Fehleisen as the *Streptococcus erysipelatis* (Fig. 55). The two organisms have much in common, but much difference of opinion exists upon the subject of their identity. It may seem unwise to omit the Streptococcus erysipelatis as a major topic for discussion, but the similarity of the organism to that just described has caused us to consider them in the same connection.

The streptococci of erysipelas can be obtained in almost pure culture from the serum which oozes from a puncture made in the margin of an erysipelatous patch. They are

---

[1] *Centralbl. für Bakt. und Parasitenk.*, Bd. xxii., Nos. 14 and 15, p. 376.
[2] *Journal of Experimental Medicine*, vol. i., No. 3, 1896.

small cocci, forming long chains—generally from six to ten individuals, but sometimes reaching a hundred in

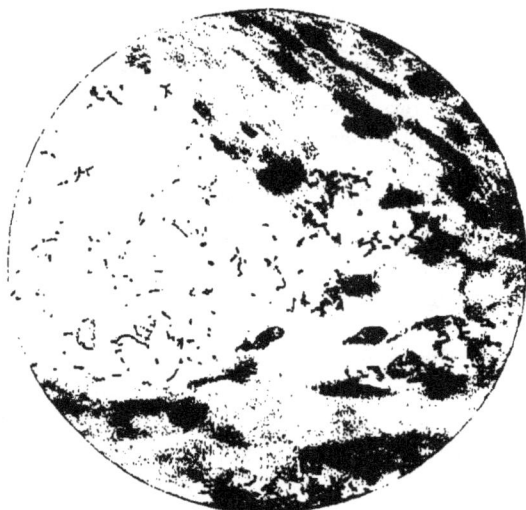

FIG. 55.—Streptococcus erysipelatis, seen in a section through human skin; × 500 (Fränkel and Pfeiffer).

number. Occasionally the chains can be found collected in tangled masses. They can be cultivated at the room-temperature, but grow much better at 30–37° C. They are not particularly sensitive to the absence of oxygen, but develop a little more rapidly in its presence.

The erysipelas cocci, like the Streptococcus pyogenes, are not motile, form no spores, and are destroyed by a low degree of heat. They stain well with aqueous solutions of anilin dyes and also by Gram's method.

The colonies upon gelatin and the development in gelatin tubes, upon agar-agar, and upon blood-serum are identical with the descriptions of the Streptococcus pyogenes. No growth occurs on potato.

The growth in bouillon is generally luxuriant, and in a short time causes the medium to be filled with chains of the cocci. As the growth progresses these chains gather in clusters and fall to the bottom as a whitish granular precipitate, above which the liquid remains clear.

When injected into animals Fehleisen's coccus behaves exactly like the Streptococcus pyogenes.

Observation has shown that dire results may follow the entrance of this organism into exposed wounds, and that it causes not only local suppuration, but sometimes a general infection.

The empiric experience that the occasional accidental infection of malignant tumors with erysipelas cocci was followed by sloughing and subsequent disappearance of the tumor, suggested inoculation with the Streptococcus erysipelatis as a therapeutic measure. The dangerous character of the remedy, however, caused many to refrain from its use, for when one inoculated the living erysipelas germs into the tissues he never could estimate the exact amount of disturbance that would follow. The difficulty seems to have been overcome by Coley, who recommends the toxin instead of the living coccus for injection. A virulent culture is obtained, inoculated into small flasks of slightly acid bouillon, allowed to grow for three weeks, then reinoculated with Bacillus prodigiosus, allowed to grow for ten or twelve days at the room-temperature, well shaken up, poured into bottles of about f℥ss capacity, and rendered perfectly sterile by an exposure to from 50–60° C. for an hour. It is claimed that the combined toxins of erysipelas and prodigiosus are much stronger than the simple erysipelas toxin. The best effects are found in cases of sarcoma, where the toxin causes a rapid necrosis of the tumor tissue, which can be scraped out with an appropriate instrument. Numerous cases are on record in which this treatment has been most efficacious; but, although Coley recommends it and Czerny still upholds it, the majority of surgeons have failed to secure the desired results.

Recently (1895) considerable attention has been bestowed upon the anti-streptococcus serum of Marmorek, which is said to act specifically upon cases of streptococcus-infection, both general and local. Numerous cases are upon record in which the serum seemed to exert a beneficial action.

It would seem as if an antiphlogistic serum should occupy an important place in the future of medicine.

The serum is prepared upon the same plan as that of Behring, except that living virulent streptococci instead of the sterile toxin are injected into the horse.

BACILLUS PYOCYANEUS.

In some cases the pus evacuated from wounds exhibits a peculiar bluish or greenish color, from the presence of

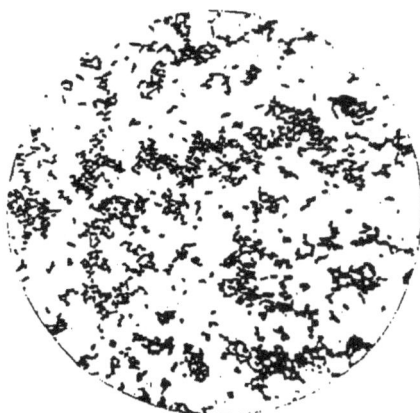

FIG. 56.—Bacillus pyocyaneus, from an agar-agar culture; × 1000 (Itzerott and Niemann).

the *Bacillus pyocyaneus* (Figs. 56, 57). This is a short, delicate bacillus of small size, measuring 0.3 : 1–2 μ, according to Flügge, frequently united in chains of four or six. It has round ends, is actively motile, has one terminal flagellum, does not form spores, and can exist with or without oxygen, though it is an almost purely aërobic organism.

It stains well with the ordinary solutions, but does not retain the color by Gram's method.

The superficial colonies upon gelatin plates form small, irregular, ill-defined collections, which produce a fluorescence of the neighboring gelatin. The gelatin softens gradually, and about five days elapse before liquefaction is complete.

The microscope shows the colonies to be round, coarsely-granulated masses

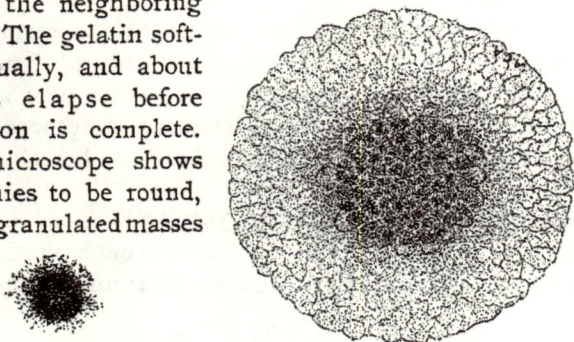

FIG. 57.—Bacillus pyocyaneus: colonies upon gelatin (Abbott).

with notched or filamentous borders. They have a yellow-green color. Upon the surface they form a delicate clump with a smooth surface, finely granular, distinctly green in the middle and pale at the edges. The colonies sink into the gelatin as the liquefaction progresses.

In gelatin puncture-cultures most of the development occurs at the upper part of the tube, where a deep saucer of liquefaction forms. The growth slowly descends into the medium, and is the point of origin of a beautiful fluorescence. The bacterial growth sinks to the bottom as it ages. At times a delicate mycoderma forms on the surface.

Upon agar-agar the growth is at first bright green, developing all along the line of inoculation. The green pigment (fluorescin) is soluble, and soon saturates the culture-medium and makes it very characteristic. As the culture ages, or if the medium upon which it grows contains much peptone, a second pigment (pyocyanin) is developed, and the bright green fades to a deep blue-green, dark-blue, or in some few cases to a deep reddish-brown.

A well-known feature of the growth upon fresh agar-agar, upon which much stress has recently been laid by

Martin is the formation of crystals in fresh cultures. Crystal-formation in cultures of other bacteria usually takes place in old, partially dried agar-agar cultures. The bacillus pyocyaneus, however, produces crystals in a few days upon fresh media. In my experience freshly isolated bacilli manifest this capability more markedly than those which have been for some time part of the laboratory stock of cultures, and subject to frequent transplantation.[1]

Upon potato a luxuriant greenish or brownish, smeary layer is produced. Milk is coagulated and peptonized.

This bacillus is highly pathogenic for laboratory animals. About 1 c.cm. of a fresh bouillon culture, if injected into the subcutaneous tissue of a guinea-pig or a rabbit, causes a rapid edema, a suppurative inflammation, and death in a short time (twenty-four hours). Sometimes the animal lives for a week or more, then dies. There is a marked hemorrhagic subcutaneous edema at the seat of inoculation. The bacilli can be found in the blood and in most of the tissues.

When the dose is too small to prove fatal, suppuration occurs in many cases.

When sterilized cultures are injected, the same results follow, a relatively larger quantity, of course, being required.

Intraperitoneal injections cause suppurative peritonitis.

The organism has been found in the human being in the pus in cases of middle-ear disease (often in pure culture), panophthalmia, bronchopneumonia, inflammations of the nasal fossæ, meningitis, etc. Escaping from such local lesions into the blood it sometimes causes nephritis.

It may, however, be stated that ordinarily the bacillus is harmless for human beings, the above-mentioned examples of pathogenic activity being marked exceptions.

It is interesting to observe, in passing, that this pathogeny can be set aside by the immunity which develops after a few inoculations with sterilized cultures. These

[1] See *Centralbl. f. Bakt.*, xxi., April 6, 1897, p. 473.

are easily prepared, as the thermal death-point determined by Sternberg is 56° C.

The bacillus appears to be rather common as a saprophyte, and, as it has been found in the perspiration, probably is not uncommon upon the skin.

Before leaving the subject of suppuration attention must be called to several rather common bacteria which may at times be the cause of troublesome suppuration. Among these are the pneumococcus of Fränkel and Weichselbaum, the typhoid bacillus, and the Bacillus coli communis (*q. v.*).

The *pneumococcus* has not infrequently been discovered most unexpectedly in abscesses of the brain and other deep-seated organs, and seems to have powerful chemotactic powers. For a careful consideration of it the reader must be referred to the chapter upon Pneumonia, where it is considered in full.

The *Bacillus coli communis*, which is always present in the intestine, seems at times to enter the blood- or lymph-channels and stimulate suppuration, and numerous cases are on record showing this. The points most frequently attacked seem to be the bile-ducts and the vermiform appendix, though the significance of the organism in appendicitis has no doubt been overrated. It has also been found in the kidney in scarlatinal nephritis, and is thought to be the exciting cause of some cases. It was originally described by Passet as the *Bacillus pyogenes fœtidus.* For a more particular study of this organism the reader is referred to the chapter devoted to its consideration.

The *Bacillus typhosus* is probably less frequently a cause of suppuration than either of the others, yet it seems to be the occasional cause of the purulent sequelæ of typhoid fever. A case has recently been reported by Flexner in which metastatic abscesses were found to be caused by it.

The *Micrococcus tetragenus* has also been found in the pus of acute abscesses: it is quite common in the cavities of pulmonary tuberculosis, and may aid in the destructive processes involved in the general phthisical infection.

MICROCOCCUS GONORRHŒÆ.

All authorities now accept the "gonococcus" to be the cause of gonorrhea. It was first observed in the urethral and conjunctival secretions of gonorrhea and purulent ophthalmia by Neisser in 1879. The organisms are of hemispherical shape, arranged in pairs, so that the inner surfaces are separated from each other by a narrow interval. Sometimes, instead of pairs of cocci, fours are seen, the group no doubt resulting from the division of a pair.

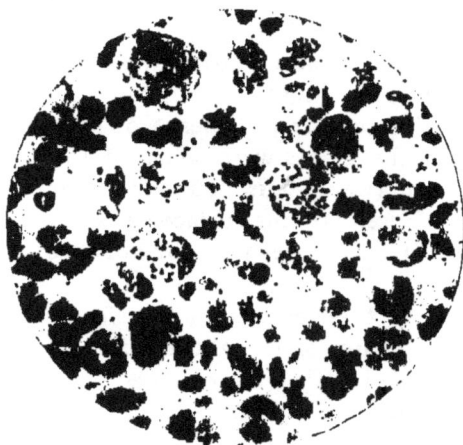

FIG. 58.—Gonococcus in urethral pus; × 1000 (Fränkel and Pfeiffer).

The described hemispherical shape is not exactly correct, for a good lens generally shows the approximated surfaces to be somewhat concave rather than flat. The Germans see in the organism a resemblance to their popular biscuit called a "semmel."

The gonococcus is small, is not motile, like other cocci, is not provided with flagella, and does not have spores. It stains readily with all the aqueous anilin dyes—best with rather weak solutions—but not by Gram's method. It can be found in the urethral discharges of gonorrhea from the beginning until the end of the disease, though in the later days its numbers may be outweighed by other organisms. Wertheim cultivated the gonococcus from a case of chronic urethritis of two years' standing, and proved its virulence by producing with it gonorrhea in a human being. The organisms are generally found within the pus-cells (Fig. 58) or attached to the surface of epithelial cells, and should always be sought for as diagnostic of gonorrhea, especially as urethritis sometimes is caused by other organisms, as the Bacillus coli communis[1] and the Staphylococcus pyogenes.

The cultivation of the gonococcus is not an easy task, but one which requires considerable bacteriologic skill. Wertheim accomplished it by diluting a drop of the pus in a little liquid *human blood-serum*, then mixing this with an equal part of melted 2 per cent. agar-agar at 40° C., and pouring into Petri dishes. As soon as the media became firm the dishes were stood in the incubator at 37° C., and in twenty-four hours the colonies could be observed. Those upon the surface showed a dark centre, around which a delicate granular zone could be made out.

When one of these colonies is transferred to a tube of human blood-serum or the above mixture obliquely coagulated, isolated little gray colonies occur; later these become confluent and produce a delicate smeary layer upon the medium. The main growth is surrounded by a thin, veil-like extension which gradually fades away into the medium. A slight growth occurs upon the water of condensation.

Turro says that the gonococci may also be cultivated upon acid gelatin, upon gelatin containing acid urine, and also in acid urine itself, in which the gonococci grow near the surface, while the pus-cocci which may be mixed with them sink deeper into the medium. His work has not been confirmed by other investigators.

Heiman,[2] who made an extensive series of culture-ex-

[1] Van der Pluyn and Loag: *Centralbl. f. Bakt. u. Parasitenk.*, Bd. xvii., Nos. 7, 8, Feb. 28, 1895, p. 233.
[2] *Med. Record*, Dec. 19, 1886.

periments, finds that the gonococcus grows best in a mixture of 1 part of pleuritic fluid and 2 parts of 2 per cent. agar. Wright[1] prefers a mixture of urine, blood-serum, peptone, and agar-agar.

It is ordinarily presumed that gonorrhea cannot be communicated to animals, but Turro asserts that the gonococci when grown upon acid gelatin readily communicate urethritis to dogs, and that no *lœsio continui* is necessary, the simple introduction of the organisms into the meatus sufficing to produce the disease.

The injection of gonococci into the subcutaneous tissue does not produce abscess.

There is no doubt that the gonococcus causes gonorrhea, as it has on several occasions been intentionally inoculated into the human urethra with resulting typical gonorrhea. It is constantly present in the disease, and very frequently also in the sequelæ—endometritis, salpingitis, oöphoritis, cystitis, peritonitis, arthritis, conjunctivitis, endocarditis, etc.—and, so far as can at present be determined, is never found under normal conditions.

In the beginning of their activities the cocci grow in the superficial epithelial cells, but soon penetrate between the cells to the deeper layers, where they continue their irritation as the superficial cells desquamate. Authorities differ as to whether the gonococci can penetrate squamous and columnar epithelium with equal facility.

The periurethral abscesses that occur in the course of gonorrhea are generally due to the Staphylococci aureus and albus, not directly to the gonococcus.

In certain of the remote secondary inflammations the gonococci disappear after a time, and either the inflammation subsides or is maintained by other bacteria. In synovitis this does not seem to be true, and the inflammation excited may last for months.

As long as the gonococci persist the patient may spread contagion. It must be pointed out that after apparent recovery from the disease the cocci sometimes remain

[1] *Jour. of the Amer. Med. Assoc.*, Feb., 1895.

latent in the urethra, and cause a relapse if the patient partake of some substance, as alcohol, irritating to the mucous membranes. Bearing this in mind, patients should not too soon be discharged as cured.

The gonococci are not easily killed, but withstand drying very well. Kratter was able to demonstrate their presence upon washed clothing six months after the original soiling, and also found that they still stained well.

Bumm found cocci similar to the gonococcus in the urethra, and points out that neither the shape nor the position in the cells is positively characteristic, but that, in addition, there must be refusal to stain by Gram's method before we can say with certainty that cocci found in urethral pus are gonococci.

All of the urethral inflammations do not depend upon the gonococcus, and in true gonorrhea all of the inflammatory symptoms do not depend upon the gonococcus, as the epithelial denudation following the disease permits the entrance of the common pus cocci of the urethra into the peri-urethral tissues. The peri-urethral abscesses and salpingitis, etc., not infrequently depend upon the ordinary pus cocci, and I have seen a case of gonorrhea with double orchitis and general septic infection, with endocarditis, in which the gonococci had no *rôle* in the sepsis, which was caused by a large dumbbell-coccus that stained beautifully by Gram's method.

MUMPS, OR EPIDEMIC PAROTITIS.

This epidemic, infectious disease of childhood, characterized by enlargement of the parotid and submaxillary glands, and rarely of the testicles, ovaries, and mammae, has not been proved to have a specific micro-organism.

Pasteur thought the disease due to bacilli which he found in the blood. Capitan and Charrin[1] and Olivier found in the blood, urine, and saliva both cocci and bacilli, but their studies are too early, and hence too crude to be of any value.

[1] *Comptes Rendu Soc. de Bioc. de Paris,* May 28, 1881.

Bouchard, Boisnet, and Bordas also found micro-organisms in the blood and saliva.

Netter, Laveran, Catrin, Mecray, and Walsh have all studied cases and isolated a diplococcus thought to be specific. The organism is described as occurring in pairs and in fours, sometimes in zoöglea. It grows slowly in the ordinary media, clouding bouillin in twenty-four hours, and appearing on gelatin after forty-eight hours as small white punctiform colonies which develop very slowly and liquefy some considerable time after coalescence. It grows on potato, and has a whitish appearance not easy to detect. Laveran and Catrin found the organism in 67 out of 72 cases examined. In their method a few drops of exudate are withdrawn from the inflamed gland with a hypodermic needle, some of the negative results being due to the fact that the needle withdrew no exudate. The blood gave pure cultures in 10 out of 15 trials.

Mecray and Walsh report that by disinfecting the mouths of patients, suffering from mumps, with a saturated boric acid solution, and cleansing Stensen's duct, by careful massage expressing its secretion, and then allowing a piece of cotton saturated with a boric acid solution to remain for five minutes between the orifice of the duct and the jaw, they were able to secure from the interior of the duct upon a bougie of sterile catgut a micrococcus identical with that Laveran had found. Of tubes inoculated with the contents of Stensen's duct 6 gave a mixed growth. All, however, showed the diplococcus. Out of 8 carefully made blood examinations, 3 gave pure cultures of the coccus and 3 mixed cultures; 2 were negative.

From Stensen's duct in healthy children they obtained the various oral bacteria, but not the diplococcus found in the cases of mumps. The experimenters do not think it possible that this diplococcus is the Staphylococcus epidermidis albus, as its growth is slower and the liquefaction of gelatin is accomplished only after a longer time than is required by the staphylococcus. They did not succeed in producing mumps in animals. In their experience a dog was encountered which suffered from swelling of the parotids, malaise, etc., after playing with a child suffering from mumps.

Concerning the diplococcus, it appeared in twos and fours; rarely in larger groups. Each was regularly rounded and about the size of the pus cocci. The colonies are small, white, glistening, distinctly defined, regularly circular spots, at first discrete and of slow growth, gradually coalescing. The slow growth is characteristic. In studying pure cultures, some gelatin tubes three days after inoculation were set aside, no growth being noted; three days later the small white colonies became distinctly visible. At ordinary temperatures gelatin is not liquefied until ten or twelve days, and the liquefaction proceeds slowly. A faint white streak appears on potato on the third day, and spreads as a delicate whitish film. The growth upon blood-serum is more rapid than on other media, but the colony is not so distinctly white in color. Litmus milk is changed to pink on the third day and is coagulated. Milk is thought to be an excellent nutrient medium, and a possible ready means of spreading contagion.

In the paper of Mecray and Walsh no mention is made of the relation of the cocci to pus cells or other organized constituents of the secretion from which they were obtained; no animal inoculations were done and nothing is said about the reaction to Gram's method of staining or possible motility the cocci might possess.

Michaelis and Bein,[1] of Leyden's clinic, found a diplococcus (previously observed by Leyden in the sputum), which occurred chiefly in the pus cells. In severe cases of the disease, which they studied by culture and microscopic section, the organism was not only secured from Stensen's duct, but in 2 cases from the pus of an abscess (parotid?) and in 1 case from the blood.

[1] *Deutsche med. Wochenschrift,* May 13, 1897.

In spite of the small number of cases studied, they were of the opinion that their coccus is the specific one. It is about 1 μ in size and resembles the gonococcus, though it is smaller. The cocci generally lie in the cells, sometimes 8 or 10 in one pus cell, and are occasionally distributed throughout the pus in long chains or strings. They stain readily with the usual anilin dyes, especially with Löffler's methylene-blue, and can be decolorized by the Gram method. They grow slowly upon the ordinary media, forming living, transparent, dew-like points on agar-agar. These little drops do not coalesce. In peptone-bouillon they form white, rather granular than flocculent deposit, the bouillon itself remaining clear. The growth is said to be more rapid in strongly than feebly alkaline media. The cocci are said to grow upon ascites-fluid and upon milk, the latter coagulating in the course of forty-eight hours. They are capable of slight movement. Numerous inoculation experiments were made, only one animal, a white mouse, succumbing. Control-experiments failed to disclose the same organisms in the healthy human parotid or its secretion.

All the observers agree in finding in the secretions of the gland and in the blood diplococci that grow slowly, produce small colonies, and coagulate milk. No one has shown their specificity by inoculation, evidence of course necessary before the claim of real importance can be accepted.

CHAPTER I.

TUBERCULOSIS.

TUBERCULOSIS is one of the most dreadful and, unfortunately, most common diseases of mankind. It affects alike the young and the old, the rich and the poor, the male and the female, the enlightened and the savage. Nor do its ravages cease with human beings, for it is common among animals, occurring with great frequency among cattle, less frequently among goats and hogs, and sometimes, though rarely, among sheep, horses, dogs, and cats.

Wild animals under natural conditions seem to escape the disease, but when caged and kept in zoölogical gardens even the most resistant of them—lions, tigers, etc.—are said at times to succumb to it, while it is the most common cause of death among captive monkeys.

The disease is not even limited to mammals, but occurs in a somewhat modified form in birds, and, it is said, even at times affects reptiles.

It is not a disease of modern times, but one which has persisted through centuries; and though, before the advent of the microscope, not always clearly separated from cancer, it has not only left unmistakable signs of its existence in the early literature of medicine, but has also imprinted itself upon the statute-books of some countries, as Naples, where its ravages were great and the means taken for its prevention radical.

While the great men of the early days of pathology clearly saw that the time must come when the parasitic

208

nature of this disease would be proved, and some, as
Klebs, Villemin, and Cohnheim, were "within an ace"
of the discovery, it remained for Robert Koch to succeed
in demonstrating and isolating the specific bacillus, now
so well known, and to write so accurate a description of
the organism and the lesions it produces as to render it
almost unparalleled in medical literature.

The tubercle bacillus (Fig. 59) is a rod-shaped organ-

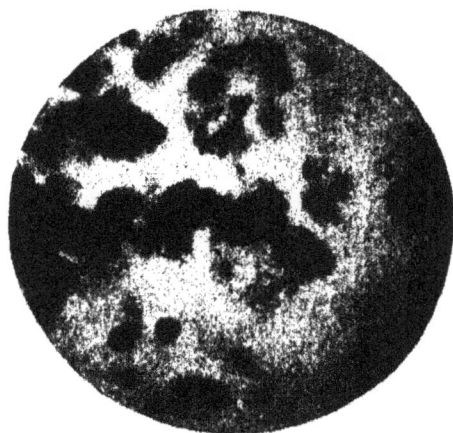

FIG. 59.—Section of a peritoneal tubercle from a cow, showing the tubercle
bacilli; × 500 (Fränkel and Pfeiffer).

ism with rounded ends and a slight curve, measuring
from 1.5–3.5 μ in length and from 0.2–0.5 μ in breadth.
It very commonly occurs in pairs, which may be associated end to end, but generally overlap somewhat and
are not attached to each other. It is very common to
observe a peculiar beaded appearance in organisms found
in pus and sputum (Fig. 60), due to the contraction of
fragmented protoplasm within the resisting capsule (?).
By some these fragmentations are thought to be bacilli
in the stage of sporulation (see Fig. 61). Koch originally held this view himself, but researches have not been
able to substantiate the opinion, and at present the evi-

14

dences *pro* and *con* point more strongly in the negative than in the positive direction.

The fragments do not look like the spores of any other organisms. When spores occur in the continuity of

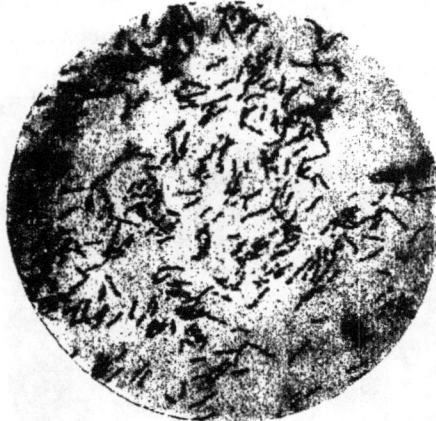

FIG. 60.—Tubercle bacillus in sputum (Fränkel and Pfeiffer).

bacilli, they are generally discrete oval refracting bodies easily recognized. The fragments seen in the tubercle bacillus are irregular and biconcave instead of oval, have

FIG. 61.—Tubercle bacilli: 1, forms suggesting sporulation; 2, forms described as beaded; the open spaces in the fragmented rods are sometimes mistaken for spores.

ragged surfaces, and are without the refraction peculiar to the ordinary spore.

The spaces between the bacillary fragments cannot be made to stain like the spores of other species. Finally, all known spores resist heat more strongly than the fully-developed bacilli, but experimentation has shown that these degenerative forms are no more capable of resisting heat than the tubercle bacilli themselves.

The organism is not motile, and does not possess flagella.

The tubercle bacillus is peculiar in its reaction to the anilin dyes. It is rather difficult to stain, requiring that the dye used shall contain a mordant (Koch), but it is also very tenacious of the color once assumed, resisting the decolorizing power of strong mineral acids (Ehrlich).

These peculiarities delayed the discovery of the bacillus for a considerable time, but now that we are familiar with them they give us a most valuable diagnostic character, for with the exception of the bacillus of lepra no known bacillus reacts in exactly the same way.

Koch first stained the bacillus with an aqueous solution of a basic anilin dye to which some potassium hydrate was added, subsequently washing with water and counter-staining with vesuvin. Ehrlich subsequently modified Koch's method, showing that pure anilin was a better mordant than potassium hydrate, and that the use of a strong mineral acid would remove the color from everything but the tubercle bacillus. This modification of Koch's method given us by Ehrlich is at the present time acknowledged to be the best method of staining the bacillus. Many other methods have been suggested, all of them, perhaps, more convenient than Ehrlich's, but none so good.

As being that most frequently performed by the physician, we will first describe the method of seeking the bacillus in sputum.

If one desires to be very exact in his examination, it may be well to have the patient cleanse the mouth thoroughly upon waking in the morning, and after the first fit of coughing expectorate into a clean wide-mouthed bottle. The object of this is to avoid the presence of fragments of food in the sputum.

The physician will secure a better result if the examination be made on the same day than if he wait a number of days, because if the bacilli are few they occur most plentifully in the small caseous flakes to be described farther on, which are easily found at first, but which break up and become part of a granular sediment that always forms in decomposed sputum.

The fresh sputum when held over a black surface generally shows a number of grayish-yellow, irregular, translucent granules somewhat smaller than the head of a pin. These consist principally of the caseous material from tuberculous tissue, and are the most valuable part of the sputum for examination. One of the granules is picked up with a pointed match-stick and spread over the surface of a perfectly clean cover-glass. If no such fragment can be found, the purulent part is next best for examination. The mucus itself rarely contains bacilli when free from scraps of tissue and pus.

In cases in which this ordinary procedure fails to reveal bacilli whose presence is strongly indicated by the clinical signs, the exact method of searching for them is to partially digest the sputum with caustic potash, and then collect the solid matter with a centrifugal apparatus. If a very few bacilli are present in the sputum, this method will often secure them.

The material spread upon the cover-glasses should not be too small in amount. Of course a massive, thick layer will become opaque in staining, but should the layer spread be, as is often advised, "as thin as possible," there may be too few bacilli upon the glass to enable one to make a satisfactory diagnosis.

As usual, the material is allowed to dry thoroughly, and is then passed three times through the flame for purposes of fixation.

Ehrlich's Method, or the Koch-Ehrlich Method.—The cover-glasses thus prepared are floated, smeared side down, upon, or immersed, smeared side up, in, a small dish of Ehrlich's anilin-water gentian-violet solution :

Anilin, 4,
Saturated alcoholic solution of gentian violet, 11,
Water, 100,

and placed in an incubator or a paraffin oven, and kept for twenty-four hours at about the temperature of the body. When removed from the stain they are washed momentarily in water, and then alternately in 25–33 per cent. nitric acid and 60 per cent. alcohol, until the blue color of the gentian violet is almost entirely lost. It must be remembered that the action of the strong acid is a powerful one, and that too long a time must not be allowed for its application. A total immersion of thirty seconds is quite enough in most cases. After final thorough washing in 60 per cent. alcohol the specimen is counter-stained in a dilute aqueous solution of Bismarck brown or vesuvin. The excess of stain is then washed off in water, and the specimen is dried and mounted in balsam. The tubercle bacilli will appear of a fine dark blue, while the pus-corpuscles, epithelial cells, and other bacteria, having been decolorized by the acid, will be colored brown by the counter-stain.

This method, requiring twenty-four hours for its completion, is naturally one which has fallen into disuse for practitioners who desire in the briefest possible time to know simply whether bacilli are present in the sputum or not.

Among clinicians Ziehl's method with carbol-fuchsin has met with great favor. After having been spread, dried, and fired, the cover-glass is held in the bite of an appropriate forceps (cover-glass forceps), and the stain[1] dropped upon it from a pipette. As soon as the entire cover-glass is covered with stain it is held over the flame of a spirit-lamp or a Bunsen burner until the stain begins to volatilize a little, as indicated by a white vapor. When

---

[1] Carbol-fuchsin (see p. 86):

    Fuchsin, 1;
    Alcohol, 10;

this is observed, the heating is sufficient, and the temperature can be subsequently maintained by intermittent heating.

If evaporation is allowed to take place, a ring of incrustation occurs at the edge of the area covered by the stain and prevents the proper action of the acid. To prevent this more stain should now and then be added. The staining is complete in from three to five minutes, after which the specimen is washed off with water, the excess of water absorbed with paper, and 3 per cent. hydrochloric acid in 70 per cent. alcohol, 25 per cent. aqueous sulphuric, or 33 per cent. aqueous nitric acid solution dropped upon it for thirty seconds, or until the red color is just extinguished. The acid is washed off with water, and the specimen is dried and mounted in Canada balsam. Nothing will be colored except the tubercle bacilli, which will appear red.

Gabbett modified the staining by adding methylene blue to the acid solution, which he makes according to this formula:

Methyl blue,	2 ;
Sulphuric acid,	25 ;
Water,	75.

In Gabbett's method, after staining with carbol-fuchsin the specimen is washed with water, acted upon by the methylene-blue solution for exactly thirty seconds, washed with water until only a very faint blue remains, dried, and finally mounted in Canada balsam. By this method the tubercle bacilli are colored red, and the pus-corpuscles, epithelial cells, and the unimportant bacteria blue.

The possible relation that the number of bacilli in the expectoration of consumptives might bear to the progress or treatment of the case has been elaborately investigated by Nuttall.[1] The total quantity of sputum expectorated in twenty-four hours was caught in covered, scrupulously

[1] *Bull. of the Johns Hopkins Hospital*, May and June, 1891, ii., 13.

clean conical glasses and measured therein. The proportion of muco-purulent to fluid matter was noted. Depending upon the degree of viscidity and number of bacilli present in the sputum, a varying amount of 5 per cent. caustic potash solution was added to it (from one-sixth to an equal volume), and after the caustic potash had rendered the sputum perfectly fluid more or less water was added to dilute the mixture. The sputum, having been measured, was poured into a perfectly clean wide-mouthed bottle containing fine sterilized gravel or broken glass. Rinsings of a measured amount of the caustic potash solution were used to free the conical glass from what matter might remain and were added to the sputum. The contents of the bottle were agitated in a shaking machine for five minutes, and allowed to stand until the caustic potash solution had had time to act. As soon as the sputum had become homogeneous an equal volume of water was added, and the whole shaken again. The sputum thus treated was of a pale-green or yellowish-brown color, and contained only small fragments of elastic tissue. It was allowed to stand two to four hours, and then shaken again for five to ten minutes.

By means of a burette of original design drops of exactly equal size were secured and caught upon clean sterile cover-glasses. The drops were subsequently spread into an even film by a very fine platinum wire, while the cover-glass was rotated upon a "turn-table." After spreading, the cover-glasses were laid upon a level brass plate slightly warmed to facilitate drying. After drying, the cover-glasses were coated with a serum film by spraying, and the temperature raised to 80°–90° C. to coagulate the serum and retain the bacteria in place, after which they were carefully stained with carbol-fuchsin and decolorized with a solution of 150 parts of water, 50 parts of alcohol, and 20–30 drops of pure sulphuric acid. Prior to this the cover-glass was washed in three alcohols and subsequently in water, and if necessary in acid and alcohol again.

A special arrangement of the microscope was devised for the purpose, and the number of bacilli in each drop estimated with extreme care. The number varied from 472 to 240,000. To estimate the number of bacilli in a given quantity the number of drops to a cubic centimeter is multiplied by the number of bacilli in the drop, and then by the number of cubic centimeters to be estimated.

The method is an ingenious one, but a glance down the columns of figures in the original article will be sufficient to show that the counting of the bacilli is devoid of any particular value.

This is only to be expected when one considers the pathology of the disease and remembers that accidents, such as unusually violent cough one day, modified by the use of sedatives the next, may cause wide variations in the quality if not in the quantity of the sputum.

When the tubercle bacilli are to be sought for in sections of tissue, considerable difficulty is at once encountered, partly because of the thickness of the section and partly because of the presence of nuclei which color intensely.

Again, Ehrlich's method must be recommended as the most certain and best method of staining a large number of bacilli.

The sections of tissue, if imbedded in celloidin or paraffin, should be freed from the foreign substances. Like the cover-glasses, they are placed in the stain for twelve to twenty-four hours at a temperature of 37° C. Upon removal they are allowed to lie in water for about ten minutes to wash away the excess of stain and to soften the tissue, which often shrinks and becomes brittle. The washing in nitric acid (20 per cent.) which follows may have to be continued for as long as two minutes. Thorough washing in 60 per cent. alcohol follows, after which the sections can be counter-stained, washed, dehydrated in 95 per cent. and absolute alcohol, cleared in xylol, and mounted in Canada balsam.

A method which has attained great and deserved praise is Unna's. It is as follows: The sections are placed in a dish of twenty-four-hours-old, newly-filtered Ehrlich's solution, and allowed to remain twelve to twenty-four hours at the room-temperature or one to two hours in the incubator. From the stain they are placed in water, where they remain for about ten minutes to wash. They are next immersed in acid (20 per cent. nitric acid) for about two minutes, and become greenish-black. From the acid they are placed in absolute alcohol, and are gently moved to and fro until the pale-blue color returns. They are then washed in three or four changes of clean water until they become almost colorless, and are then removed to the slide by means of a section-lifter. The water is absorbed with filter-paper, and then the slide is heated over a Bunsen burner until the section becomes shining, when it receives a drop of xylol balsam and a cover-glass.

It is said that sections stained in this manner do not fade as quickly as those stained by Ehrlich's method.

The tubercle bacillus also stains well by Gram's method, but as this is a general method by which many different bacteria are colored, it is ill adapted for purposes of differentiation, especially when the prosecution of the characteristic methods is not more difficult.

So far as is known, the tubercle bacillus is a purely parasitic organism. It has never been found except in the bodies and excretions of animals affected with tuberculosis, and in dusts of which these are component parts. This purely parasitic nature greatly interferes with the isolation of the organism, which cannot be grown upon the ordinary culture-media. Koch first achieved its artificial cultivation by the use of blood-serum. When planted upon this medium the bacilli are first apparent to the naked eye in about two weeks, and occur in the form of small dry, whitish flakes, not unlike fragments of chalk. These slowly increase at the edges, and gradually form scale-like masses of small size, which under the microscope are seen to consist of tangled masses of bacilli, many of which are in a condition of involution.

The best method of obtaining a culture is to inoculate a guinea-pig with tuberculous material, allow an artificial tuberculosis to develop, kill the animal after a couple of months, and make the cultures from the centre of one of the tuberculous glands.

Of course many technical difficulties must be overcome. The tuberculous material used for inoculation may be sputum, injected beneath the skin by a hypodermic syringe. The animal is allowed to live for a month or six weeks, then killed. The autopsy is performed according to directions already given. A large lymphatic gland with softened contents or a nodule in the spleen being selected for the culture, an incision is made into it with a sterile knife, or a rigid sterile platinum wire is introduced; some of the contents are removed and planted upon blood-serum. After receiving the inoculated material the tubes are closed, either by a rubber cap placed over the cotton stopper, which is cut off and pushed in, or by a rubber cork above the cotton, the idea of this rubber corking being simply to prevent evaporation. The tubes must be kept in an incubator at the temperature of 37–38° C.

Kitasato has published a method by which Koch has been able to secure the tubercle bacillus in pure culture from sputum. After carefully cleansing the mouth the patient is allowed to expectorate into a sterile Petri dish. By this method the contaminating bacteria from the mouth and the receptacle are excluded, and the expectorated material is made to contain only such bacteria as were present in the lungs. The material is carefully washed a great many times in renewed distilled sterile water until all bacteria not enclosed in the muco-purulent material are removed; it is then carefully opened with sterile instruments, and the culture-medium—glycerin agar-agar or blood-serum—is inoculated from the centre. Kitasato has been able by this method to demonstrate that many of the bacilli ordinarily present in tubercular sputum are dead, although they continue to stain well.

Kitasato's method of washing the sputum has been modified and simplified by Czaplewski and Hensel.[1] In their studies of whooping-cough, instead of washing the flakes in water in dishes, they shook them in peptone water in test-tubes. The shaking in the test-tube being so much more thorough than the washing in dishes, fewer ·changes are necessary, three or four washings being sufficient.

In 1887, Nocard and Roux gave a great impetus to investigations upon tuberculosis by their discovery that the addition of 4–8 per cent. of glycerin to bouillon and agar-agar made them suitable for the development of the bacillus, and that a much more luxuriant development could be obtained upon these media than upon blood-serum. The growth upon such "glycerin agar-agar" (Fig. 62) very much resembles that upon blood-serum. The growth upon bouillon with 4 per cent. of glycerin is also luxuriant. As tubercle bacilli require considerable oxygen for their proper development, they grow only upon the surface of the bouillon, where a rather thick mycoderma forms. The surface-

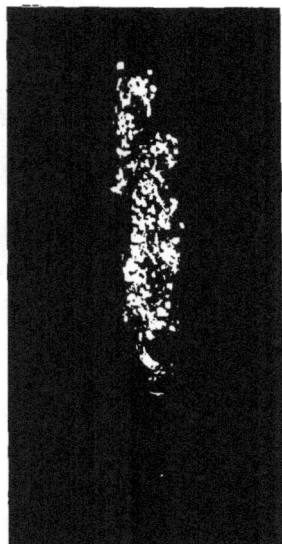

FIG. 62.—Bacillus tuberculosis on "glycerin agar-agar."

growth is rather brittle, and after a time gradually subsides fragment by fragment.

The tubercle bacillus can be grown in gelatin to which glycerin has been added, but as its development takes place only at 37°–38° C., a temperature at which gelatin is always liquid, its use for the purpose is disadvantageous.

[1] *Centralbl. f. Bakt. u. Parasitenk.*, xxii., Nos. 22 and 23, p. 643.

Pawlowski was able to cultivate the bacillus upon potato, but Sander, who found that it could be readily grown upon various vegetable compounds, especially upon acid potato mixed with glycerin, also found that upon such compounds its virulence was constantly lost.

It has also been shown that the continued cultivation of the tubercle bacillus upon such culture-media as are appropriate so lessens its parasitic nature that in the

Fig. 63.—Bacillus tuberculosis : adhesive cover-glass preparation from a fourteen-day-old blood-serum culture; × 100 (Fränkel and Pfeiffer).

course of time it can be induced to grow feebly upon the ordinary agar-agar.

It is really surprising to note the extremely simple compounds in which the tubercle bacillus can be grown. Instead of requiring the most concentrated albuminous media, as was once supposed, Proskauer and Beck have shown that the organism can grow in non-albuminous media containing asparagin, and that it can even be induced to grow upon a mixture of commercial ammonium carbonate, 0.35 per cent.; primary potassium phosphate, 0.15 per cent.; magnesium sulphate, 0.25 per cent.; glycerin, 1.5 per cent. It was even found that tuberculin was produced in this inorganic mixture.

The tubercle bacillus seems to require a considerable amount of oxygen for its development. It is also peculiarly sensitive to temperatures, not growing at a temperature below 29° C. or above 42° C. Temperatures above 75° C. kill it after a short exposure.

The tubercle bacillus does not develop well in the light, and when its virulence is to be maintained should always be kept in the dark. Sunlight kills it in from a few minutes to several hours, according to the thickness of the mass exposed to its influence.

The widespread character of tuberculosis at one time suggested the idea that tubercle bacilli were ubiquitous in the atmosphere, that we all inhaled them, and that it was only our *vital resistance* that prevented us all from becoming its victims.

Cornet must be given the credit of having shown that such an idea is untrue, and that tubercle bacilli only exist in the atmospheres frequented by consumptives. His experiments were made by collecting dusts from numerous places—streets, sidewalks, houses, rooms, walls, etc. Injecting them into guinea-pigs, whose constant susceptibility to the disease makes them a very delicate reagent for its detection, Cornet showed the bacilli to be present only in the dust with which pulverized sputum was mixed, and found such infectious dust to be most common where the greatest carelessness in respect to cleanliness prevailed.

Our present knowledge of the life-history of the tubercle bacillus, by showing its indisposition to multiply outside the bodies of animals, the deleterious influence of sunlight upon it, the absence of positive permanent forms, and its sensitivity to temperatures beyond a certain range, confirms all that Cornet has pointed out, and shows us why the expectoration of millions of consumptives has not rendered our atmospheres pestilential.

As long as tuberculosis exists among men or cattle, it shows that the existing hygienic precautions are insufficient. While not so radical as to suggest the unreasonable isolation of patients and destruction of property once practised in the kingdom of Naples, the author would favor the registration of all tuberculous cases as a means of collecting accurate data concerning their origin, would insist upon domestic sterilization and disinfection, and would have special hospitals for as many, especially of the poorer classes, among whom hygienic measures are almost always opposed, as could be persuaded to occupy them.

It has already been declared the duty of the physician to use every means in his power to prevent the spread of infection in the households in his care, and no disease is more deserving of attention than this neglected one. Patients should cease to kiss the members of their family and friends; their individual knives, forks, spoons, cups, etc. should be carefully kept apart—secretly if the patient be sensitive upon the subject—from those of the family, and scalded after each meal; the napkins and handkerchiefs, as well as whatever clothing or bed-clothing is soiled by the discharges, should be kept apart from the common wash, and boiled; and of course the expectoration should be carefully attended to, received in a suitable receptacle, sterilized or disinfected, and never allowed to dry, for it has been shown that the tubercle bacillus can remain vital in dried sputum for as long as nine months. A very neat arrangement for collecting and disposing of the expectoration is recommended by some boards of health. It consists of a metal case into which a pasteboard box is fitted. When the box is to be emptied the whole of the pasteboard portion is removed, and, together with the expectoration, burned. The metal part is disinfected, provided with a new pasteboard box, and is again ready for use. (See Fig. 20, page 120.) The physician should also give directions for disinfecting the bedroom occupied by a consumptive before it becomes the chamber of a healthy person.

Boards of health are now becoming more and more in-

terested in tuberculosis, and, though exceedingly slow
and conservative in their movements, are disseminating
literature among doctors for distribution to their patients,
with the hope of achieving by volition that which they
would otherwise regard as cruel compulsion.

The channels by which the tubercle bacillus enters the
organism are varied. A few cases are on record where
the micro-organisms have passed through the *placenta*,
so that a tuberculous mother was able to infect her
unborn child. It is not impossible that the passage of
bacilli in this manner through the placenta causes the
development of tuberculosis in infants after birth, the
disease having remained latent during fetal life, for
Birch-Hirschfeld has shown that fragments of a fetus,
itself showing no tubercular lesions, but coming from a
tuberculous woman, were fatal to guinea-pigs into which
they were inoculated.

The most frequent channel of infection is the *respiratory tract*, into which the finely-pulverized dust of rooms
and streets enters. Probably all of us at some time in
our lives inhale living virulent tubercle bacilli, yet not
all of us suffer from tuberculosis. Personal predisposition seems of great importance, for it has been shown
that without the formation of tubercles virulent bacilli
may be present for considerable lengths of time in the
bronchial lymphatic glands—the dumping-ground of the
pulmonary phagocytes.

In order that infection shall occur it does not seem
necessary that the least abrasion or laceration shall exist
in the mucous lining of the respiratory tract. The
tubercle bacillus is a foreign body of irritating properties, and, lodging upon a cell, is soon engulfed in its
protoplasm, or, arrested by a leucocyte, is dragged off to
some other region in whose narrow passages a most hos-
tile strife doubtless takes place.

Infection also commonly takes place through the *gastro-intestinal tract* by infected food. At present an over-
whelming weight of evidence points to the presence of

bacilli in the milk of cattle affected with tuberculosis. It does not seem necessary that tuberculous ulcers shall be present in the udders; indeed, the bacilli have been demonstrated in considerable numbers in milk from udders without tubercular lesions discoverable to the naked eye.

The meat from tuberculous animals is less dangerous than the milk, because the meat is nearly always cooked before being eaten, while the milk is generally taken uncooked. The bacilli enter the intestinal lymphatics, sometimes produce lesions immediately beneath the mucous membrane, and lead later on to the formation of ulcers ; but generally they first involve the mesenteric lymphatic glands. The thoracic duct is sometimes affected, and from such a lesion it is easy to understand the development of a general miliary tuberculosis. The occasional absorption of tubercle bacilli by the lacteals, and their entrance into the systemic 'circulation and subsequent deposition in the brain, bones, joints, etc., are supposed to explain primary lesions of these tissues.

Infection is said also to take place occasionally through the *sexual apparatus.* In sexual intercourse tubercle bacilli from tuberculous testicles may be discharged into the female organs, with resulting tuberculous lesions. The infection in this way generally is from the male to the female, primary tuberculosis of the testicle being much more common than primary tuberculosis of the uterus or ovaries.

While most probably rare, in comparison with the preceding, *wounds* also are avenues of entrance for the tubercle bacilli. Anatomical tubercles are not uncommon upon the hands of anatomists and pathologists, most of these growths being tuberculous in character. An interesting fact concerning these dermal lesions is the exceedingly small number of bacilli which they contain.

The macroscopic lesions of tuberculosis are too familiar to require a description of any considerable length. They

Tuberculosis of the lung: the upper lobe shows advanced cheesy consolidation with cavity-formation, bronchiectasis, and fibroid changes; the lower lobe retains its spongy texture, but is occupied by numerous miliary tubercles.

consist in nodes, nodules, or collections of agminated nodules, called tubercles, scattered irregularly through the tissues, which are devitalized or disorganized by their presence. When tubercle bacilli are introduced beneath the skin of a guinea-pig, the animal shows no sign of disease for a week or two; it then begins to lose appetite and gradually to diminish in flesh and weight. Examination at this time will show a nodule at the point of injection and enlargement of the neighboring lymphatic glands. The atrophy increases, the animal shows a febrile reaction, and at the end of a varying period of time, averaging about twelve weeks, dies. Post-mortem examination shows a cluster of tubercles at the point of inoculation, enlargement of lymphatic glands both near and remote from the primary lesion (due to the presence of tubercles), and a widespread invasion of the lungs, liver, kidneys, peritoneum, and other organs and tissues, with tuberculous tissue in a more or less advanced condition of necrosis. Sometimes there are no tubercles discoverable at the point of inoculation. There is no regularity in the distribution of the disease. Tubercle bacilli are demonstrable in immense numbers in all the diseased tissues. The disease as seen in the guinea-pig is more extended than in other animals because of its greater susceptibility, and the death of the animal is more rapid than in other species for the same reason. In rabbits the lesion runs a longer course with similar lesions. In bovines and sheep the infection is generally first seen in, and is principally confined to, the alimentary apparatus and the associated organs, though pulmonary disease also occurs. In man the disease is chiefly pulmonary, though gastro-intestinal and general miliary forms are also common. The development of the lesions in whatever tissue or animal always depends upon the distribution of the bacilli by the lymph or the blood, and is first inflammatory, then degenerative, in type.

The experiments of Koch, Prudden and Hodenphyl, and others have shown that when dead tubercle bacilli

15

are injected into the subcutaneous tissues of rabbits small local abscesses develop in the course of a couple of weeks, showing that the tubercle bacilli are chemotactically potent.

While it is extremely interesting to observe that this chemotactic property exists, it seems to be by some other irritant that most of the lesions of tuberculosis are caused. When the dead tubercle bacilli, instead of being injected *en masse* into the areolar tissue, are so introduced into the body—as by intravenous injection—as to disseminate themselves or remain in small groups, the result is quite different, and much more closely resembles that of the action of the living organism.

Baumgarten, whose researches were made upon minute tubercles of the iris, has shown that the first manifestation of the irritation caused by the bacillus is not the attraction of leucocytes, but the stimulation of the fixed connective-tissue cells of the part affected. These cells increase in number by karyokinesis, and form about the irritating bacterium a minute focus which is the primitive tubercle.

The leucocytes are of secondary advent, and are no doubt attracted both by the substance shown by Prudden and Hodenphyl to exist in the bodies of the dead bacilli and by the necrotic changes which already affect the primary cells. For reasons not understood, the amount of chemotaxis varies greatly in different cases. Sometimes the tubercles will be sufficiently purulent in type almost to justify the name "tubercular abscess;" sometimes there will be a marked absence of cellular elements derived from the blood.

The important toxic substance produced by the bacillus is evidently not associated with chemotaxis, for when the leucocytes are absent the necrosis which is so characteristic persists.

The groups of cells constituting the primitive tubercle have scarcely reached microscopic proportions before a distinct coagulation-necrosis is observable. The protoplasm of the cells affected takes on a hyaline character, and seems abnormally viscid, so that contiguous cells have a tendency to become partially confluent. The chromatin of their nuclei becomes dissolved in the nuclear juice and gives stained nuclei a pale but homogeneous appearance. Sometimes this nuclear change is only observed very late. As the necrosis advances the contiguous cells flow together and form large protoplasmic masses—giant-cells—which contain as many nuclei as there were component cells. It may be that these nuclei multiply by karyokinesis after the protoplasmic coalescence, but only one observer, Baumgarten, has found signs of this process in giant-cells. While these changes are in progress in the cells of the primary focus, the leucocytes may collect in such numbers as to obscure them and make themselves appear to constitute the primitive cells. When the irritant substance is produced in considerable quantities, the most delicate cells die first; and it is not infrequent to find a tubercle rich in leucocytes suddenly showing degeneration of these cells, with recurring prominence of the original epithelioid cells.

It has been taught by some that the giant-cells are produced by the union of the leucocytes, but a careful observation of the rôle played by these cells will convince one that such an origin for these monstrous cells must be very rare.

Giant-cells are not always produced, for sometimes the necrotic changes are so violent and widespread as to convert the whole cellular mass into a granular detritus of unrecognizable fragments.

Tubercles are constantly avascular, as would be expected of a process which is a combination of progressive irritation and necrosis. The avascularity may be a factor in the necrosis of the larger tuberculous masses, but it plays no part in the degeneration of the smallest tubercles, which is purely toxic.

Tubercles may be developed in any tissue and in any organ. In whatever situation they occur, space is occu-

pied at the expense of the tissue, whose component cells
are pushed aside or else included in the nodule. In miliary tuberculosis of the kidney it is not unusual to find a
tubercle including a whole glomerule, and resolving its
component thrombosed capillaries and epithelium into
necrotic fragments.

As almost all tissues contain a supporting tissue-frame-
work of connective-tissue fragments, some of these must
be embodied in the new growth. The fibres which possess little vitality are more resistant than cells, and, after
all the cells of a tubercle have been destroyed, will be
distinctly visible among the granules, so that the tubercle
has a reticulated appearance.

As a rule, tubercles steadily increase in size by the invasion of fresh tissue. The tubercle bacillus does not seem
to find the necrotic centres of the tubercles adapted to its
growth, and completes its life-cycle with the tissue-cells.
It is unusual to find healthy-looking bacilli in the necrotic
areas, most of them being observed at the edges of the
tubercle, where the nutrition is good. From such edges
the bacilli are occasionally picked up by leucocytes and
transported through the lymph-spaces, until the phago-
cyte falls a prey to its prisoner, dies, and sows the seed
of a new tubercle. However, for the spread of tubercle
bacilli from place to place phagocytes are not always
necessary, for the bacilli seem capable of transportation
by streams of lymph alone.

Notwithstanding the steady advance which takes place
in most observed cases of tuberculosis, and the thoroughly
comprehensible microscopic explanation of it, many cases
of tuberculosis make quite perfect recoveries.

The periphery of every tubercle is a zone of reaction,
with a marked tendency to granulation and organization.
If the vital condition is such that through inappropriate nutriment or through unusually active phagocytosis the activity of the bacilli is checked or their
death is brought about, this tendency to cicatrization is
allowed to progress unmolested, and the necrosed mass is

soon surrounded with a zone of newly-formed contracting fibrillar tissue, by which it is perfectly isolated. In such isolated masses lime-salts are commonly deposited. Sometimes this process is perfected without the destruction of the bacilli, but with their incarceration and inhibition. Such a condition is called *latent tuberculosis,* and may at any time be the starting-point of a new infection and lead to a fatal termination.

In 1890, Koch announced some observations upon toxic products of the tubercle bacillus and their relation to the diagnosis and treatment of tuberculosis, which at once aroused an enormous but, unfortunately, a transitory enthusiasm.

These observations, however, are of capital importance. Koch observed that when guinea-pigs are inoculated with a mixture containing tubercle bacilli the wound ordinarily heals readily, and soon all signs of local disturbance other than enlargement of the lymphatic glands of the neighborhood disappear. In about two weeks there occurs at the point of inoculation a slight induration which develops into a hard nodule, then ulcerates, and remains until the death of the animal. If, however, in the course of a short time the animals are reinoculated, the course of the process is altogether changed, for, instead of healing, the wound and the tissue surrounding it assume a dark color and become obviously necrotic, and ultimately slough away, leaving an ulcer which rapidly and permanently heals without enlargement of the lymph-glands.

Having made this observation with injected cultures of the living bacillus, Koch next observed that the same change occurred when the secondary inoculation was made with pure cultures of the dead bacilli.

It was also observed that if the material used for the secondary injection was not too concentrated and not too often repeated (only every six to forty-eight hours), the animals thus treated improved in condition, and, instead of dying of the tuberculosis induced by the primary injection in from six to ten weeks, continued to live, sometimes (Pfuhl) as long as nineteen weeks.

Koch also discovered that a 50 per cent. glycerin extract of cultures of the tubercle bacillus produced the same effect as the dead cultures originally used, and gave this substance, *tuberculin*, to the scientific world for experimental purposes, in the hope that the prolongation of life observed in the guinea-pig might be true in the case of man.

The active substance of the "tuberculin" seems to be an albuminous derivative insoluble in absolute alcohol. It is not a toxalbumin.

The action of the tuberculin upon the animal organism is peculiar, but readily understandable. *It does not exert the slightest influence upon the tubercle bacillus,* but acts upon the living tuberculous tissue. In the description of the tissue-changes already given it has been shown that the tubercle bacillus effects the coagulation-necrosis of the cells, but does not derive its nutriment from the dead tissue. As the cells die and are incorporated in the necrotic mass, the bacilli find the conditions of life unfavorable, and likewise seem to die. The active bacilli, therefore, are always found at the margins of the tuberculous tissues, where the cells are fairly active. The necrosis is due to bacillary poisons. When tuberculin is injected into the organism the result is to double the amount of poisonous influence upon the cells surrounding the bacilli, to destroy their vitality, to remove the favorable conditions of growth from the organism, and to leave it for a time checkmated.

Virchow, who well understood the action of the tuberculin, soon showed that as a diagnostic and therapeutic agent in man its use was attended with great danger. The destroyed tissue was absorbed, and with it the bacilli were likewise absorbed and transported to new areas, where a rapid invasion occurred. Old tuberculous lesions which had been encapsulated were softened, broken down, and became sources of dangerous infection to the individual, so that, a short time after its enthusiastic reception as a "gift of the gods," tuberculin was placed upon its proper footing as a diagnostic agent valuable in veterinary practice, but dangerous in human medicine, except in cases of lupus and other external forms of the disease where the destroyed tissue could be discharged from the surface of the body.

The method of preparation of tuberculin is rather simple. Small flasks exposing a considerable surface of liquid are filled with about 25 c.cm. of bouillon containing about 4 per cent. of glycerin. The bouillon is preferably made with calf- instead of ox-meat. When thoroughly sterile the surfaces are inoculated with pure cultures of the tubercle bacillus and are stood in an incubator. In the course of two weeks a slight surface growth is apparent, which in the course of time develops into a pretty firm pellicle and gradually subsides. At the end of four or six weeks development ceases and the pellicle sinks. The contents of a number of flasks are then collected in an appropriate vessel and evaporated over a water-bath to one-tenth their volume, then filtered through a Pasteur-Chamberland filter. This is crude tuberculin.

When such a product is injected in doses of a fraction of a cubic centimeter an inflammatory and febrile reaction occurs. The inflammation sometimes causes superficial tuberculous lesions (lupus) to ulcerate and slough away, and for this reason is of some value in therapeutics, although attended with the dangers mentioned above. The fever is sufficiently characteristic to be of diagnostic value, though the tuberculin can only be used as a diagnostic agent in practice upon animals.

A recent important work upon tuberculin has been done by Koch.[1]

In his experience the attempts made to produce immunity to the tubercle bacillus by the injection into animals of attenuated cultures proved failures, because

[1] *Deutsche med. Wochenschrift,* 1897, No. 14.

abscesses invariably followed their introduction, whether dead or alive, and nodular growths in the lungs were constant sequelæ of their injection into the circulation. In such nodules the bacilli could be found unabsorbed and unaltered. It seemed as if the fluids of the body could not effect solution of the bacteria. The ineffectual attempts at immunization, with the results given, probably depend upon the inability of the tissues to take up from the bacilli whatever immunizing substances they might contain, first, because of the impossibility of dissolving them, and, second, because the irritating powers they possess interfere with the direct action of normal fluids and uninjured body-cells, and always subject the bacteria to semi-pathological conditions.

From these data, which he carefully studied out, Koch concluded that it would be necessary to bring about some artificial condition advantageous to the absorption of the bacilli, and for the purpose tried the action of diluted mineral acids and alkalies. The chemical change brought about in this manner facilitated absorption, but the absorption of bacilli in this altered condition was not followed by immunity, probably because the chemical composition of tubercle-toxin (or whatever one may name the poisonous products of the bacillus) was changed by the reagents used.

Tuberculin, with which Koch performed many experiments, was found to produce immunity only to tuberculin, not to bacillary infection.

Pursuing the idea of fragmenting the bacilli, or in some way treating them chemically in order to increase their solubility, Koch found that a 10 per cent. sodium hydrate solution yielded an alkaline extract of the bacillus, which, when injected into animals, produced effects similar to those following the administration of tuberculin, except that they were briefer in duration and more constant in result. The marked disadvantage of abscess-formation following the injections, however, remained. This fluid, when filtered, possessed the properties of tuberculin.

The mechanical fragmentation of the bacilli had been used by Klebs in the studies of antiphthisin and tuberculocidin. Koch now used it with advantage in his studies, and pulverized living, fresh, virulent, but perfectly dry bacteria in an agate mortar, in order to liberate the bacillary substance from its protecting envelope of fatty acid. In the trituration only a very small quantity of the bacteria could be handled at a time, and Koch seemed thoroughly aware of the risk incurred from inhalation of the finely pulverized bacillary mass.

Having reduced the bacilli to fragments, they were removed from the mortar in distilled water, and collected by centrifugation, in a small glass tube, as a muddy residuum at the bottom of an opalescent, clear fluid. For convenience he named the clear fluid TO, the sediment TR. TO was found to contain tuberculin. In order to separate the essential poison of the bacteria as perfectly as possible from the irritating tuberculin, the TR or fragments were dried perfectly, triturated once more, re-collected in fresh distilled water and re-centrifugated. After the second centrifugation microscopic examination showed that the bacillary fragments had not been resolved into a uniform mass, for when TO was subjected to staining with carbol-fuchsin and methyl-blue it was found to exhibit a blue reaction, while in TR a cloudy violet reaction was obtained.

The addition of 50 per cent. of glycerin had no effect upon TO, but caused a cloudy white deposit to be thrown down from TR. This last reaction showed that TR contained fragments of the bacilli which are insoluble in glycerin.

Experiment showed that TR had decided immunizing powers. Injected into tuberculous animals in too large dose it produced a reaction, but its effects were entirely independent of the reaction. Koch's aim in using this substance in therapeutics was to produce immunity in the patient without reactions, by gradual but rapid increase of the dose. In so large a number of cases did

Koch produce immunity to tuberculosis by the administration of TR, that he thinks it proved beyond a doubt that the observations are correct.

In making the TR preparation Koch advises the use of a fresh, highly virulent culture not too old. It must be perfectly dried in a vacuum exsiccator, and the trituration, in order to be thorough, should not be done upon more than 100 mg. of the bacilli at a time. A satisfactory separation of the TR from TO is said only to occur when the perfectly clear TO takes up at least 50 per cent. of the solid substance, as otherwise the quantity of TO in the final preparation is so great as to produce undesirable reactions.

The fluid is best preserved by the addition of 20 per cent. of glycerin, which does not injure and prevents decomposition of the TR.

The finished fluid contains 10 mg. of solid constituents to the c. cm., and before administration should be diluted with physiological salt solution (not solutions of carbolic acid). When administering the remedy to man the injections are made with a hypodermic syringe into the tissues of the back. The beginning dose is $\frac{1}{500}$ of a mg., rapidly increased to 20 mg., the injections being made daily.

In speaking of the results of experiments upon guinea-pigs, Koch says:

"I have, in general, got the impression in these experiments that full immunization sets in two or three weeks after the use of large doses. A cure in tuberculous guinea-pigs, animals in which the disease runs, as is well known, a very rapid course, may, therefore, take place only when the treatment is introduced early—as early as one or two weeks after the infection with tuberculosis.

"This rule avails also for tuberculous human beings, whose treatment must not be begun too late. . . . A patient who has but a few months to live cannot expect any value from the use of the remedy, and it will be of little value to treat patients who suffer chiefly from secondary infection, especially with the streptococcus, and in whom the septic process has put the tuberculosis entirely in the background.''

By proper administration of the TR Koch was able to render guinea-pigs so completely immune that they were able to withstand inoculations of virulent bacilli. The point of inoculation presents no changes when the remedy is administered, and the neighboring lymph-glands are generally normal, or when slightly swollen contain no bacilli.

One very important objection found by Trudeau and Baldwin against commercially prepared TR is that it is possible for it to contain unpulverized, and hence live, virulent tubercle bacilli. In the preparation of the remedy it will be remembered that no antiseptic or germicide was added to the solutions, by which the effects of accidental failure to crush every bacillus could be overcome, Koch having specially deprecated such additions as producing destructive changes in the TR. Until this objection can be removed, and our confidence that our attempts to cure patients will not cause their death be restored, it becomes a question whether TR can find a place in human medicine at all, or must remain an interesting scientific laboratory demonstration.

Probably the most interesting use to which the TR-tuberculin has thus far been put is found in the experiments of Fisch,[1] who immunized a horse with it, hoping to produce an antitoxin that might be useful in treating tuberculosis. His experiment seems to have met with remarkable success, for the serum thus secured, which he calls "Antiphthisic Serum, TR," is found to thoroughly immunize guinea-pigs to tuberculosis, to cure tuberculous guinea-pigs in the early stages of the disease, and to neutralize the effects of tuberculin upon tuberculous animals.

Upon human beings it is too early to make a positive

[1] *Jour. of the Amer. Med. Assoc.*, Oct. 30, 1897.

report, but Fisch's cases have shown remarkable improvement. The subject is pregnant with interest and deserves attention.

Hirshfelder [1] claims to have cured a large number of cases of tuberculosis by the use of a preparation known as *oxytuberculin.* It consists of a 4 per cent. glycerin-bouillon culture of very virulent tubercle bacilli, which after being sterilized for one hour, and filtered, receives the addition of 8–10 volumes of hydrogen peroxid, and is then sterilized for ninety-six hours in a steam apparatus. During the sterilization the fluid is kept in a glass vessel, plugged with cotton wool. The peroxid of hydrogen is renewed every twelve hours.

From the fluid obtained in this way the excess of the peroxid is removed by alkalinization. Before being employed in human medicine the remedy is tested upon guinea-pigs. The dose may gradually be increased to 20 c.cm. The theory of action is based upon a claimed destruction of the toxic property of the tuberculin by the oxidation of the peroxid of hydrogen, which leaves a harmless but potent immunizing substance in the fluid.

Paterson [2] has suggested, for the production of immunity to tuberculosis, the use of gradually increasing doses of the serum of a fowl immunized to avian tuberculosis by gradually increased doses of sterilized, attenuated, and virulent cultures of the bacillus of avian tuberculosis. Curative results were observed in fowls thus treated, and in mammals similarly treated, and the inference drawn is that men treated in the same manner can be similarly benefited. The dose recommended is 2 c.cm.

The theory depends upon the supposed identity or near relationship of the bacilli of avian and mammalian tuberculosis.

Klebs has claimed much advantage from the treatment of tuberculosis by *antiphthisin.* According to the ex-

---

[1] *Deutsche med. Wochenschrift,* 1897, No. 19, and *Jour. of the Amer. Med. Assoc.,* 1897.

[2] *Amer. Medico-Surg. Bull.,* Jan. 25, 1898.

perimental studies of Trudeau and Baldwin, however, antiphthisin is only much diluted tuberculin, and exerts no demonstrable influence upon the tubercle bacillus *in vitro*, does not cure tuberculosis in guinea-pigs, and probably inhibits the growth of the tubercle bacillus upon culture-media to'which it has been added, only by its acid reaction.

On the other hand, Ambler has used antiphthisin with excellent results in the treatment of human tuberculosis.

Numerous experimenters, prominent among whom are Tizzoni, Cattani, Bernheim, and Paquin, have experimented with the tubercle bacillus and tuberculin, hoping that the principles of serum-therapy might be applicable to the disease. Nothing positive has, however, been achieved. The first-named observers claim to have immunized guinea-pigs, in whose blood an antitoxin formed; the last-named thinks the serum of immunized horses a specific for tuberculosis. The field of experimentation is an inviting one, though the chronic course of the disease lessens the certainty with which the results can be estimated.

Babes and Proca, in an experimental research upon the action of the antituberculous serum, claim for it a decided specific action, and demonstrate experimentally that animals inoculated with tubercle bacilli and injected with the serum are protected from the spread of the disease.

Mafucci and diVestra found that by injecting guinea-pigs with serum from sheep immunized by injections first of dead, then of living cultures of tubercli bacilli, although no cures were brought about, the vitality of the animals was maintained longer. Unprotected animals died in fifty to fifty-three days. Those injected after infection, seventy-four days, and those injected before infection, ninety-one days.

The author[1] made an elaborate study of the so-called *antituberculin*, suggested by Viquerat, and widely praised

[1] *Jour. of the Amer. Med. Assoc.*, Aug. 21, 1897.

by Paquin. For a long period, donkeys were injected with increasing doses of tuberculin, in order that an antitoxin—antituberculin—might be generated in their blood. Experiments upon guinea-pigs showed that the serum was powerless to immunize against the tubercle bacillus, or to cure established tuberculosis. The serum, however, had the power of annulling the effects of tuberculin upon tuberculous animals. While a failure experimentally, certain clinicians claim that in practice it exerts a beneficial action upon patients. Indeed, presuming that an antituberculin is formed, it is but natural that it should do good in all cases in which it is probable that the patient is poisoned by tuberculin or a similar product.

Rather nearer the desideratum are the experiments of De Schweinitz,[1] who injected cows and horses with increasing quantities of bouillon cultures of a greatly attenuated tubercle bacillus, and subsequently found that the serum possessed the property of rendering guinea-pigs immune to the virulent bacilli.

The Bacillus of Fowl-tuberculosis (*Tuberculosis gallinarum*).—The cases of tuberculosis which occasionally occur spontaneously in chickens, parrots, ducks, and other birds were originally attributed to the Bacillus tuberculosis hominis, but the recent works of Rivolta, Mafucci, Cadio, Gilbert, Roget, and others have shown that, while very similar in many respects to the Bacillus tuberculosis, the organism found in the disease of birds has distinct peculiarities which stamp it a different variety, but not a separate species. Cadio, Gilbert, and Roger succeeded in infecting fowls by feeding them upon food containing tubercle bacilli, and keeping them in cages in which dust containing tubercle bacilli was placed. The infection was aided by lowering the temperature with antipyrin and lessening vitality by starvation. Morphologically, the organisms are similar, the bacillus of fowl-tuberculosis being a little longer and more slender than its ally.

[1] *Centralbl. f. Bakt. und Parasitenk.*, Sept. 15, 1897, Bd. xxii., Nos. 8 and 9.

Upon culture-media a distinct rapidity of growth is observable, and we find that, instead of growing only where glycerin is present, the Bacillus tuberculosis gallinarum will grow upon blood-serum, agar-agar, and bouillon as ordinarily prepared. It will not grow upon potato. The bacillus will grow at 42–43° C. quite as well as at 37° C., while the growth of the tubercle bacillus ceases at 42° C. Moreover, the temperature of 43° C. does not attenuate its virulence. The thermal death-point is 70° C. Upon culture-media it can retain its virulence for two years.

The growth upon artificial culture-media is luxuriant, and lacks the dry quality characteristic of ordinary tubercle-bacillus cultures. As it becomes old a culture of fowl-tuberculosis turns slightly yellow.

Birds are the most susceptible animals for experimental inoculation, the embryos and young more so than the adults; guinea-pigs are quite immune. Artificial inoculation can only be made in the subcutaneous tissue, never through the intestine. The chief seat of the disease is the liver, where cellular nodes, lacking the central coagulation and the giant-cells of mammalian tuberculosis, and enormously rich in bacilli, are found. The disease never begins in the lungs, and the fowls which are diseased never show bacilli in the sputum or the dung.

Rabbits are easily infected, an abscess forming at the seat of inoculation, and later nodules forming in the lung, so that the distribution is quite different from that seen in birds.

The bacillus stains like the tubercle bacillus, but takes the stain rather more easily. The resistance to acids is about the same.

Pseudo-tuberculosis.—Eberth, Chantemesse, Charrin, and Roger have reported certain cases of so-called pseudo-tuberculosis. The disease occurred spontaneously in guinea-pigs, and was characterized by the formation of cellular nodules in the liver and kidneys much resembling miliary tubercles. Cultures made from them showed the

presence of a small motile bacillus which could easily be
stained by ordinary methods (Fig. 64).　When introduced

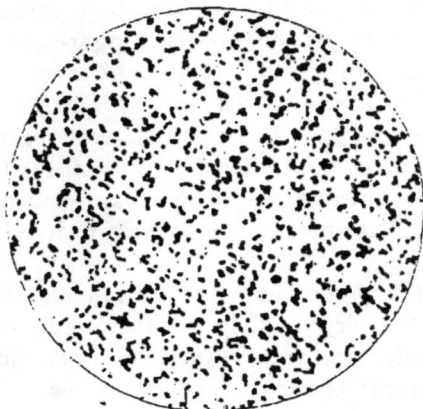

FIG. 64.—Bacillus pseudo-tuberculosis from agar-agar; × 1000 (Itzerott and
Niemann).

subcutaneously into guinea-pigs the original disease was
produced.

Pseudo-tuberculosis seems to be an indefinite affection
of which we have very little knowledge, and which is
certainly in no way connected with or related to true
tuberculosis.

CHAPTER II.

LEPROSY.

LEPROSY is a disease of great antiquity, and very early received much attention and study. In giving the laws to Israel, Moses included a large number of rules for its recognition, the isolation of the sufferers, the determination of recovery, and observances to be fulfilled before the convalescent could once more mingle with his people. The Bible is replete with accounts of miracles wrought upon lepers, and during the times of biblical tradition it must have been an exceedingly common and malignant disease.

At the present time, although we in the Northern United States hear very little about it, leprosy is still a widespread disease. It exists in much the same form as two thousand years ago in Palestine, Syria, Egypt, and the adjacent countries. It is exceedingly common in China, Siam, and parts of India. Cape Colony has many cases. In Europe, Norway, Sweden, and parts of the Mediterranean coast furnish a considerable number of cases. Certain islands, especially the Sandwich Islands, are regular hot-beds for its maintenance. The United States is not exempt, the Gulf coast being chiefly affected.

At one time the view was prevalent that the disease was spread only by contagion, at another that it was miasmatic. At present the tendency is to view it as contagious to a degree rather less than tuberculosis. Sometimes it is hereditary.

The cause of leprosy is now pretty certainly determined to be the lepra bacillus (Fig. 65), which was dis-

16   241

covered by Hansen, and subsequently clearly described by Neisser.

The bacillus is almost the same size as the tubercle bacillus—perhaps a little shorter—but lacks the curve which is so constant in the latter. It stains in very much the same way as the tubercle bacillus, but permits of a rather more rapid penetration of the stain, so that

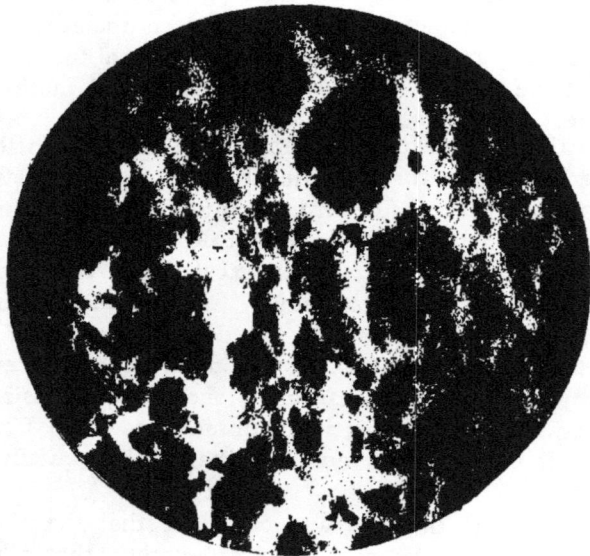

FIG. 65.—Bacillus lepræ, seen in a section through a subcutaneous node; × 500 (Fränkel and Pfeiffer).

the ordinary aqueous solutions of the anilin dyes color it quite readily. It stains well by Gram's method, by which beautiful tissue specimens can be prepared. The peculiar property of retaining the color in the presence of the mineral acids which characterizes the tubercle bacillus also characterizes the lepra bacillus, and the methods of Ehrlich, Gabbett, and Unna can be used for its detection.

Like that of the tubercle bacillus, its protoplasm often presents open spaces or fractures, which have been regarded by some as spores, but which are even less likely to be spores than the similar appearances in the tubercle bacillus.

The organism almost always occurs singly or in irregular groups, filaments being unknown. It is not motile.

Many experimenters have endeavored to grow this bacillus upon artificially prepared substances, but in spite of modern methods, improved apparatus, and refined media, few claim to have met with success.

Bordoni-Uffredozzi was able to grow upon a blood-serum-glycerin mixture a bacillus which partook of the staining peculiarities of the lepra bacillus as it appears in the tissues, but differed very much from it in its morphology. After numerous generations this bacillus was induced to grow upon ordinary culture-media. It commonly presented a club-like form, which was regarded by Baumgarten as an involution appearance. Fränkel points out that the bacillus of Bordoni is possessed of none of the essential characters of the lepra bacillus except its staining.

Czaplewski[1] offers a confirmation of the work of Bordoni-Uffredozzi, together with a description of a bacillus supposed to be the lepra bacillus, which he succeeded in cultivating from the nasal secretions of a leper.

The bacillus was first isolated upon a culture-medium consisting of glycerinized serum without the addition of salt, pepton, or sugar. The mixture was placed in flat dishes, coagulated by heat, and sterilized by the intermittent method.

The secretion, rich in lepra bacilli, was taken up with a platinum wire and inoculated upon the culture-medium by a series of linear strokes. The dishes (Petri dishes were used for the experiment) were securely closed with paraffin and stood in the incubating-oven at 37° C.

Upon the surface of the medium there grew numerous colonies of staphylococcus aureus, the bacillus of Fried-

---

[1] *Centralbl. f. Bakt. und Parasitenk.*, Jan. 31, 1898, vol. xxiii., Nos. 3 and 4, p. 97.

länder and a number of colonies consisting of fine, slender, often somewhat nodose bacilli about the size and form of the lepra bacillus.

These colonies were grayish-yellow, humped in the middle, 1–2 mm. in diameter, irregularly rounded, and irregular at the edges. They could be inverted entire with the platinum wire and were excavated on the under side. The consistence was crumbly.

When a transfer was made from one of these colonies to fresh media, in a few days the growth became apparent and assumed a band-like form, with a plateau-like elevation in the center.

The bacillus thus isolated grew with moderate rapidity upon all the ordinary culture-media except potato. Upon blood-serum the growth was more luxuriant and fluid than upon the solid media. Upon coagulated serum the growth was rather dry and elevated, and was frequently so loosely attached to the surface of the medium as to be readily lifted up by the platinum wire.

The growth was especially good upon sheep's blood-serum with the addition of 5 per cent. of glycerin. The growth upon the Löffler-mixture was excellent.

Upon agar-agar the growth is not so good as upon blood-serum ; it is more luxuriant upon glycerin agar-agar than upon plain agar-agar; it is grayish and flatter upon agar-agar than upon blood-serum. The growth never extends to the water of condensation to form a floating layer, as does that of the tubercle bacillus.

The colonies that form upon agar-agar are much like those described by Bordoni-Uffredozzi, and appear as isolated, grayish, rounded flakes, thicker in the center than at the edges, and characterized by an irregular serrated border from which a fine irregular network extends upon the medium. These projections consist of bundles of the bacilli.

Upon gelatin the bacillus develops well after it has grown artificially for a number of generations. Upon the surface of gelatin the growth is, in general, similar to that upon agar-agar. In puncture-cultures most of the growth is on the surface in the form of a whitish, or grayish, or yellowish folded layer. In the depths of the gelatin the development occurs as a granular rather thick column. The medium is not liquefied.

Bouillon is not clouded; no superficial growth occurs. The vegetation occurs only at the bottom of the tube in the form of a powdery sediment.

Czaplewski found that the bacillus stained well with Löffler's methylen-blue, and with the aqueous solutions of the anilin dyes. It also stains by Gram's method, and has the same resisting power to the decolorizing action of mineral acids and alcohol as the lepra bacillus as seen in tissue. The young bacilli color homogeneously, but older ones are invariably granular. They are usually pointed at the ends when young, but may be rounded or knobbed when older. The more rapidly the bacillus grows, the longer and more slender it appears.

All attempts to infect the lower animals with leprosy, either by the purulent matter or solid tissue from lepers, or by inoculating them with the supposed specific bacilli that have been isolated, have failed.

Ducrey seems to have cultivated the lepra bacillus in grape-sugar, agar, and in bouillon "*in vacuo*." His results need confirmation. Very few instances are recorded in which actual inoculation has produced leprosy in either men or animals. Arning was able to secure permission to experiment upon a condemned criminal in the Sandwich Islands. The man was of a family entirely free from disease. Arning introduced beneath his skin fragments of tissue freshly excised from a lepra nodule, and kept the man under observation. In the course of some months typical lesions began to develop at the points of inoculation and spread gradually, ending in general lepra in the course of about five years.

Melcher and Artmann introduced fragments of lepra nodules into the anterior chambers of the eyes of rabbits, and observed the death of the animals after some months with typical lepra lesions of all the viscera, especially the cecum.

While the lepra bacillus has much in common with the tubercle bacillus, there is not the slightest evidence of any real identity. It has already been shown that lepra bacilli do not grow upon artificial media, and that they cannot be readily transmitted by inoculation. The following description will show that the relation of the bacilli to the lesions is entirely different from that of the tubercle bacilli to the tubercles.

Like the Bacillus tuberculosis, the Bacillus lepræ probably only occurs in places frequented by persons suffering from the disease. That individuals are infected by the latter less readily than by the former bacilli probably depends upon the fact that leprous infection seems to take place most commonly by the entrance of the organisms into the individual through cracks or fissures in the skin, while the tuberculous infection occurs through the more accessible respiratory and digestive apparatus. Once established in the body, the bacillus by its growth produces chronic inflammatory nodes—the analogues of tubercles.

The nodes of lepra consist of various kinds of cells and of fibres. Unlike the tubercles, the lepra nodes are vascular, and much of the embryonal tissue completes its formative function by the production of fibres. The bacilli are not distributed through the nodes like tubercle bacilli, but are found in groups enclosed within the protoplasm of certain large cells—the "lepra cells." These cells seem to be overgrown and partly degenerated lymphoid cells. Sometimes they are anuclear, sometimes they contain several nuclei (giant-cells).

Lepra nodules do not degenerate like tubercles, and the formation of ulcers, which constitutes a large part of the disease, seems largely due to the action of external agencies upon the feebly vital pathological tissue, which is unable to recover itself when injured.

According to the recent studies of Johnston and Jamieson,[1] the bacteriological diagnosis of nodular leprosy can be made by spreading the serum obtained by scraping a leprous nodule upon a cover-glass, drying, fixing, and staining with carbol-fuchsin and Gabbet's solution as for the tubercle bacillus. In such preparations the bacilli are present in enormous numbers, thus forming a marked contrast to the tubercular skin diseases, in which very few can be found.

In that form known as anesthetic leprosy, nodules form upon the peripheral nerves, and by connective-tissue formation, as well as the entrance of the bacilli into the nerve-sheaths, cause irritation, then degeneration, of the nerves. The anesthesia which follows these peripheral nervous lesions is one of the conditions predisposing to the formation of ulcers, etc. by allowing injuries to occur without detection and to progress without observation. The ulcerations and occasional loss of phalanges that follow these lesions occur, probably, in the same manner as in syringomyelia.

The disease advances, having first manifested itself upon the face, extensor surfaces, elbows, and knees, to the lymphatics and the internal viscera. Death ultimately occurs from exhaustion, if not from the frequent intercurrent affections to which the conditions predispose.

[1] *Montreal Med. Journal,* Jan., 1897.

CHAPTER III.

GLANDERS.

GLANDERS is an infectious mycotic disease which, very fortunately, is almost confined to the lower animals. Only occasionally does it secure a victim from hostlers, drovers, soldiers, and bacteriologists, whose frequent association with and experimentation upon animals bring them in frequent contact with those which are diseased. Of all the infectious diseases studied by scientists, none has caused the havoc which glanders has wrought. Several men of prominence have succumbed to accidental infection.

Glanders was first known to us as a disease of the horse and ass characterized by the occurrence of discrete, cleanly-cut ulcers upon the mucous membrane of the nose. These ulcers are formed by the breaking down of nodules which can be detected upon the diseased membranes, and show no tendency to recover, but slowly spread and discharge a virulent pus. The edges of the ulcers are indurated and elevated, the surfaces often smooth. The disease does not progress to any great extent before the submaxillary lymphatic glands begin to enlarge. Later on these glands form large lobulated masses, which may soften, open, and become discharging ulcers. The lungs may also become infected by inspiration of the infectious material, and contain small foci not unlike tubercles in appearance. The animals ultimately die of exhaustion.

In 1882, shortly after the discovery of the tubercle bacillus, Löffler and Schütz discovered in the discharges and tissues of this disease the specific micro-organism, the glanders bacillus (*Bacillus mallei;* Fig. 66), which is its cause.

248

The glanders bacillus is somewhat shorter and distinctly thicker than the tubercle bacillus. It has rounded ends, and it generally occurs singly, though upon blood-

FIG. 66.—Bacillus mallei, from a culture upon glycerin agar-agar; × 1000 (Fränkel and Pfeiffer).

serum, and especially upon potato, several joined individuals may be found. Long threads are never formed.

The bacillus is non-motile. Various observers have claimed the discovery of spores, but although in the interior of the bacilli there have been observed irregular spaces like the similar spaces in the continuity of the tubercle bacillus not colored by the stains, they have not yet been definitely proven to be spores. The observation of Löffler that the bacilli can be cultivated after being kept in a dry state for three months makes it appear as if some permanent form (spore) occurs. No flagella have been demonstrated upon the bacillus.

Like the tubercle bacillus, the glanders bacillus does not seem to find conditions outside the animal body suitable for its existence, and probably does not occur except as a parasite.

The organism only grows between 25° and 42° C., and generally grows very slowly, so that attempts at its isolation and cultivation by the usual plate method are apt to fail, because the numerous other organisms in the material grow much more rapidly.

The best method of isolation seems to be the use of an animal reagent. It has been said that glanders principally affects horses and asses. Recent observations, however, have shown the goat, cat, hog (slightly), field-mouse, wood-mouse, marmot, rabbit, guinea-pig, and hedgehog all to be susceptible animals. Cattle, house-mice, white mice, and rats are immune.

The guinea-pig, being a highly susceptible as well as a readily procurable animal, naturally becomes the reagent for the detection and isolation of the bacillus. When a subcutaneous inoculation of some glanders pus is made, the disease can be observed in guinea-pigs by a tumefaction in from four to five days. Somewhat later this tumefaction changes to a caseous nodule, which ruptures and leaves a chronic ulcer with irregular margins. The lymph-glands speedily become involved, and in a month to five weeks signs of general infection are present. The lymph-glands suppurate, the testicles undergo the same process, and still later the joints exhibit a suppurative arthritis containing the bacilli. The animal finally dies of exhaustion. In guinea-pigs no nasal ulcers form. In field-mice, which are even more susceptible, the disease is much more rapid. No local lesions are visible. In two or three days the animal seems unwell, the breathing is hurried, it sits still with closed eyes, and without any other preliminaries tumbles over on its side, dead.

From the tissues of the inoculated animals the pure cultures are most easily made. Perhaps the best places to secure the culture are from softened nodes which have not ruptured or from the suppurating joints. Strauss has, however, given us a method which is of great use, because of the short time required. The material suspected to contain the glanders bacillus is injected into the peritoneal cavity of a male guinea-pig. In three or four days the disease becomes established. The testicles enlarge a little; the skin over them becomes red and shining. The testicles themselves begin to suppurate, and often discharge through the skin. The animal dies in about two weeks. If such an animal be killed and its testicles examined, the tunica vaginalis testis will be found to contain pus, and sometimes to be partially obliterated by inflammatory exudation. The bacilli are present in this pus, and can be secured from it in pure cultures.

The value of Strauss's method has, however, been lessened by the discovery by Kutcher,[1] that a new bacillus, which he has classified among the pseudo-tubercle bacilli, produces a similiar testicular swelling when injected into the abdominal cavity.

The purulent discharges from the noses of horses and from other lesions of large animals generally contain very few bacilli, so that their detection by the use of the guinea-pig inoculation is made much more simple.

The bacillus is an aërobic organism, and can be grown in bouillon, upon agar-agar, better upon glycerin agar-agar, very well upon blood-serum, and quite characteristically upon potato. It grows in gelatin, but this is not an appropriate medium, because the bacillus develops best at temperatures at which the gelatin is liquid.

Upon 4 per cent. glycerin agar-agar plates the colonies appear upon the second day as pale-yellow or whitish, shining round dots. Under the microscope they appear as brownish-yellow, thick granular masses with sharp borders.

The culture upon agar-agar and glycerin agar-agar occurs as a moist, shining layer not possessed of distinct peculiarities. Upon blood-serum the growth is rather characteristic. The colonies along the line of inoculation first develop as circumscribed, clear, transparent drops, which later become confluent and form a transparent layer unaccompanied by liquefaction.

[1] *Zeitschrift für Hygiene*, Bd. xxi., Heft i., Dec. 6, 1895.

The most characteristic growth is upon potato. It first appears in about forty-eight hours as a transparent, honey-like, yellowish layer, developing only at incubation-temperature and soon becoming reddish-brown. As this brown color of the colony develops, the potato for a considerable distance around it becomes greenish-brown. (See *Frontispiece.*) No other known organism produces the same appearance upon potato.

In litmus milk the growth of the glanders bacillus is associated with the production of an acid that reddens the reagent, with the formation of a firm coagulum and the subsequent separation from it of a clear reddish whey.

The organism loses its virulence if cultivated for many generations upon artificial media.

The bacillus is killed in five minutes by exposure to 55° C.

That this bacillus is the cause of glanders there is no room to doubt. Löffler and Schütz have succeeded by the inoculation of horses and asses in producing the well-known disease.

The organisms when in cultures can be stained with the watery anilin-dye solutions, but are difficult to stain in tissues. They do not stain by Gram's method.

The chief difficulty in staining the bacillus in tissues is the readiness with which it gives up the stain in the presence of decolorizing agents. Löffler at first accomplished the staining by allowing the sections to lie for some time (five minutes) in the alkaline methylene-blue solution, then transferring them to a solution of sulphuric and oxalic acids—

Concentrated sulphuric acid,	2 drops ;
5 per cent. oxalic-acid solution,	1 drop ;
Distilled water,	10 c. cm.

for five seconds, then transferring to absolute alcohol, xylol, etc. The bacilli appear dark blue upon a paler ground. This method gives very good results, but has been largely superseded by the use of Kühne's carbol-methylene blue:

Methylene blue,	1.5
Alcohol,	10.
5 per cent. aqueous phenol solution,	100.

Kühne's method of staining is to place the section in the stain for about half an hour, wash in water, decolorize carefully in hydrochloric acid (10 drops to 500 c.cm. of water), immerse at once in a solution of lithium carbonate (8 drops of a saturated solution of lithium carbonate in 10 c.cm. of water), place in a bath of distilled water for a few minutes, dip into absolute alcohol colored with a little methylene blue, dehydrate in anilin oil containing a little methylene blue in solution, wash in pure anilin oil, not colored, then in a light ethereal oil, clear in xylol, and mount in balsam.

When stained in sections of tissue the bacilli are found to occupy the interior of small inflammatory zones not unlike tubercles in appearance. These nodules can be seen with the naked eye scattered through the livers, kidneys, and spleens of animals dead of experimental glanders. The nodules consist principally of leucocytes, but also contain numerous epithelioid cells. As is the case with tubercles, the centres of the nodules are prone to degenerate, soften, and also to suppurate. The retrogressive processes upon exposed surfaces, where the breaking down of the nodules allows their contents to escape, are the sources of the typical ulcerations. At times the process is progressive, and some of the lesions heal by the formation of a stellate scar.

Baumgarten regarded the origin and course of the histological lesions of glanders to be much like those of the tubercle. In his studies epithelioid cells first accumulated, and were followed by leucocytes. Tedeschi was not able to confirm the results of Baumgarten's work, but found the primary change to be due to a necrosis of the affected tissue followed by an invasion of leucocytes. The recent researches of J. H. Wright[1] are in accord with those of Tedeschi rather than with those of Baumgarten, for Wright observed first a marked degenerative effect upon the tissue, and then an inflammatory exudation amounting in some cases to actual suppuration.

As has been mentioned, cultures of the bacillus lose their virulence more or less after four or five generations in artificial media. While this is true, attempts to attenuate fresh cultures by heat, etc. have so far failed.

Leo has pointed out that white rats, which are immune to the disease, may be made susceptible by feeding with phloridzin and causing a glycosuria.

Kalning, Preusse, Pearson, and others have prepared a substance, "mallein," from cultures of the bacillus, and suggested its employment for diagnostic purposes. It seems to be quite useful in veterinary medicine, the reaction occasioned by its injection being similar to that caused by the injection of tuberculin in tuberculous patients. The manufacture of mallein is not attended with great difficulty. The bacilli are grown in glycerin bouillon for several weeks, killed by heat, the culture filtered through porcelain and evaporated to one-tenth of its volume. It has also been prepared from potato cultures, which are said to produce a stronger toxin. A febrile reaction of more than 1.5° C. following the injection is said to be specific of the disease. Babes has asserted that the injection of this toxic product into susceptible animals will protect them from the disease.

Various experiments have been made with curative objects in view. Certain observers claim to have seen good results follow the injection of mallein in repeated small doses. Others, as Chenot and Picq, find the blood-serum from immune animals like the ox to be curative when injected into infected guinea-pigs.

[1] *Jour. of Exp. Med.*, vol. i., No. 4, p. 577.

CHAPTER IV.

SYPHILIS.

ALTHOUGH syphilis is almost as well known as it is widespread, we have not yet discovered for it a definite specific cause. Whether it is due to a protozoan parasite, or whether it is due to a bacterium, the future must decide. Numerous claims have been made by those whose studies have revealed organisms of one kind or another in syphilitic tissues, but no one has yet succeeded either in isolating, cultivating, or successfully inoculating them.

In 1884 and 1885, Lustgarten published a method for the staining of bacilli which he had found in syphilitic tissues and assumed to be the cause of the disease. The staining, which is very complicated, requires that the sections of tissue be stained in Ehrlich's anilin-water gentian-violet solution for twelve to twenty-four hours at the temperature of the room, or for two hours at 40° C.; washed for a few minutes in absolute alcohol; then immersed for about ten seconds in a 1½ per cent. permanganate-of-potassium solution, after which they are placed in an aqueous solution of sulphurous acid for one to two seconds, thoroughly washed in water, run through alcohol and oil of cloves, and finally mounted in Canada balsam dissolved in xylol.

If the bacilli are supposed to be present in pus or discharges from syphilitic lesions, the cover-glasses spread with the material are stained in the same manner, except that for the first washing distilled water instead of absolute alcohol is used.

This method undergoes a modification in the hands of De Giacomi, who prefers to stain the cover-glasses in hot anilin-water-fuchsin solution for a few moments, sections in the same solution cold for twenty-four hours; then immerse them first in a weak, then in a strong, solution of chlorid of iron. The cover-glasses are washed in water, sections in alcohol, and subsequently passed through the usual reagents for dehydration and clearing.

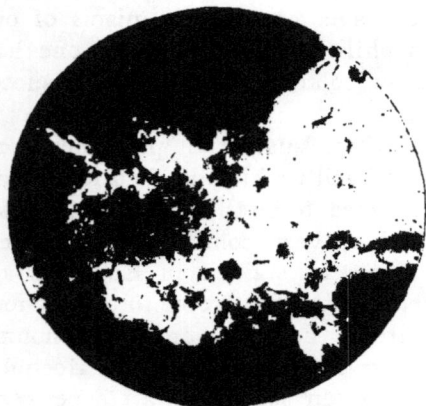

FIG. 67.—Bacillus of syphilis (Lustgarten), from a condyloma; × 1000 (Itzerott and Niemann).

In some syphilitic tissues these methods suffice to define distinct bacilli with a remarkable similarity to the tubercle bacillus. The organism is about the same size as the tubercle bacillus, and even more frequently curved, but often presents a club-like enlargement of one end (involution-form ?). The bacilli very frequently occur singly, though more often in groups, and never lie free, but are always enclosed in cells. These bacilli are not always found in syphilitic lesions, nor is their demonstration easy under the most favorable circumstances. Lustgarten emphasizes particularly that they are only demonstrable after the most painstaking technical procedures.

The probability of the specificity of this organism was considerably lessened by the observation by Matterstock, Travel, and Alvarez that in preputial smegma, and also in vulvar smegma from healthy individuals, a similar organism, identical both in morphology and staining peculiarities, could be demonstrated. Of course the occurrence of Lustgarten's bacillus in the internal organs could not but argue against the probability of its identity with the smegma bacillus ; but Lustgarten himself pointed out that the bacilli of both tuberculosis and leprosy stain by his method, and thus gave Baumgarten the right to suggest that the few cases well adapted for the demonstration of the Lustgarten bacilli might be cases of mixed infection of tuberculosis and syphilis.

The most recent research upon the bacteriology of syphilis is that of van Niessen,[1] who claims to have cultivated a syphilis bacillus from the blood of a few cases. Blood secured from a deep puncture at the end of a thoroughly disinfected finger is caught in a sterile glass, diluted with an equal quantity of distilled water and kept for from ten to fourteen days at a temperature of $10°$–$20°$ R. ($13°$–$15°$ C.). Very often the blood of syphilitics is found subject to accidental contamination by various well-known bacteria. When this is not the case, however, the serum remains almost perfectly clear and contains a large number of bacilli—syphilis bacilli. The bacillus can be transplanted to bouillon, in which it grows with the production of grayish-white shreds and floating flocculi, some of which are suspended in the liquid, while others form a membrane upon the surface.

When transplanted to obliquely solidified gelatin and kept at room temperature, in the course of forty-eight hours a very fine, grayish-white, thready mass like cloudy streaks, and having a peculiar reflecting surface, can be seen. Under a lens this is seen to consist of lines of threads which sometimes seem to penetrate into the depths of the gelatin. After a time a layer is formed upon the surface of the medium. Some liquefaction of the medium occurs and causes the growth to slide down upon

---

[1] *Centralbl. f. Bakt. und Parasitenk.*, Bd. xxiii., No. 2, Jan. 19, 1898, p. 49 ; No. 344, Jan. 31, 1898, p. 97 ; and No. 546, Feb. 11, 1898, p. 177.

17

itself so as to assume the form of a fragment of a tape-worm. Upon agar-agar after the lapse of two days the growth consists of a central pellicle along the line of inoculation, with little sprouts projecting in all directions from the edges. The growth is grayish, with an occasional yellowish tinge.

Punctures in agar-agar were unsuccessful, but in gelatin the appearance of the growth is similar to that of the cholera spirillum.

The bacillus also grows upon potato in the form of an elevated layer of exactly the same color as the potato. In the course of time the entire potato becomes colored a dark gray. It also grows in milk, urine, serum, and water.

The colonies of this bacillus are quite characteristic, but so varied in appearance as to make one suspect that the plate upon which which they grow is contaminated with various other species of bacteria. In general, the colonies may be said to appear slowly as transparent whitish drops, which become grayish and later yellowish, and finally brownish in color. The gelatin about them presents concentric, wave-like rings, depending upon the liquefaction of the medicine.

When the growth is more rapid and occurs at higher temperatures bundles of threads, somewhat resembling the early stages of a mould, are observed. Examining microscopically, one finds in the slowly growing colonies a surrounding zone of small centrifugally arranged fine threads or hairs extending in all directions, with one or two exceptionally long bundles extending beyond the others and beyond the limits of the colony. The long threads are never found to divide. Many of the colonies are highly suggestive of those of anthrax.

The bacillus is motile in very slight degree. It forms spores. It is, in general, about the size of the tubercle bacillus.

The vegetation of the organism is said to be peculiar in that the bacillary stage is of short duration and soon gives place to the formation of septate, V-shaped, and branched forms. It seems to be normally a strepto-bacillus in its early stages, but eventually becomes very pleomorphous, varying in appearance from a chain of oval cocci to the hypha of the moulds. There seems to be nothing peculiar about the staining-capacity of the bacillus. It stains with the ordinary solutions of the anilin dyes, retains the stain of Gram's method, and is decolorized by mineral acids.

Döhle [1] succeeded in staining certain protoplasmic bodies in the tissues in syphilis, which resembled the actively motile protoplasmic bodies which he had previously encountered in the discharges. They were for the most part round or oval, sometimes with irregular outlines, and were provided with flagella. The staining took place in a mixture of hematoxylon and carbol-fuchsin, subsequently treated with iodin or chromatin, and washed in alcohol.

Convinced that these bodies were the cause of syphilis, he excised small fragments from gummata and other syphilitic tissues, and placed them beneath the skin of guinea-pigs, which subsequently fell ill with a chronic marasums which ultimately caused death.

In the inoculation experiments of van Niessen there were observed as evidences of the specificity of the organism discovered by him: (1) abortion in pregnant female rabbits; (2) extra-genital primary lesions on the ears of inoculated rabbits in the form of nodes; (3) secondary ulcer and tumor formations, and irregular lesions, such as occasional thrombosis and pneumonia.

[1] *Münch. med. Wochenschrift,* 1897, No. 43.

CHAPTER V.

ACTINOMYCOSIS.

IN 1845, Langenbeck discovered that the specific disease of cattle known as actinomycosis could be communicated to man. His observations, however, were not given to the world until 1878, one year after Bollinger had discovered the cause of the disease in animals.

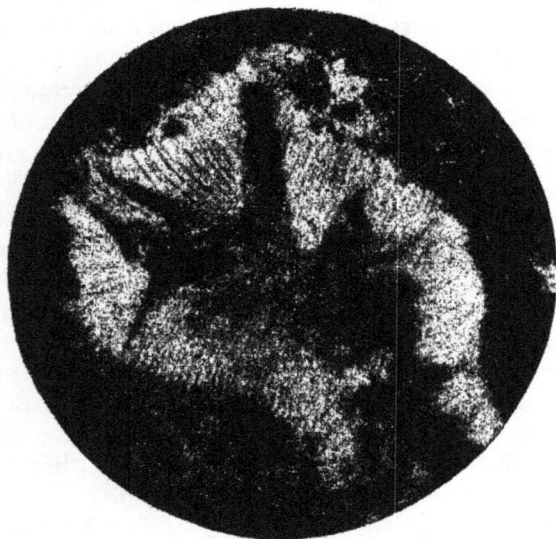

FIG. 68.—Actinomyces bovis, from the tongue of a calf; × 500 (Fränkel and Pfeiffer).

Actinomycosis is a disease almost peculiar to the bovine animals, though sometimes occurring in hogs, horses, men, and other animals.

The first manifestations of the disease are usually found either about the jaw or in the tongue, in either of which localities there are produced considerable enlargements which are sometimes dense and fibrous (wooden tongue) and sometimes suppurative. In sections of these nodular formations small yellowish granules surrounded by some pus can be found. These granules when viewed beneath the microscope exhibit a peculiar rosette-like body—the ray-fungus or actinomyces.

The fungus is of sufficient size to be detected by the naked eye. It can be colored, in sections of tissue, by the use of Gram's method, or better by Weigert's fibrin stain. Tissues pre-stained with carmin, then stained by Weigert's method, give beautiful pictures.

The entire fungus-mass consists of several distinct zones embracing entirely different elements. At the centre of the mass there is found a granular substance containing numerous bodies resembling micrococci. Extending from this centre into the neighboring tissue is a radiating, apparently branched, thickly-tangled mass of mycelial threads. These threads seem to terminate in a zone of conspicuous club-shaped radiating forms which give the colonies the rosette-like appearance. The cells of the tissues affected and a larger or smaller collection of leucocytes form the surrounding resisting tissue-zone.

The degree of chemotactic influence exerted by the organism seems to depend partly upon the tissue affected and partly upon the individuality of the animal. When the animal is but slightly susceptible, and when the tongue is the part affected, the disease is characterized by the production of enlargement due to the formation of cicatricial tissue. If, on the other hand, the animal is highly susceptible or the jaw is affected, the chief symptom is suppuration, with the formation of cavities communicating by sinuses.

Before the nature of the affection was understood it was confounded with various diseases of the bones, principally with osteosarcoma.

From the tissues primarily affected the disease spreads to the lymphatic glands, and not infrequently to the lungs. Israel has pointed out certain cases of human actinomycosis beginning in the peribronchial tissues, probably from inhalation of the fungi.

The occurrence of three distinct elements as components of the rays served to class this organism among the pleomorphous bacteria in the genus Cladothrix, where it has remained undisturbed for at least a decade. Recent researches have, however, changed the view held by some bacteriologists in regard to the actinomyces, and caused them to regard the organism as a bacillus. If it be a bacillus, the central zone of granular cocci-like elements is to be regarded as consisting of individuals in process of rapid division and spore(?)-formation, the mycelial zone as consisting of perfect individuals, and the peripheral zone, with the rosette-like, club-shaped elements, as consisting of individuals partly degenerated through the activity of the cells and tissue-juices (involution-forms).

Jones is of the opinion that the disease, if not indentical with, is closely allied to, tuberculosis, and that the occasional branched forms of tubercle bacilli prove the tendency of the individual bacillus to form a reticulum.

When the mycelial threads are carefully examined, the branchings, which appear distinct upon hasty inspection, are found to be more the effect of a peculiar relation which the threads bear to one another than actual bifurcations, so that it must be regarded as very questionable whether these threads ever so divide.

The organism may be grown upon artificial culture-media, as has been proven by Israel and Wolff.

Upon agar-agar or glycerin agar-agar it forms translucent colonies, about the size of a pin's head, of firm, almost cartilaginous, consistence. These colonies consist of bacillary individuals, sometimes seemingly branched. In bouillon similar dense globular organisms can be grown. The blood-serum colonies, which grow similarly to the agar-agar colonies, are rather more luxuriant, and slowly liquefy the medium.

When the actinomyces are grown upon artificial media their virulence is retained for a considerable length of time. If introduced into the abdominal cavities of rabbits, there are produced in the peritoneum, mesentery, and omentum typical nodules containing the actinomyces rays.

The organism can also be grown in raw eggs, into which it is carefully introduced through a small opening made under aseptic precautions. In the egg the organism forms peculiar long mycelial threads quite unlike the short forms developing upon agar-agar.

The characteristic rosettes which are constantly found in the tissues are never seen in artificial cultures.

The exact manner by which the organism enters the body is unknown. In some cases it may be by direct inoculation with pus, but there is reason to believe that the organism occurs in nature as a saprophyte, or as an epiphyte upon the hulls of certain grains, especially barley. Woodhead records a case where a primary mediastinal actinomycosis in the human subject was supposed to be traced to perforation of the posterior pharyngeal wall by a barley spikelet swallowed by the patient.

Cases of actinomycosis are fortunately of rare occurrence in human medicine, and do not always occur in those brought in contact with the lower animals. The fungi may enter the organism through the mouth and pharynx, through the respiratory tract, through the digestive tract, or through wounds.

The invasion has been known to take place at the roots of carious teeth, and is more liable to occur in the lower than in the upper jaw. Israel reported a case in which the primary lesion seemed to occur external to the bone of the lower jaw, as a tumor about the size of a cherry, with an external opening. In two cases of the disease observed by Murphy of Chicago both began with toothache and swelling of the jaw.

When inhaled, the organisms gain entrance to the deeper portions of the lung, and bring about a suppura-

tive bronchopneumonia with adhesive inflammation of
the contiguous pleura.   After the formation of the pleuritic adhesions the disease may penetrate the newly
formed tissue, extend to the chest-wall, and form external

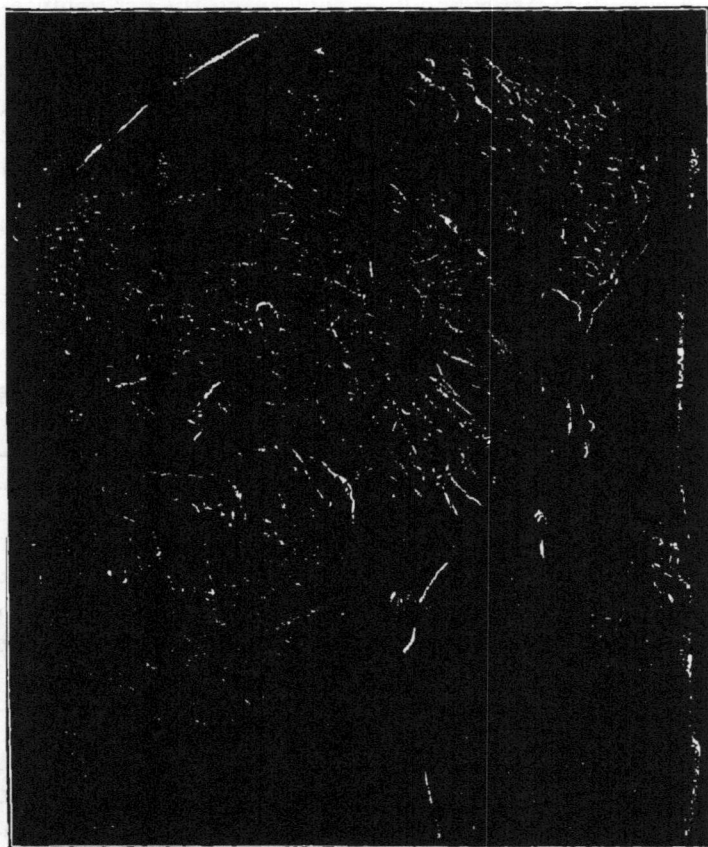

FIG. 69.—Section of liver from a case of actinomycosis in man (Crookshank).

sinuses.   Or it may penetrate the diaphragm and invade
the abdominal organs, causing an interesting and characteristic lesion in the liver and other large viscera (see
Fig. 69).

Microscopically the lesion consists chiefly of a round-cell infiltration with circumscribing granulation-tissue leading to the formation of cicatricial bands. In the form known as "wooden tongue" the disease runs an essentially chronic course, with the production of considerable amounts of connective tissue.

But few cases recover, the disease terminating by death from exhaustion or from complicating pneumonia or other organic lesions.

CHAPTER VI.

MYCETOMA, OR MADURA-FOOT.

A CURIOUS disease of not infrequent occurrence in the Indian province of Scinde is one known as mycetoma, Madura-foot, or *pied de Madura*. It almost invariably affects natives of the agriculturist class, and in most cases begins in or is referred by the patient to the prick of a thorn. It generally affects the foot, more rarely the hand, and in one instance was seen by Boyce in the shoulder and hip. It is more common in men than in women, individuals between twenty and forty years of age suffering most frequently, but persons of any age or sex may suffer from the disease. It is insidious in its onset, as has been said, generally following a slight injury, such as the prick of a thorn. No symptoms are observed in what might be called an incubation stage of a couple of weeks' duration, but after this time elapses a nodular growth gradually forms, attaining in the course of time the size of a marble. Its deep attachments are indistinct and diffuse. The skin becomes purplish, thickened, indurated, and adherent. The points most frequently invaded at the onset are the ball of the great toe and the pads under the bases of the fingers and toes.

In the course of months, although progressing slowly, the lesions attain very perceptible size, distinct tumors being present. Later, sometimes not until after a year or two, the nodes begin to soften, break down, discharge their purulent contents, and originate ulcers and communicating sinuses. The discharge at this stage is a thin sero-pus, and is always mixed with a number of fine round black or pink bodies, described, when black, as resembling gunpowder; when pink, as resembling fish-roe. It is the detection of these particles upon which the diagnosis rests, and upon which the division of the disease into the *melanoid* and *pale* varieties depends.

The progress of the disease causes an enormous size and a peculiar deformity of the affected foot or hand. The malady is generally painless.

The micro-organismal nature of the disease was early suspected. In spite of the confusion caused by some who confounded the disease with and described it as "Guinea-worm," Carter held that it was due to some indigenous fungus as early as 1874. Boyce and Surveyor believe that the black particles of the melanoid variety represent a curious metamorphosis of a large branching septate fungus, and that the white particles of the other variety are the remains of a lowly-organized fungus and of caseous particles.

Kanthack tried to prove the identity of the fungus with the well-known actinomyces, but there seems to be considerable doubt about the correctness of his view.

Vincent succeeded in isolating the micro-organism by puncturing one of the nodes with a sterile pipette, and has cultivated it upon artificial media. Acid vegetable infusions seem suitable to its growth. It develops scantily in bouillon at the room-temperature, better at 37° C.—in from four to five days. In twenty to thirty days the colony attains the size of a little pea.

In the liquid media the colonies which cling to the glass, and thus remain near the surface of the medium, develop a rose- or bright-red color.

Cultures in gelatin are not very abundant, are colorless, and are unaccompanied by liquefaction.

Upon the surface of agar-agar strikingly beautiful rounded, glazed colonies are formed. They are at first colorless, but later become rose-colored or bright red. The majority of the clusters remain isolated, some of them attaining the size of a small pea. They are generally umbilicated like a variola pustule, and present a curious appearance when the central part is pale and the periphery red. As the colony ages the red color is lost and it becomes dull white. The colonies are very adherent to the surface of the medium, and are said to be of cartilaginous consistence. The organism also grows in milk without coagulation.

Upon potato the development is meagre, slow, and with very little tendency to chromogenesis. The color-production is more marked if the potato be acid in reac-

FIG. 70.—Streptothrix Maduræ in a section of diseased tissue (Vincent).

tion. Some of the colonies upon agar-agar and potato have a powdery surface, no doubt from the occurrence of spores. It is, of course, an aërobic organism.

Under the microscope the organism is found by Vincent to be a streptothrix—a true branched fungus consisting of long bacillary branching threads in a tangled mass. In many of the threads spores could be made out.

Vincent was unable to communicate the disease to animals by inoculation.

Microscopic study of the diseased tissues in cases of mycetoma is not without interest. The healthy tissue is said to be sharply separated from the diseased masses. The latter appear as large degenerated tubercles, except that they are extremely vascular. The mycelial or filamentous fungous mass occupies the centre of the degeneration, where its long filaments can be beautifully demonstrated by the use of appropriate stains, Gram's method being excellent for the purpose. The tissue surrounding the disease-nodes is infiltrated with small round cells. The youngest nodules are seen to consist of granulation-tissue, which in its organization is checked by the coagulation-necrosis which is sure to overtake it. Giant-cells are few.

Not infrequently small hemorrhages occur from the ulcers and sinuses of the diseased tissues ; the hemorrhages can be explained from the abundance of small blood-vessels in the diseased tissue.

Although the disease has been described as occurring in Scinde, it is not limited to that province, having been met with in Madura, Hissar, Bicanir, Dehli, Bombay, Baratpur, Morocco, Algeria, one case by Bastini and Campana in Italy, and one by Kempner in America.

CHAPTER VII.

FARCIN DU BŒUF.

THE peculiar disease which sometimes affects numbers of cattle in Guadeloupe, and which was described by the older writers as *farcin du bœuf*, has been carefully studied by Nocard. It is a disease of cattle characterized by a superficial lymphangitis and lymphadenitis, affecting the tracheal, axillary, prescapular, and other glands. The affected glands enlarge, suppurate, and discharge a creamy, sometimes a grumous, pus. The internal organs are often affected with a pseudo-tuberculosis whose central areas undergo a purulent or caseous degeneration.

In the researches of Nocard it was discovered, by staining by Gram's and by Kühne's methods, that in the centres of the tubercles micro-organisms could be defined. They resembled long delicate filaments rather intricately woven, characterized by distinct ramifications which made clear the proper classification of the organism as a *streptothrix*. The organism was successfully cultivated by Nocard upon various culture-media at the temperature of the body. It is aërobic.

In bouillon the organism develops in the form of colorless masses irregular in size and shape, some of which float upon the surface, others of which sink to the bottom of the liquid. Sometimes the surface is covered by an irregular fenestrated pellicle of a gray color.

Upon agar-agar the growth develops in small, rather discrete, irregularly rounded, opaque masses of a yellowish-white color. The surfaces of the colonies are tuberculated, and an appearance somewhat like a lichen is observed (see Fig. 71).

270

Upon potato very dry scales of a pale-yellow color rapidly develop.

The growth upon blood-serum is less luxuriant, but similar to that upon agar-agar.

In milk the organism produces no coagulation by its growth, and does not alter the reaction.

Microscopic study always reveals the organism as the same tangled mass of filaments seen in the tissues. The old cultures are rich in spores, which are very small and

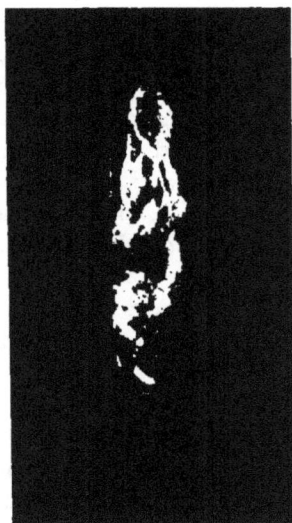

Fig. 71.—Streptothrix of farcin du bœuf growing on glycerin agar-agar.

develop upon the most superficial portions of the growth. These spores resist the penetration of stains to a rather unusual extent.

Cultures retain their virulence for a long time: Nocard found one virulent after it had been kept for four months in an incubating oven at 40° C.

The streptothrix of *farcin du bœuf* is pathogenic for guinea-pigs, cattle, and sheep; dogs, rabbits, horses, and asses are immune.

When the culture or some pus containing the micro-organism is injected subcutaneously into a guinea-pig, a voluminous abscess results.   Not long afterward the lymphatic vessels and glands of the region are the seat of swelling and induration, and extensive phlegmons form, which rupture externally and discharge considerable pus.   The animal, of course, becomes extremely ill and seems about to die ; instead, it slowly recovers its normal condition.

In other animals, as the cow and the sheep, the subcutaneous inoculation results in an abscess relatively less extensive.   This ulcerates, then indurates, and seems to disappear, but after the lapse of several weeks or months opens again in the form of a new abscess.

In animals which are immune or nearly immune, like the horse, the ass, the dog, and the rabbit, the subcutaneous inoculation is followed by the formation of a small abscess which speedily cicatrizes.

Intraperitoneal inoculation in the guinea-pig gives rise to an appearance resembling tuberculosis.   The omentum may be extensively involved and full of softened nodes. The liver, spleen, and kidneys appear full of tubercles, but careful examination will satisfy the observer that the tubercles are only upon the peritoneal surfaces, not in the organs.

Intravenous introduction of the cultures produces a condition much resembling general miliary tuberculosis. All the organs contain the pseudo-tubercles in considerable numbers.

CHAPTER VIII.

RHINOSCLEROMA.

In Austria, Hungary, Italy, and some parts of Germany there sometimes occurs a peculiar disease of the anterior nares, characterized by the occurrence of circumscribed tumors, known as rhinoscleroma. The tumor-masses are somewhat flattened, isolated or coalescent, grow with great slowness, and recur if excised. The disease commences in the mucous membrane and the adjoining skin, and spreads to the skin in the neighborhood by a slow invasion, involving the upper lip, jaw, hard palate, and sometimes the pharynx. The growths are without evidences of inflammation, do not ulcerate, and consist microscopically of infiltration of the papilla and corium of the skin, with round cells which change in part to fibrillar tissue. The tumors possess a well-developed lymph-vascular system. Sometimes the cells undergo hyaline degeneration.

In these little tumors the researches of Von Frisch discovered little bacilli much resembling both in morphology and vegetation the pneumo-bacilli of Friedländer, and, like them, surrounded by capsules. The only marked difference between the so-called bacillus of rhinoscleroma and the Bacillus pneumoniæ of Friedländer is that the former stains well by Gram's method, while the latter does not, and that the former is rather more distinctly rod-shaped than the latter, and more often shows its capsule in culture-media.

The bacillus can easily be cultivated, and in all media resembles the bacillus of Friedländer too closely to be distinguished from it. Even when inoculated into animals the bacillus behaves much like Friedländer's bacillus.

Inoculation has, so far, failed to produce the disease either in men or in the lower animals.

18   273

B. THE TOXIC DISEASES.

CHAPTER I.

TETANUS.

ONE of the most exquisitely toxic bacteria of which we have any knowledge is the bacillus discovered in 1884 by Nicolaier, obtained in pure culture by Kitasato in 1889, and now universally recognized as the cause of tetanus. It is a peculiar organism, whose striking feature is a considerable enlargement of one end, in which a bright round spore is seen (Fig. 72). The bacilli which

FIG. 72.—Bacillus tetani; × 1000 (Fränkel and Pfeiffer).

are not sporiferous, are long, rather slender, have rounded ends, seldom unite in chains or pairs, are motile, and have no flagella. The bacilli stain readily with ordinary aqueous solution of the anilin dyes, and also very readily by Gram's method.

The tetanus bacillus is a common saprophytic organism which can be found in most garden-earth, in dust, in manure, and sometimes in the intestinal discharges of animals. It is extremely difficult to isolate and cultivate, because it will not grow where the smallest amount of oxygen is present.

The method now generally employed for the isolation of this bacillus is that originated by Kitasato, and based upon his observation that its spores can resist high temper-

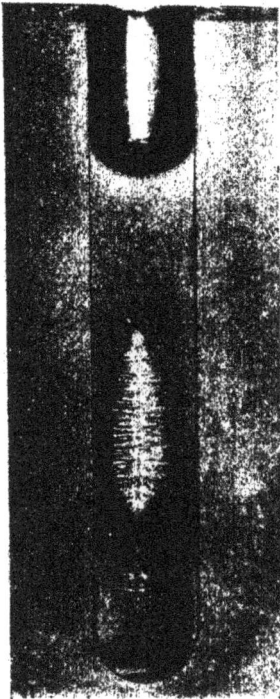

FIG. 73.—Bacillus tetani : six-days-old puncture-culture in glucose-gelatin (Fränkel and Pfeiffer).

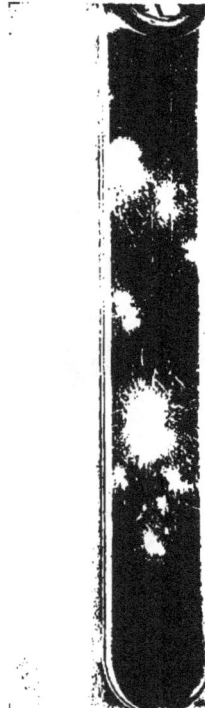

FIG. 74.—Bacillus tetani : culture four days old in glucose-gelatin (Fränkel and Pfeiffer).

atures. After finding that the typical bacilli are present in earth or pus, or whatever the material to be investigated was, Kitasato exposed a portion of it for an hour to a temperature of 80° C. By this heating all the fully-developed bacteria, tetanus as well as the others, and the great majority of the spores except those of tetanus, were destroyed, and, as little other than tetanus spores remained, their cultivation was made comparatively easy.

The resistance which the tetanus bacilli manifest toward heat is only part of a great general resisting power of which they are possessed. It is said that they can retain their vitality in the dried condition for months. Sternberg says they can resist 5 per cent. carbolic solutions for ten hours, but will not grow after fifteen hours' immersion. 5 per cent. carbolic acid, to which 0.5 per cent.

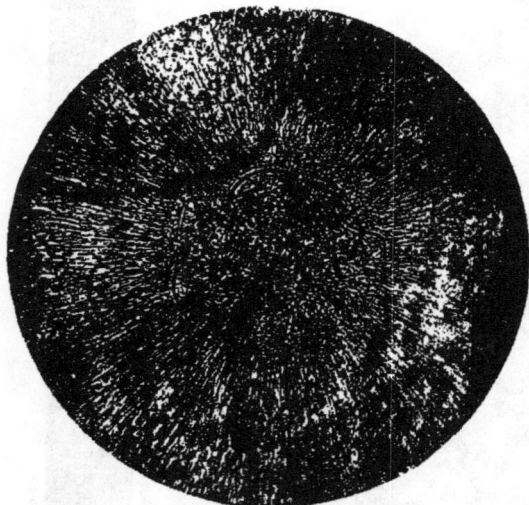

Fig. 75.—Bacillus tetani: five-days-old colony upon gelatin containing glucose; × 1000 (Fränkel and Pfeiffer).

of hydrochloric acid has been added, destroys them in two hours. They are also destroyed in three hours by 1 : 1000 bichlorid-of-mercury solution ; but when to such a solution 0.5 per cent. of hydrochloric acid is added, its activity is so increased that the spores are destroyed in thirty minutes. The resistance to heat is only within certain limits, for exposure to passing steam for from five to eight minutes is certain to kill the spores.

The colonies of the tetanus bacillus, when grown in

an atmosphere of hydrogen upon gelatin plates, somewhat resemble those of the well-known hay bacillus. There is a dense rather opaque central mass from which a more transparent zone is readily separable. The margins of this outer zone are made up of a radiating fringe of projecting bacilli (Fig. 75). The liquefaction that occurs is much slower than that caused by bacillus subtilis.

When grown in gelatin puncture-cultures the development occurs deep in the puncture, and consists of multitudes of short-pointed processes radiating from the puncture, somewhat resembling a fir tree (Fig. 73). Liquefaction begins in the second week and causes the disappearance of the radiating processes. The liquefaction spreads slowly, but may involve the entire mass of gelatin and resolve it into a grayish-white syrupy liquid, at the bottom of which the bacilli accumulate. The growth in gelatin containing glucose is much more rapid ; that in agar-agar punctures is much slower, but similar to the gelatin cultures except for the absence of liquefaction. The organism can also be grown in bouillon, and attains its maximum development at a temperature of 37° C. Much gas is given off from the cultures.

Cultures of the tetanus bacillus in all media give off a peculiar characteristic odor—a burnt-onion smell, with a suggestion of putrefaction about it.

The methods for excluding the oxygen from the cultures and replacing it by hydrogen, as well as other methods suggested for the cultivation of the strictly anaërobic organisms, are given under the appropriate heading (Anaërobic Cultures), and need not be repeated here.

A very simple method of cultivating the bacillus in bouillon for the purpose of securing a large amount of toxin has been suggested by the author.[1] An ordinary bottle is filled with bouillon to the mouth, and closed by a perforated rubber stopper containing a glass tube

[1] *Centralbl. f. Bakt. u. Parasitenk.*, xix., Nos. 14 and 15, April 25, 1896, p. 550.

a couple of inches long. Connected with this glass tube, by means of a short piece of rubber tubing, is the bulb of a broken pipette, the other end of which is plugged with cotton (Fig. 76). When the steam sterilization takes place the expanding fluid ascends to the reservoir represented by the pipette bulb, descending again as the fluid cools. When the sterilization is completed the reservoir is detached, the inoculation made by passing a very fine pipette into the bottle, the projecting glass tube drawn out to a fine tube, and the bottle stood in hot water until the expanding fluid ascends to the apex of the pointed glass tube. The tube is now sealed in a flame and the bottle and its contents allowed to cool. In cooling the retracting fluid leaves a vacuum which at once draws up any minute bubbles of air remaining, and allows the tetanus bacillus to grow in a condition of very fair anaërobiosis.

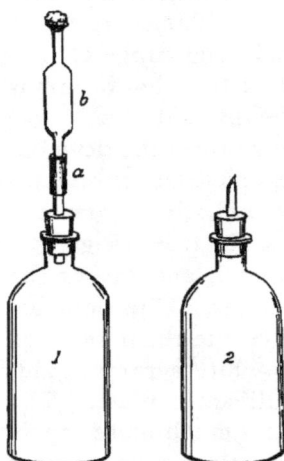

FIG. 76.—Tetanus bottle.

Tetanus bacilli exist in nature as widely distributed saprophytes. They are quite common in the soil, and the fact that they are most plentiful in manured ground has suggested that they originate in the intestines of horses and reach the earth from their excrement. Le Dentu has, however, shown that the tetanus bacillus is a common organism in New Hebrides, where there are no horses. In these islands the natives poison their arrows by dipping them into a clay rich in tetanus bacteria.

The work of Kitasato has given us a very exact knowledge of the tetanus bacillus and completely establishes its specific nature.

The organisms generally enter the animal body through a wound caused by some implement which has been in contact with the soil, or enter abrasions from the soil directly. Doubtless many of the wounds are so small that their existence is overlooked, and this, together with the fact that the period of incubation of the disease, especially in man, is of considerable duration, and at times permits the wound to heal before any symptoms of intoxication occur, serves to explain to us at least some of the reported cases in which no wound is said to have existed.

It would seem that in some rare cases tetanus can occur without the previous existence of a wound. Such a case has been reported by Kamen, who found that the intestine of a person dead of the disease was rich in the Bacillus tetani. Kamen is of the opinion that the bacilli can grow in the intestine and be absorbed, especially where there are imperfections in the mucosa. It is not impossible, though he does not think it probable, that the bacteria growing in the intestine could elaborate enough toxin to produce the disease by absorption.

All animals are not alike susceptible to the disease. Men, horses, mice, rabbits, and guinea-pigs are all susceptible ; dogs are much less so. Most birds are scarcely at all susceptible either to the bacilli or to the poison. Amphibians are immune, though it is said that frogs can be made susceptible by elevation of their body-temperature.

When a white mouse is inoculated with an almost infinitesimal amount of bouillon or solid culture, or is inoculated with garden-earth containing the tetanus bacillus, the disease is almost certain to follow, the first symptoms coming on in from one to two days. The mouse develops typical tetanic convulsions, which begin first in the neighborhood of the inoculation, but soon become general. Death follows sometimes in a very few hours. In rabbits the period of incubation is nearly two weeks, and in man may be three weeks.

The conditions in the animal body are not favorable for the development of the bacilli, because of the free supply of oxygen contained in the blood, and we find that they grow with great slowness, remain localized at the seat of inoculation, and never enter the blood- or lymph-circulation. Doubtless most cases of tetanus are cases of mixed infection in which the bacillus enters with bacteria, which greatly aid its growth by using up the oxygen in their neighborhood. The amount of poison produced must be exceedingly small and its power tremendous, else so few bacilli growing under adverse conditions could not produce fatal toxemia. The poison is produced rapidly, for Kitasato found that if mice were inoculated at the root of the tail, and afterward the skin and the subcutaneous tissues around the inoculation were either excised or burned out, this treatment would not save the animal unless the operation were performed *within an hour after the inoculation.*

Some incline to the view that the toxin is a ferment, and the experiments of Nocard, quoted before the Académie de Médecine, October 22, 1895, might be adduced in support of the theory. He says: "Take three sheep with normal tails, and insert under the skin at the end of each tail a splinter of wood covered with the dried spores of the tetanus bacillus; watch these animals carefully for the first symptoms of tetanus, then amputate the tails of two of them 20 cm. above the point of inoculation, . . . the three animals succumb to the disease without showing any sensible difference."

The circulating blood of diseased animals is fatal to susceptible animals because of the toxin which it contains; and the fact that the urine is also toxic to mice proves that the toxin is excreted by the kidneys.

From pure cultures of tetanus bacilli grown in various media, and from the blood and tissues of animals affected with the disease, Brieger succeeded in separating two alkaloidal substances—"tetanin" and "tetano-toxin," both very poisonous and productive of tonic convulsions; and Brieger and Fränkel later isolated an extremely poisonous toxalbumin.

The pathology of the disease is of much interest because of its purely toxic nature. There is generally a small wound with a slight amount of suppuration. At the autopsy the organs of the body are normal in appearance, except the nervous system, which bears the greatest insult. It, however, shows little else than congestion either macroscopically or microscopically.

An interesting fact contributed to our knowledge of the disease has been presented by Vaillard and Rouget, who found that if the tetanus bacilli were introduced into the body freed from their poison, they were unable to produce any signs of disease because of the promptness with which the phagocytes took them up. If, however, their poison was not removed, or if the body-cells were injured by the simultaneous introduction of lactic acid or other chemical agents, the bacilli would immediately begin to manufacture the toxin and produce symptoms of the disease.

The toxin is easily prepared, being readily soluble in water. The most ready method of preparation is to grow the bacilli in bouillon, keeping the culture-medium at a temperature of 37° C., and allowing it to remain undisturbed for from two to four weeks, by which time it will have attained a toxicity so great that 0.000005 c.cm. will cause the death of a mouse. The toxin is very rapidly destroyed by heat, and cannot bear any temperature above 60°–65° C. It is also decomposed by light. The best method of keeping it is to add 0.5 per cent. of phenol, and then store it in a cool, dark place. It will not keep its strength very long under the best conditions.

The purified toxin of Brieger and Cohn was surely fatal to mice in doses of 0.00000005 gram. Lambert,[1] in his comprehensive review of the use of tetanus antitoxin, points out that this is the most poisonous substance that has ever been discovered.

By the gradual introduction of such a toxin into animals Behring and Kitasato have been able to produce in

[1] *New York Med. Jour.*, June 5, 1897.

their blood a distinctly potent and valuable antitoxic substance.

The method for the production of this tetanus antitoxic serum is very much like that for the diphtheria antitoxic serum (*q. v.*), except that a much longer time is required for its production, that the doses of toxin are of necessity smaller because its toxicity is greater, and that trichlorid of iodin or Gram's solution will probably need to be added to the toxin to prevent too powerful a local reaction. Horses, dogs, and goats may be used.

As tetanus cases are not very common, and the antitoxic serum when produced is not very stable in its properties, Tizzoni and Cattani have successfully prepared it in a solid form, in which, it is claimed, it can be kept indefinitely, shipped any distance, and used after simple solution in water. Their method is to precipitate the antitoxin from the blood of immunized dogs with alcohol. Numerous cases of the beneficial action of this antitoxin are on record.

The strength of the serum is generally expressed 1 : 1,000,000, 1 : 10,000,000, etc., which indicates that 1 c.cm. of the serum is capable of protecting 1,000,000 or 10,000,000 grams of guinea-pig from infection.

The experiments of Alexander Lambert show that a protective power of 1 : 800,000,000 can be attained.

As Welch has pointed out, the antitoxin of tetanus has proved to be rather a disappointment in human medicine, and also for the treatment of large animals, such as the horse. The results following its injection, in combination with the sterile toxin, into mice, guinea-pigs, and rabbits are highly satisfactory, but the amount needed, in proportion to the body-weight, to save the animal from the toxin being manufactured in its body by bacilli increases so enormously with the day or hour of the disease as to make the dosage, which increases millions of times where that of diphtheria antitoxin increases but tenfold, a matter of difficulty and uncertainty. Nocard also calls attention to the fact that the existence of tetanus is unknown until there is sufficient toxemia to produce spasms, and that therefore it is impossible to attack the disease in its inception; we are obliged to meet it upon the same grounds as diphtheria in the later days of the disease— a time when it is well known that the chances of improvement are greatly lessened.

Of course, as there is no other remedy that combats the disease at all, the antitoxin is one which, when obtainable, should always be employed.

An interesting observation has been recently made by Wasserman,[1] who, assuming that the destruction of nerve-cells in the cerebrum and cord during tetanus toxemia might have something to do with immunity, believed it possible to obtain from these cells an immunizing substance. Investigating the subject, he found that when fresh brain or spinal cord was rubbed up in a mortar with physiological salt solution, and injected into animals, the mixture had the power not only to confer upon them an immunity lasting for twenty-four hours, but also was potent enough to neutralize the effects of an injection of tetanus toxin ten times as large as that necessary to kill the animal in doses of 1 c.cm.

These observations may offer a possible solution of the difficult problem laid before us by Montesano and Montesson,[2] who unexpectedly found the tetanus bacillus in pure culture in the cerebro-spinal fluid of a case of paralytic dementia that died without a tetanic symptom.

[1] *Berlin. klin. Wochenschrift,* 1898, No. 1.
[2] *Centralbl. f. Bakt., u. Parasitenk.,* Bd. xxii., Nos. 22, 23, p. 663. Dec., 1897.

CHAPTER II.

DIPHTHERIA.

In 1883, Klebs pointed out the existence of a bacillus in the pseudo-membranes upon the fauces of patients suffering from diphtheria, but it was not until 1884 that Löffler succeeded in isolating and cultivating the organism, which is now known by both their names—the Klebs-Löffler bacillus.

The bacillus as described by Löffler is about the length of the tubercle bacillus, about twice its diameter, has a

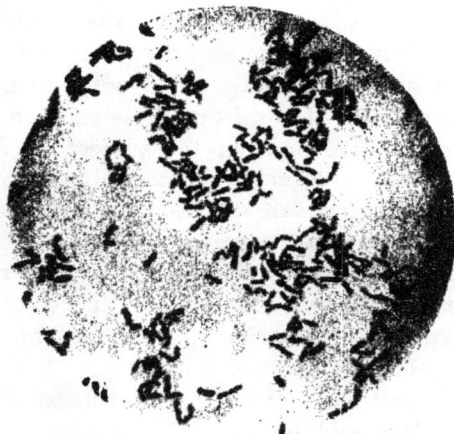

FIG. 77.—Bacillus diphtheriæ, from a culture upon blood-serum; × 1000 (Fränkel and Pfeiffer).

curve similar to that which characterizes the tubercle bacillus, and has rounded ends (Fig. 77). It does not form chains, though two, three, and rarely four individuals may be found joined; generally the individuals are all separate from one another. The morphology of the bacillus is peculiar in its considerable irregularity, for among the well-formed individuals which abound in fresh cultures a large number of peculiar organisms are to be found, some much larger than normal, some with one end enlarged to a club-shape, some greatly elongated, with both ends expanded into club-shaped enlargements. These bizarre forms seem to represent an involution-form of the organism, for, while present in perfectly fresh cultures, they are so abundant in old cultures that scarcely a single well-formed bacillus can be found. It not infrequently happens that in unstained bacilli distinct granules can be defined at the ends—polar granules—thus giving the organism somewhat the appearance of a diplococcus.

The bacillus can be readily stained by aqueous solutions of the anilin colors, but more beautifully and characteristically with Löffler's alkaline methylene blue :

Saturated alcoholic solution of methylene blue,   30 ;
    1 : 10,000 aqueous solution of caustic potash,   100 ;

and an aqueous solution of dahlia, as recommended by Roux.

When cover-glass preparations are stained with these solutions, the bizarre forms already mentioned are much more obvious than in the unstained individuals, and the contrast between the polar granules, which color intensely, and the remainder of the bacillus, which tinges slightly, is marked. Through good lenses the organisms are always distinct bacilli, notwithstanding the fact that the ends stain more deeply than the centres, and it is only through poor lenses that the organisms can be mistaken for diplococci. The bacilli stain well by Gram's method, this being a good method to employ for their definition in sections of tissue, though Welch and Abbott assert that Weigert's fibrin method and picro-carmin give the most beautiful results.

The diphtheria bacillus does not form spores, and is delicate in its thermal range. Löffler found that it could not endure a temperature of 60° C., and Abbott has shown that a temperature of 58° C. for ten minutes is fatal to it. Notwithstanding this susceptibility, the organism can be kept alive for several weeks after being dried upon shreds of silk or when surrounded by dried diphtheria membrane.

No flagella have been demonstrated upon the bacillus. It is non-motile.

Fernbach has shown that when the organisms are grown in a medium exposed to a passing current of air, the luxuriance of their development is increased, though their life-cycle is shorter. The growth can also take place when the air is excluded, so that the bacillus must be classed among the optional anaërobic organisms.

The diphtheria bacillus grows readily upon all the ordinary media, and is a very easy organism to obtain in pure culture. Löffler has shown that the addition of a small amount of glucose to the culture-medium increases the rapidity of the growth, and suggests a special medium which bears his name—Löffler's blood-serum mixture:

Blood-serum, 3 ;
Ordinary bouillon + 1 per cent. of glucose, 1.

This mixture is filled into tubes, coagulated, and sterilized like blood-serum, and is one of the best-known media in connection with the study of diphtheria.

The studies of Michel[1] have shown that the development of the culture is much more luxuriant and rapid when horse serum instead of beef or calves' blood is used. Horse's blood can easily be secured by the introduction of a trocar into the jugular vein; 5 liters of it can be withdrawn without causing the animal any inconvenience or producing symptoms.

The impossibility of clinically making an accurate diagnosis of diphtheria without a bacteriologic examination has caused many private physicians and many medical societies and boards of health to equip laboratories where

[1] *Centralbl. f. Bakt. u. Parasitenk.*, Sept. 24, 1897, Bd. xxii., Nos. 10 and 11.

accurate examinations can be made. The method requires some apparatus, though a competent bacteriologist can often make shift with a bake-oven, a wash-boiler, and other household furniture instead of the regular sterilizers and incubators, which are expensive.

When it is desired to make a bacteriologic diagnosis of a suspected case of diphtheria or to secure the bacillus in pure culture, a sterile platinum wire having a small loop at the end, or a swab made by wrapping a little cotton around the end of a piece of wire and carefully sterilizing in a test-tube, is introduced into the throat and touched to the false membrane, after which it is smeared carefully over the surface of at least three of the blood-serum-mixture tubes, without either again touching the throat or being sterilized. The tubes thus inoculated are stood away in an incubating oven at the temperature of 37° C. for twelve hours, then examined. If the diphtheria bacillus is present upon the first and second tubes, there will be a smeary yellowish-white layer, with outlying colonies on the second tube, while the third tube will show rather large isolated whitish or slightly yellowish colonies, smooth in appearance, but rather irregular in outline. Very often the colonies are china-white in appearance. These colonies, *if found by microscopic examination to be made up of diphtheria bacilli*, will confirm the diagnosis of diphtheria, and will at the same time give pure cultures when transplanted. There are very few other bacilli which grow so rapidly upon Löffler's mixture, and scarcely one other which is found in the throat.

Ohlmacher recommends the microscopic examination of the still invisible growth in five hours. A platinum loop is rubbed over the inoculated surface; the material secured is then mixed with distilled water, dried on a cover-glass, stained with methylene blue, and examined. This method, if reliable, will be very valuable in making an early diagnosis preparatory to the use of the antitoxin.

The presence of diphtheria bacilli in material taken

from the throat does not necessarily prove the patient to
be diseased.   Virulent bacilli can often be discovered in
the throats of healthy persons who have knowingly or
unknowingly come in contact with the disease.   The
bacteriologic examination is only an adjunct to the
clinical diagnosis, and must never be taken as positive
in itself.

The bacillus grows similarly upon blood-serum and
Löffler's mixture.   Upon glycerin agar-agar and agar-agar
the colonies are much larger, more translucent, always

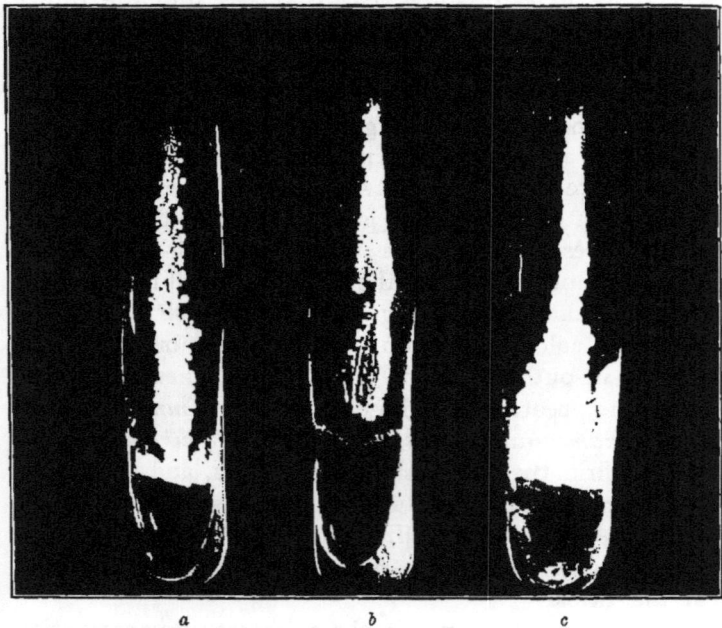

FIG. 78.—Diphtheria bacilli (from photographs taken by Prof. E. K. Dunham, Carnegie Laboratory, New York): *a*, pseudo-bacillus; *b*, true bacillus;
c, pseudo-bacillus.

appearance. It must be remarked that when sudden transplantations are made from blood-serum to agar-agar the growth resulting is meagre, but the oftener this growth is transplanted to fresh agar-agar the more luxuriant it becomes.

The growth in gelatin puncture-cultures is characterized by small spherical colonies which develop along the entire length of the needle-track. The gelatin is not liquefied.

Upon the surface of gelatin plates the colonies that develop do not attain anything like the size of the colonies upon Löffler's mixture. They appear to the naked

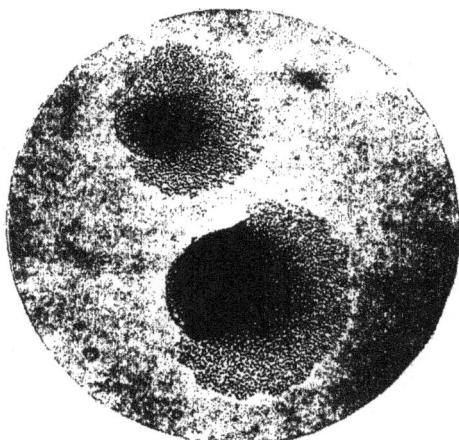

Fig. 79.—Bacillus diphtheriæ, colony twenty-four hours old upon agar-agar; × 100 (Fränkel and Pfeiffer).

eye as whitish points with smooth contents and regular though sometimes indented borders. Under the microscope they appear as granular, yellowish-brown colonies with irregular borders (Fig. 79).

When planted in bouillon the organism causes a diffuse cloudiness at first, but, not being motile, soon settles to the bottom in the form of a rather flocculent precipitate which has a tendency to cling to the sides of the glass. Sometimes a delicate irregular mycoderma forms upon

19

the surface, especially when the cultivation is made by the method of Fernbach with a passing current of air. This mycoderma, which may appear quite regular when the flask is undisturbed, is so brittle that it at once falls to pieces if the flask be moved.

Spronck has recently determined that the characteristics of the growth of the diphtheria bacillus in bouillon, as well as the amount of toxin-production, vary according to the amount of glucose in the bouillon. He divides the cultures into three types :

Type A. The reaction of the bouillon becomes acid and remains acid, the acidity increasing. The bacilli accumulate at the bottom of the clear liquid. The toxin-production is meagre.

Type B. There is no change from alkalinity to acidity, but the original alkalinity of the bouillon steadily increases. The culture is very rich, the bottom of the flask shows a considerable sediment, the liquid is cloudy, and a delicate growth occupies the surface. The toxicity is very great.

Type C. In a few days the reaction of the culture becomes acid, and then later on changes to alkaline. During the acid period the liquid is clear, with a white surface-growth. When the alkalinity returns the bouillon clouds and the surface-growth increases in thickness. Sediment accumulates at the bottom of the flask. The toxicity of the culture is much less than in Type B.

Spronck regards the varying reaction as due to the fermentation of the glucose, and asserts that the most luxuriant and toxic cultures are those in which no glucose is present. To exclude as much of the undesirable sugar as possible, he makes the bouillon from the stalest meat obtainable, preferring it when just about to putrefy. Of the meats experimented with, beef was found to be the best.

In large cities meat is ordinarily kept sufficiently long before being offered for sale to meet Spronck's requirement.

Upon potato the bacillus develops only when the reaction is alkaline. The potato-growth is not characteristic. Welch and Abbott always secured a growth of the organism when planted upon potato, but do not mention the reaction of the medium they employed.

Milk is an excellent medium for the cultivation of the Bacillus diphtheriæ, and is possibly at times a means of infection. Litmus milk is an excellent medium for observing the changes of reaction brought about by the growth of the bacillus. At first the alkalinity, which is always favorable to the development of the bacillus, is destroyed by the production of an acid. When the culture is old the acid is replaced by a strong alkaline reaction.

Palmirski and Orlowski[1] assert that the bacillus produces indol, but only after the third week. Smith, however, came to a contrary result, and found that when diphtheria bacillus grew in the dextrose-free bouillon that he recommends no indol was produced.[2]

Diphtheria as it occurs in man is generally a disease characterized by the formation of a pseudomembrane upon the fauces. It is a local infection, due to the presence and development of the bacilli in the pseudomembrane, but is accompanied by a general toxemia resulting from the absorption of a violently poisonous substance produced by the bacilli. The bacilli are found only in the membranous exudation, and most plentifully in its older portions. As a rule, they do not distribute themselves through the circulation of the animal, though at times they may be found in the heart's blood.

The most malignant cases of the disease are thought to be due to pure infection by the diphtheria bacillus, though such cases are more rare than those in which the Streptococcus pyogenes and Staphylococcus aureus and albus are found in association with it.

In a series of 234 cases carefully and statistically studied

[1] *Centralbl. f. Bakt. u. Parasitenk.*, Mar., 1895.
[2] *Jour. of Exper. Med.*, vol. ii., No. 5, Sept., 1897, p. 546.

by Blasi and Russo-Travali it was found that in 26 cases of pseudomembranous angina due to streptococci, staphylococci, colon bacilli, and pneumococci, 2 patients died, the mortality being 3.84 per cent. In 102 cases of pure diphtheria 28 died, a mortality of 27.45 per cent. Seventy-six cases showed diphtheria bacilli and staphylococci; of these, 25, or 32.89 per cent., died. Twenty cases showed the diphtheria bacilli and Streptococcus pyogenes, with 6 deaths—30 per cent. In 7 cases, of which 3, or 43 per cent., were fatal, the diphtheria bacillus was in combination with streptococci and pneumococci. The most dangerous forms met were 3 cases, all fatal, in which the diphtheria bacillus was found in combination with the Bacillus coli.

It may be well to remark that all pseudomembranous diseases of the throat are not diphtheria, but that some of them, exactly similar in clinical picture, result from the activity of the pyogenic organisms alone, and are neither diphtheria nor contagious.

Diphtheritic inflammations of the throat are not always accompanied by the formation of the usual pseudomembrane, it rarely but occasionally happening that in the larynx a rapid inflammatory edema without a fibrinous surface-coating causes a fatal suffocation. Only a bacteriological examination will reveal the nature of the disease in such cases.

Herman Biggs,[1] in an interesting discussion of the occurrence of the diphtheria bacillus and its relation to diphtheria, comes to the following conclusions:

1. "When the diphtheria bacillus is found in healthy throats investigation almost always shows that the individuals have been in contact with cases of diphtheria. The presence of the bacillus in the throat, without any lesion, does not, of course, indicate the existence of the disease.

2. "The simple anginas in which virulent diphtheria bacilli are found are to be regarded from a sanitary stand-

[1] *Amer. Jour. of the Med. Sciences,* Oct., 1896, vol. xxii., No. 4, p. 411.

point in exactly the same way as the cases of true diphtheria.

3. "Cases of diphtheria present the ordinary clinical features of diphtheria, and show the Klebs-Löffler bacilli.

4. "Cases of angina associated with the production of membrane in which no diphtheria bacilli are found might be regarded from a clinical standpoint as diphtheria, but bacteriological examination shows that some other organism than the Klebs-Löffler bacillus is the cause of the process."

No more convincing proof of the existence of a powerful poison in diphtheria could be desired than the evidences of general toxemia resulting from the absorption of material from a comparatively small number of bacilli situated upon a little patch of mucous membrane.

In animals artificially inoculated the lesions produced are not identical with those seen in the human subject, yet they have the same general features of local infection with general toxemia.

Guinea-pigs, kittens, and young pups are susceptible animals. When half a cubic centimeter of a twenty-four-hour-old bouillon culture is injected beneath the skin of such an animal, the bacilli multiply at the point of inoculation, with the production of a patch of inflammation associated with a distinct fibrinous exudation and the presence of an extensive edema. The animal dies in twenty-four to thirty-six hours. The liver is enlarged, and sometimes shows minute whitish points, which in microscopic sections prove to be necrotic areas in which the cells are completely degenerated and the chromatin of their nuclei is scattered about in granular form. Similar necrotic foci, to which attention was first called by Oertel, are present in nearly all the organs in cases of death from the toxin. The bacilli are constantly absent from these lesions. Welch and Flexner[1] have shown these foci to be common to numerous irritant poisonings, and not peculiar to diphtheria.

[1] *Bull. of the Johns Hopkins Hospital*, Aug., 1891.

The lymphatic glands are usually enlarged ; the adrenals are also enlarged, and, in cases into which the live bacilli have been injected, are hemorrhagic.

Sometimes the bacilli themselves are present in the internal organs, and even in the blood, but generally this is not the case.

It might be argued, from the different clinical pictures presented by the disease as it occurs in man and in animals, that they were not expressions of the same thing. A careful study, however, together with the evidences adduced by Roux and Yersin, who found that when the bacilli were introduced into the trachea of animals opened by operation a typical false membrane was produced, and that diphtheritic palsy often followed, and of hundreds of investigators, who find the bacilli constantly present in the disease as it occurs in man, must satisfy us that the doubt of the etiological rôle of the bacillus rests on a very slight foundation.

All possible skepticism of the specificity of the bacillus on my part was dispelled by an accidental infection that kept me housed for three weeks during the busiest season of the year. Without having been exposed to any known diphtheria contagion, while experimenting in the laboratory, a living virulent culture of the diphtheria bacillus was drawn into a pipette and accidentally entered my mouth. Through carelessness no precautions were taken to prevent serious consequences, and as a result of the accident, two days later, my throat was full of typical pseudomembrane which private and Health Board bacteriological examinations showed contained pure cultures of the Klebs-Löffler bacilli.

One reason for skepticism in this particular is the supposed existence of a *pseudodiphtheria bacillus*, which has so many points in common with the real diphtheria bacillus that it is difficult to distinguish between them. We have, however, come to regard this pseudobacillus as an attenuated form of the real bacillus. The chief points of difference between the bacilli are that the pseudobacillus seems to be shorter than the diphtheria bacillus when grown upon blood-serum; that the cultures in bouillon seem to progress much more rapidly at a temperature of from 20°–22° C. than those of the true bacillus; and that the pseudobacillus is not pathogenic for animals. These slight distinctions are all exactly what should be expected of an organism whose virulence had been lost, and whose vegetative powers had been altered, by persistent manipulation or by unfavorable surroundings.

Park[1] carefully studied this subject, and found that all bacilli with the exact morphology of the diphtheria bacillus, found in the human throat, are virulent Klebs-Löffler bacilli, while forms found in the throat closely resembling them, but more uniform in size and shape, shorter in length, and of more equal staining properties with Löffler's alkaline methylene blue solution, can be, with reasonable safety, regarded as pseudodiphtheria bacilli, especially if it be found that they produce an alkaline rather than an acid reaction by their growth in bouillon. The pseudodiphtheria bacilli were found in about 1 per cent. of throats examined in New York; they seem to have no relationship to diphtheria. They are never virulent.

A difference of possibly much greater importance is that observed by Martini,[2] that the diphtheria bacillus will not grow in fluid antitoxic serum in which the pseudodiphtheria bacillus thrives. Both the real and the pseudobacilli flourish upon coagulated antitoxic serum. If this difference in the behavior of the bacilli bears any relation to the so-called "specific immunity-reaction" of cholera and typhoid fever, it may be of great future importance.

The diphtheria bacilli are always present in the throats of patients suffering from diphtheria, and constitute the element of contagion by being accidentally discharged by the nose or mouth by coughing, sneezing, vomiting,

---

[1] *Scientific Bulletin No. 1,* Health Department, City of New York, 1895.

[2] *Centralbl. f. Bakt. u. Parasitenk.,* Bd. xxi., No. 3, Jan. 30, 1897.

etc. Whoever comes in contact with such materials is in danger of infection.

It is of great interest to notice the remarkable results obtained by Biggs, Park, and Beebe in New York, where the bacteriological examinations conducted in connection with diphtheria show that the virulent bacilli may be found in the throats of convalescents as long as five weeks after the discharge of the membrane and the commencement of recovery, and that they exist not only in the throats of the patients themselves, but also in the throats of their care-takers, who, while not themselves infected, may be the means of conveying the disease from the sick-room to the outer world. Even more extraordinary are the observations of Hewlett and Nolen,[1] who found the bacilli in the throats of patients seven, nine, and in one case *twenty-three weeks* after convalescence. The importance of this observation must be apparent to all readers, and serves as further evidence why most thorough isolation should be practised in connection with this dreadful disease.

Park[2] found virulent diphtheria bacilli in about 1 per cent. of the healthy throats examined in New York City. Diphtheria was, however, prevalent in the city at the time. Most of the persons in whose throats they existed had been in direct contact with cases of diphtheria. Very many of those whose throats contained the virulent bacilli did not develop diphtheria. He concludes that the members of a household in which a case of diphtheria exists should be regarded as sources of danger, unless cultures from their throats show the absence of diphtheria bacilli.

In connection with the contagiousness of diphtheria the recent experiments of Reyes are interesting. He has demonstrated that in absolutely dried air distributed diphtheria bacilli die in a few hours. Under ordinary

---

[1] *Brit. Med. Jour.*, Feb. 1, 1896.

[2] Report on Bacteriological Investigations and Diagnosis of Diphtheria, from May 4, 1893, to May 4, 1894, *Scientific Bulletin No. 1*, Health Department, City of New York.

conditions their vitality, when dried on paper, silk, etc., continues for a few days. In air that is moist the duration of vitality is prolonged to about a week. In sand exposed to a dry atmosphere they die in five days in the light; in sixteen to eighteen days in the dark. When the sand is exposed to a moist atmosphere the duration of vitality is doubled. In fine earth they remained alive seventy-five to one hundred and five days in dry air, and one hundred and twenty days in moist air.

From time to time reference has been made to the toxin elaborated by the diphtheria bacillus. Roux and Yersin first demonstrated the existence of this substance in cultures passed through a Pasteur porcelain filter. The toxin is intensely poisonous; it is not an albuminous substance, and can be elaborated by the bacilli when grown in non-albuminous urine, or, as suggested by Uschinsky, in non-albuminous solutions whose principal ingredient is asparagin. The toxic value of the cultures is greatest in the second or third week.

In addition to the toxin, a toxalbumin has been isolated by Brieger and Fränkel.

Behring discovered that the blood of animals rendered immune to diphtheria by inoculation, first with attenuated and then with virulent organisms, contained a neutralizing substance which was capable of annulling the effects of the bacilli or the toxin when simultaneously or subsequently inoculated into non-protected animals. This substance, in solution in the blood-serum of the immunized animals, is the *diphtheria antitoxin.*

The preparation of the antitoxin for therapeutic purposes received a further elaboration in the hands of Roux. The subject is one of great interest, but must be considered briefly in a work of this kind.

The antitoxin is manufactured commercially at present, the method being the immunization of large animals to great quantities of the toxin, and the withdrawal of their antitoxic blood when the proper degree of immunity has been attained. The details are as follows:

The Preparation of the Toxin.—The method employed by Roux and others at the present time was first suggested by Fernbach, and consists in growing the most virulent bacilli obtainable in alkaline bouillon exposed in a thin layer to the passage of a current of air.

The cultures are allowed to grow for three or four weeks at a temperature of 37° C., with a stream of moist air constantly passing over them. After the given time has passed, it will be found that the acidity primarily produced by the bacillus gives place to a much more intense alkalinity than originally existed. The acme of the toxin-production seems to keep pace with this alkaline production. When "ripe," 0.4 per cent. of trikresol is added to the cultures, which are then filtered through porcelain. If the toxin must be kept before using, it is best to preserve it unfiltered, as it deteriorates more rapidly after filtration. Unfiltered toxin causes too much local irritation. If the bacillus employed was virulent and the conditions of culture were favorable, the filtered culture should be so toxic that 0.1 c.cm. would be fatal to a 500-gram guinea-pig in twenty-four hours (Roux). Even under the most favorable circumstances it is difficult to obtain a toxin which will kill in less than thirty hours.

The experience of the author with Fernbach's apparatus has not been satisfactory. The passing current of air is a frequent source of contamination to the culture, and requires great care. In the end it is questionable whether the toxin thus produced is better than that obtained from an ordinary flask exposing a large surface to the air.

Park and Williams did an elaborate work upon the production of diphtheria toxin.[1] They found that "toxin of sufficient strength to kill a 400-gram guinea-pig in three days and a half in a dose of 0.025 c.cm., developed in suitable bouillon, contained in ordinary Erlenmeyer flasks, within a period of twenty-four hours. In such bouillon the toxin reached its greatest strength in four to seven days (0.005 c.cm. killing a 500-gram

[1] *Jour. of Exper. Med.*, vol. i., No. 1, Jan., 1896, p. 164.

guinea-pig in three days). This period of time covered that of the greatest growth of the bacilli, as shown both by the appearance of the culture and by the number of colonies developing on agar plates.''

''The bodies of the diphtheria bacilli did not at any time contain toxin in considerable amounts.'' ''The type of growth of the bacilli and the rapidity and extent of the production of toxin depended more on the reaction of the bouillon than upon any other single factor.'' '' The best results were obtained in bouillon which, after being neutralized to litmus, had about 7 c.cm. of normal soda solution added to each liter. An excessive amount of either acid or alkali prevented the development of toxin.'' ''Strong toxin was produced in bouillon containing peptone ranging from 1 to 10 per cent.'' ''The strength of toxin averaged greater in the 2 and 4 per cent. peptone solution than in the 1 per cent.''

''When the stage of acid reaction was brief and the degree of acidity probably slight, strong toxin developed while the culture bouillon was still acid; but when the stage of acid reaction was prolonged little if any toxin was produced until just before the fluid became alkaline.''

''Glucose is deleterious to the growth of the diphtheria bacillus and to the production of toxin when it is present in sufficient amounts to cause by its disintegration too great a degree of acidity in the culture-fluid. When the acid resulting from the decomposition of glucose is neutralized by the addition of an alkali the diphtheria bacillus again grows abundantly.''

The Immunization of the Animal.—The animals chosen to furnish the antitoxic serum should be animals which present a distinct natural immunity to ordinary doses of the toxin, and should be sufficiently large to furnish large quantities of the finished serum. Behring originally employed dogs and sheep ; Aronson at first preferred the goat ; but Roux introduced the horse, which is more easily immunized than the other animals mentioned, and, being large enough to furnish a considerable quantity of serum, recommends itself strongly for the purpose.

The animal chosen should be free from tuberculosis and glanders, as tested by tuberculin and mallein, but need not be expensive. A horse with a disabled foot will answer well. Rheumatic horses should be rejected. In the beginning a small dose of the toxin—about 1 c.cm.—should be given hypodermically to detect individual susceptibility. Horses vary much in this particular, as Roux has pointed out. The author found light-colored horses to be distinctly more susceptible than dark-colored ones, a fact which has some substantiation in the clinical observation that blonde children suffer more severely from diphtheria than dark-complexioned ones.

If well borne, the preliminary injection is followed in about six days by a larger dose, in six days more by a still larger one, and the increase is continued every six days or so, according to the condition of the animal, until enormous quantities—500–1,000 c.cm.—are introduced at a time.

As the expression of quantity alone is very misleading, and to know exactly what strength the horse is receiving, the author has devised a special nomenclature by which to express it. Instead of stating that the animal received 10, 50, or 100 c.cm. of toxin, we now say it receives 10, 50, or 100 *factors*, the term factor being used to express 100 times the least certainly fatal dose of toxin per 100 grams of guinea-pig. The number of factors in a given quantity of antitoxin naturally varies with its strength, and it will at once be seen that it is advantageous to express strength regardless of quantity.

The toxin causes some local reaction—at first a distinct inflammation, later a painful edema and a febrile reaction. The amount of local irritation is much less marked when the injections are made slowly; and a gravity apparatus, which is filled with the amount of serum to be injected, suspended from the ceiling of the

stable so that the toxin is allowed to take its own time
to enter the tissues, can be recommended. Sometimes
it takes an hour to inject 500 c.cm. in this manner.

The amount of local reaction, edema, etc., the appetite
and general condition, the temperature-curve, and the
stability of the body-weight, must all be taken into consideration, so that the administration shall not be too
rapid and the animal be thrown into a condition of
cachexia with toxic instead of antitoxic blood.

One of the principal things to be avoided is haste.
Too frequent or too large dosage is almost certain to kill
the animal.

Behring found that mixing the toxin with trichlorid
of iodin lessened the irritant effect upon susceptible animals. I prefer not to use susceptible horses.

The suggestion of Prof. Pearson, that the large doses
of toxin might with readiness be introduced into the
trachea when the absorption is good, has been successfully accomplished by the author. The absorption seems
to take place without any change in the toxin, and to be
as rapid as from the subcutaneous tissue.

As the antitoxin protects the horse perfectly against
the toxin, a preliminary dose will enable one to omit all
the small preliminary doses of toxin, and render the
horse immune at once. Thus, I have frequently administered 100 c.cm. of antitoxin of about 100 units strength
to a horse one day and 500 c.cm. of strong toxin (500
factors) the next. This is just 500 times as much toxin
as has twice killed a horse in the laboratory. After the
lapse of a few days the same quantity can be administered
again, and in a week a third time. In this rapid way
antitoxin can often be secured at short notice. It is yet
a question, however, whether this method, modified from
Pawlowski, is as good and certain as the slow way suggested by Behring.

The possibility of producing serum rapidly may depend
upon the method, but the production of strong serums depends chiefly upon the *horse* and not upon its treatment.

The Preparation of the Serum for Therapeutic Pur-poses.—When, because of the tolerance to large quanti-ties of toxin, the horse seems to possess antitoxic blood, a "twitch" is applied to the upper lip, the eyes are blindfolded, a small incision is made through the skin, a trocar thrust into the jugular vein, and the blood al-lowed to flow through a cannulated tube into sterile bottles. It is allowed to coagulate, and kept upon ice for two days or so, that the clear serum may be pi-petted off. This serum is the *antitoxic serum.* It does not always materialize according to the desires of the experimenter, sometimes proving surprisingly strong in a short time, sometimes very weak after months of patient preparation.

The serums are preserved by Roux with camphor, by Behring with carbolic acid (0.5 per cent.), and by Aron-son with trikresol (0.4 per cent.). I prefer to use tri-kresol, as it is not poisonous, is a reliable antiseptic, and has a very pronounced local anesthetic action. Formalin has been tried, but it gelatinizes the serum and causes much local pain when injected beneath the skin.

Dried antitoxic serum has also been placed upon the market under the impression that it will keep longer and bear shipment better than any other. This is not, how-ever, shown to be the case, and as the dried serum dis-solves with difficulty it is much less convenient than the usual preparations. It is also less likely to be sterile than the liquid forms.

The strength of the serum is expressed in what are known as *immunizing units.* This denomination origin-ated with Behring and Ehrlich, whose *normal serum* was of such strength that 0.1 c.cm. of it would protect against ten times the least certainly fatal dose of toxin when simultaneously injected into guinea-pigs. Each cubic centimeter of this normal serum they called an *immunizing unit.* Later it was shown that the strength of the serum could easily be increased tenfold, so that 0.01 c.cm. of the serum would protect the guinea-pig against the ten-times fatal dose. Each cubic centimeter of this stronger serum was described as an antitoxic unit, and, of course, contained ten immunizing units. Still later it was shown that the limits of strength were by no means reached, and he succeeded in making serums three hundred times the normal strength, each cubic centimeter of which contained 300 immunizing units, or 30 antitoxic units.

In the course of the development of strength in the serum the exact meaning of "immunizing unit" gradually became obscured, until it is at present an expression of strength rather than one of quantity.

While it is difficult to define an immunizing unit, it is not at all difficult for one skilled in laboratory technique to determine the number present in a sample of serum. There are three rules of practice:

1. Determine accurately the least certainly fatal dose of a sterile diphtheria toxin for a standard guinea-pig.

2. Determine accurately the least quantity of the serum that will protect a guinea-pig against *ten times* the determined least certainly fatal dose of toxin.

3. Express the required dose of antitoxic serum as a fraction of a cubic centimeter and multiply it by ten. There will then be as many immunizing units in 1 c.cm. of the serum as there are parts in the resulting fraction.

Example: It is found that 0.01 c.cm. of toxin kills at least 9 out of 10 guinea-pigs. It is then regarded as the least certainly fatal dose. Guinea-pigs receive ten times this dose (0.1 c.cm.) and varying quantities of the serum, measured by dilution, say, $\frac{1}{1000}$ c.cm., $\frac{1}{2500}$ c.cm., $\frac{1}{5000}$ c.cm. The first two live. The fraction $\frac{1}{2500}$ is now multiplied by 10; $\frac{1}{2500} \times 10 = \frac{1}{250}$, and we find that there are as many units per cubic centimeter in the serum as there are parts in the result—*i. e.*, 250.

The most accurate definition of an immunizing unit is: ten times the least amount of antitoxic serum that will protect a standard (300-gram) guinea-pig against ten times the least certainly fatal dose of diphtheria toxin.

The strongest serum ever obtained by the author contained 1400 units per cubic centimeter.

As the quantity to be injected at each dose diminishes according to the number of units per cubic centimeter the serum contains, it is of the highest importance that the serums be as strong as possible. Various methods of concentration have been suggested, such as the partial evaporation of the serum *in vacuo*, but none have proved satisfactory. The latest suggestion comes from Bujwid,[1] who finds that when an antitoxic serum is frozen and then thawed, it separates into two layers, an upper watery stratum and a lower yellowish one; the antitoxic value of the yellowish layer is about three times that of the original serum.

Ehrlich asserts that 500 units are valueless: 2000 units are probably an average dose, and, as the remedy seems harmless, it is better to err on the side of too much than on that of too little. Fourteen thousand units have been administered in one case with beneficial results.

The largest collection of statistics upon the results of antitoxic treatment in diphtheria in the hospitals of the world are probably those collected by Prof. Welch, who, excluding every possible error in the calculations, "shows an apparent reduction of case-mortality of 55.8 per cent."

One of the most important things in the treatment is to begin it early enough. Welch's statistics show that 1115 cases of diphtheria treated in the first three days of the disease yielded a fatality of 8.5 per cent., whereas 546 cases in which the antitoxin was first injected after the third day of the disease yielded a fatality of 27.8 per cent.

After the toxin has set up destructive organic lesions in various organs and tissues of the body, no amount of neutralization will restore the integrity of the parts, so that the antitoxin must fail in these cases.

The urticaria which sometimes follows the injection

---

[1] *Centralbl. f. Bakt. u. Parasitenk.*, Sept., 1897, Bd. xxii., Nos. 10 and 11, p. 287.

of antitoxic serum seems to bear a distinct relation to the age of the serum, fresh serums being more liable to produce it than those which have been kept for a month or two.

I have found that the "keeping" qualities of the serums, when properly preserved, are of long duration. Samples examined two years after having been exposed for sale in the markets have been found unchanged. The serums most prone to deteriorate seem to be those of highest potency, but even here the good qualities are unchanged for months.

Freezing is without effect and ordinary temperature-changes are harmless to the serum. The antitoxic power is destroyed at 72° C., the point at which the serum coagulates.

The erythemata are probably in some way associated with the globulicidal action of the blood. Keeping the serum "until it is ripe" lessens this effect. The serums from different horses probably vary much in both their irritant and globulicidal properties, so that antitoxins prepared by mixing the serums from a number of horses are probably preferable to those from single horses.

Dried serums are much less active than fresh ones.

For purposes of immunization smaller doses than those used for treatment suffice. According to Biggs, 2 cubic centimeters are sufficient to give complete protection. The immunity that results from the injection is of a month or six weeks' duration.

The transitory nature of this immunity is probably dependent upon the fact that the antitoxin is slowly excreted through the kidneys.

20

CHAPTER III.

No micro-organism of hydrophobia has as yet been discovered, yet the peculiarities of the disease are such as to leave no doubt in the mind of a bacteriologist that one must exist. To find it is now the desideratum.

Although many men have labored upon hydrophobia, no name is so well known or so justly honored as that of the great pioneer in bacteriology, Pasteur. The profession and laity are alike familiar with his name and work, and although at times the newspapers of our country and certain members of the profession have opposed the methods of treatment which he has suggested as the result of his experimentation, we cannot but feel that this skepticism and opposition are due either to ignorance of the principles upon which Pasteur reasoned or to a culpable conservatism. The most vehement opponent that Pasteur has in America seems to disbelieve the existence of rabies. It is impossible to argue with him.

Hydrophobia, or rabies, is a specific toxemia to which dogs, wolves, skunks, and cats are highly susceptible, and which can, through their saliva, be communicated to men, horses, cows, and other animals. The means of communication is almost invariably a bite, hence the inference that the specific organism is present in the saliva.

The animals that are infected manifest no symptoms during a varying incubation-period in which the wound generally heals kindly. This period may last for as long a time as twelve months, but in rare cases may be only some days. An average duration of the period of incubation might be stated as about six weeks.

As the incubation-period comes to an end there is an observable alteration in the wound, which becomes reddened, sometimes may suppurate a little, and is painful. The victim, if a man, is much alarmed and has a sensation of horrible dread. The period of dread passes into one of excitement, which in many cases amounts to a wild delirium and ends in a final stage of convulsion and palsy. The convulsions are tonic, rarely clonic, and subsequently cause death by interfering with the respiration, as do those of tetanus and strychnia.

During the convulsive period much difficulty is experienced in swallowing liquids, and it is supposed that the popular term "hydrophobia" arose from the reluctance of the diseased to take water because of the inconvenience and occasional spasms which the attempt causes.

This description, brief and imperfect as it is, will illustrate the parallelism existing between hydrophobia and tetanus. In the latter we observe the entrance of infectious material through a wound, which, like the bite in hydrophobia, sometimes heals, but often suppurates a little. We see in both affections an incubation-period of varying duration, though in hydrophobia it is much longer than in tetanus, and convulsions of tonic character causing death by asphyxia.

It is maintained by some that the stage of excitement argues against the specific nature of the disease, but these subjective symptoms are like the mental condition of tuberculosis, which leads the patient to make a hopeful prognosis of his case, and the mental condition of anthrax, in which it is stated that no matter how dangerous his condition the patient is seldom much alarmed about it.

Pasteur and his co-workers found that in animals that die of rabies the salivary glands, the pancreas, and the nervous system contain the infection, and are more appropriate for experimental purposes than the saliva, which is invariably contaminated with accidental pathogenic bacteria.

The introduction of a fragment of the medulla oblongata of a dog dead of rabies beneath the dura mater of a rabbit causes the development of rabies in the rabbit in a couple of weeks. The medulla of this rabbit introduced beneath the dura mater of a second rabbit produced a more violent form of the disease in a shorter time, and by frequently repeated implantations Pasteur found that an extremely virulent material could be obtained.

Inasmuch as the toxins of diphtheria and tetanus circulate in the blood, and not infrequently saturate the nervous systems of animals affected, it might be concluded that the material with which Pasteur worked was a toxin. This is readily disproven, however, not only by the fact that the toxin would weaken instead of strengthen by the method of transfer from animal to animal, it not being a vital entity, but also by the discovery that when an emulsion of the nervous system of an affected animal is filtered through porcelain, or when it is heated for a few moments to 100° C., or exposed for a considerable time to a temperature of 75° or 80° C., its virulence is entirely lost. This would seem to prove that that which is in the nervous system and communicates the disease is a living, active body—a parasite, and in all probability a bacterium. However, all endeavors to discover, isolate, or cultivate this organism have failed.

Pasteur noted that the virulence of the poison was less in animals that had been dead for some time than in the nervous systems of those just killed, and by experimentation showed that when the nervous system was dried in a sterile atmosphere the virulence was attenuated in proportion to the length of time it had been dry. This attenuation of virulence of course suggested to Pasteur the idea of a protective vaccination, and by inoculating a dog with much attenuated, then with less attenuated, then with moderately strong, and finally with strong, virus, the dog developed an immunity which enabled it to resist the infection of an amount of virulent material that would certainly kill an unprotected animal.

It is remarkable that this thought, which was a theory based upon a broad knowledge, but experience with comparatively few bacteria, should every day find more and more grounds for confirmation as our knowledge of immunity, of toxins, and of antitoxins progresses. What Pasteur did with rabies is what we now do in producing the antitoxin of diphtheria—*i. e.* gradually accommodate the animal to the poison until its body-cells are able to neutralize or resist it. As the poison cannot be secured outside of the body because the bacilli, micrococci, or whatever they may be cannot be secured outside of the body, he does what Behring originally did in diphtheria—introduces attenuated poison-producers—bacilli crippled by heat or drying, and capable of producing only a little poison—accustoms the animal to these, and then to stronger and stronger ones until immunity is established.

The genius of Pasteur did not cease with the production of immunity, but, we rejoice to add, extended to the kindred subject of therapy, and has now given us a *cure for hydrophobia.*

For the production of a cure in infected cases very much the same treatment is followed as has been described for the production of immunity. The patient must come under observation early. The treatment consists of the subcutaneous injection of about 2 grams of an emulsion of a rabbit's spinal cord which had been dried for from seven to ten days. This beginning dose is not increased in size, but each day the emulsion used is from a cord which has not been dried so long, until, when the twenty-fifth day of treatment is reached, the patient receives 2 grams of emulsion of spinal cord dried only three days, and is considered immune or cured.

It will be observed that this treatment is really no more than the immunization of the individual during the incubation stadium, and the generation of a vital force— shall we call it an antitoxin?—in the blood of the animal in advance of the time when the organism is expected to saturate the body with its toxic products.

This, in brief, is the theory and practice of Pasteur's system of treating hydrophobia. It is exactly in keeping with the ideas of the present, and is most extraordinary in its reasonings and details when we remember that the first application of the method to human medicine was made October 26, 1885, nearly ten years before the time we began to understand the production and use of antitoxins.

CHAPTER IV.

CHOLERA AND SPIRILLA RESEMBLING THE CHOLERA SPIRILLUM.

CHOLERA is a disease from which certain parts of India are never free. The areas in which it is endemic are the foci from which the great epidemics of the world, as well as the constant smaller epidemics of India, probably spread. No one knows when cholera was first introduced into India, and the probabilities are that it is indigenous to that country, as yellow fever is to Cuba. Very early mention of it is made in the letters of travellers, in books and papers on medicine of a century ago, and in the governmental statistics, yet we find that little is said about the disease except in a general way, most attention being directed to the effect upon the armies, native and European, of India and adjacent countries. The opening up of India by Great Britain in the last half century has made possible much accurate scientific observation of the disease and the relation which its epidemics bear to the manners and customs of the people.

The filthy habits of the people of India, their poverty, their crowded condition, and their religious customs, all serve to aid in the distribution of the disease. We are told that the city of Benares drains into the Ganges River by a most imperfect system, which distributes the greater part of the sewage immediately below the banks upon which the city is built. It is a matter of religious observance for every zealot who makes a pilgrimage to the "sacred city" to take a bath in and drink a large quantity of this sacred but polluted water, and, as may be imagined, the number of pious Hindoos who leave Benares with comma bacilli in their intestines or upon their clothes is great, for there are few months in the

311

year when there are not at least some cases of cholera in the city.

The frequent pilgrimages and great festivals of the Hindoos and Moslems, by bringing together an enormous number of people who crowd in close quarters where filth and bad diet are common, cause a rapid increase in the number of cases during these periods and the dispersion of the disease when the festivals break up. The disease extends readily along the regular lines of travel, visiting town after town, until from Asia it has frequently extended into Europe, and by the steamships plying on foreign waters has been several times carried to our own continent and to the islands of the seas. Many cases are on record which show conclusively how a single ship, having a few cholera cases on board, may be the cause of an outbreak of the disease in the port at which it arrives.

It seems strange to us now, with the light of present information illuminating the pages of the past, to observe how the distinctly infectious nature of such a disease could be overlooked in the search for some atmospheric or climatic cause, some miasm, which was to account for it.

The discovery of the organism which seems to be the specific cause of cholera was made by Koch, who was appointed one of a German cholera-commission to study the disease in Egypt and India in 1883–84. Since his discovery, but a decade ago, the works upon cholera and the published investigations to which the spirillum has been subjected have produced an immense literature, a large part of which was stimulated by the Hamburg epidemic of a few years ago.

The micro-organism described by Koch, and now generally accepted to be the cause of cholera, is a short individual about half the length of a tubercle bacillus, considerably stouter, and distinctly curved, so that the original name by which it was known was the "comma bacillus" (Figs. 80, 81).

A study of the growth of the organism and the forms which it assumes upon different culture-media soon convinces us that we have to do with an organism in no way related to the bacilli. If the conditions of nutrition are

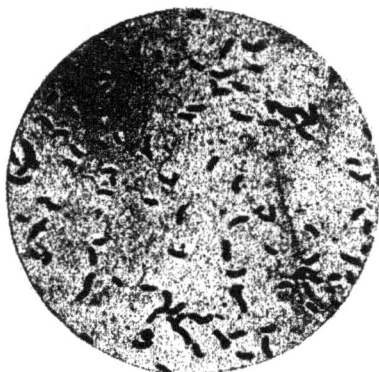

FIG. 80.—Spirillum of Asiatic cholera, showing the flagella; × 1000 (Günther).

diminished so that the multiplication of the bacteria by simple division does not progress with the usual rapidity, we find a distinct tendency toward—and in some cases, as upon potato, a luxuriant development of—long spiral threads with numerous windings—unmistakable spirilla. Fränkel has found that the exposure of cultures to unusually high temperatures, the addition of small amounts of alcohol to the culture-media, etc., will so vary the growth of the organism as to favor the production of spirals instead of commas. One of the most common of the numerous forms observed is that in which two short curved individuals are so joined as to produce an S-shaped curve.

The cholera spirilla are exceedingly active in their movements, and in hanging-drop cultures can be seen to swim about with great rapidity. Not only do the comma-shaped organisms move, but when distinct spirals exist, they, too, move with the rapid rotary motion so common among the spirilla.

The presence of flagella upon the cholera spirillum can be demonstrated without difficulty by Löffler's method (*q. v.*).   Each spirillum possesses a single flagellum attached to one end.

Inoculation-forms of most bizarre appearance are very common in old cultures of the spirillum, and very often

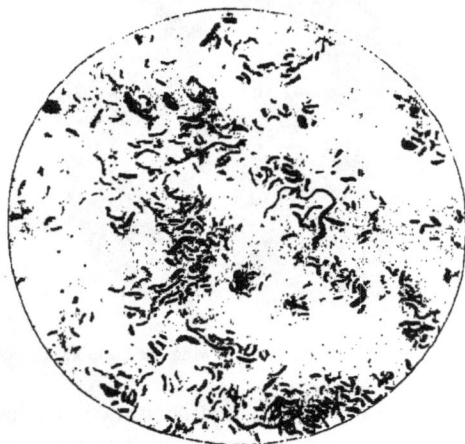

FIG. 81.—Spirillum of Asiatic cholera, from a bouillon culture three weeks old, showing numbers of long spirals; × 1000 (Fränkel and Pfeiffer).

there can be found in fresh cultures many individuals which show by granular protoplasm and irregular outline that they are partly degenerated.   Cholera spirilla from various sources seem to differ in this particular, some of the forms being as pronounced in their involution as the diphtheria bacilli.

In partially degenerated cultures in which long spirals are numerous Hüppe observed, by examination in the "hanging drop," in the continuity of the elongate members, certain large spherical bodies which he described as spores.   These bodies were not enclosed in the organisms like the spores of anthrax, but seemed to exemplify the form of sporulation in which an entire individual transforms itself into a spore (arthrospore).   Koch, and indeed all other observers, failed to find signs of fructification in the cholera organism, and the true nature of the bodies described by Hüppe must be regarded as doubtful. Most bacteriologists disagree with Hüppe in believing that arthrospores exist at all, and the fact (which will be pointed out later on) that there is very little permanence about cholera cultures throws additional doubt upon the accuracy of Hüppe's conclusion.

The cholera spirillum stains well with the ordinary aqueous solutions of the anilin dyes; fuchsin seems particularly appropriate. At times the staining must be continued for from five to ten minutes to secure homogeneity. The cholera spirillum does not stain by Gram's method. It may be colored and examined while alive; thus Cornil and Babes, in demonstrating it in the rice-water discharges, "spread out one of the white mucous fragments upon a glass slide and allow it to dry partially; a small quantity of an exceedingly weak solution of methyl violet in distilled water is then flowed over it, and it is flattened out by pressing down on it a cover-glass, over which is placed a fragment of filter-paper, which absorbs any excess of fluid at the margin of the cover-glass. Comma bacilli so prepared and examined with an oil-immersion lens (× 700–800) may then be seen: their characters are the more readily made out because of the slight stain which they take up, and because they still retain their power of vigorous movement, which would be entirely lost if the specimen were dried, stained, and mounted in the ordinary fashion."

The colonies of the spirillum when grown upon gelatin plates are highly characteristic. They appear in the lower strata of the gelatin as small white dots, gradually grow out to the surface, effect a gradual liquefaction of the medium, and then appear to be situated in little pits with sloping sides (Fig. 82). This peculiar appearance, which gives one the suggestion that the plate is full of little holes or air-bubbles, is due to the evaporation of the liquefied gelatin.

One of the best methods of securing pure cultures of

the cholera spirillum, and also of making a diagnosis of the disease in a suspected case, is probably that of Schottelius. The method is very simple : A small quantity of the fecal matter is mixed with bouillon and stood in an incubating oven for twenty-four hours. If the

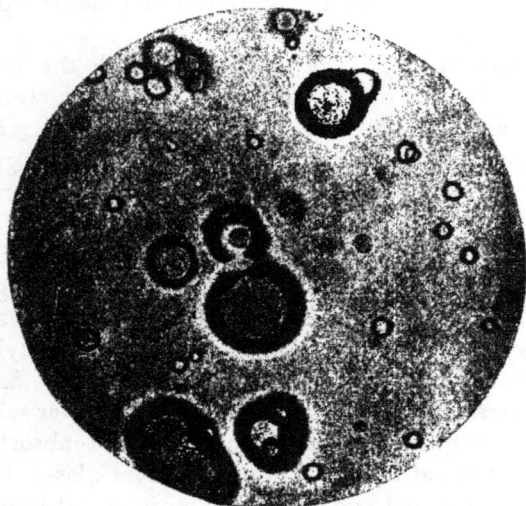

FIG. 82.—Spirillum of Asiatic cholera: colonies two days old upon a gelatin plate; × 35 (Heim).

cholera spirilla are present, they will grow most rapidly at the surface of the liquid when the supply of air is good. A pellicle will be formed, a drop from which, diluted in melted gelatin and poured upon plates, will show typical colonies.

Under the microscope the principal characteristics can be made out. The colony of the cholera spirillum scarcely resembles that of any other organism. The little colonies which have not yet reached the surface of the gelatin begin very soon to show a pale-yellow color and to exhibit irregularities of contour, so that they are almost never smooth and round. They are coarsely granular, and have the largest granules in the centre. As the colony increases in size the granules also increase in size, and attain a peculiar transparent character which is suggestive of powdered glass. The commencement of liquefaction causes the colony to be surrounded with a transparent halo. When this occurs the colony begins to sink, from the digestion and evaporation of the medium, and also to take on a peculiar rosy color.

In puncture-cultures in gelatin the growth is again so characteristic that it is quite diagnostic (Fig. 83). The

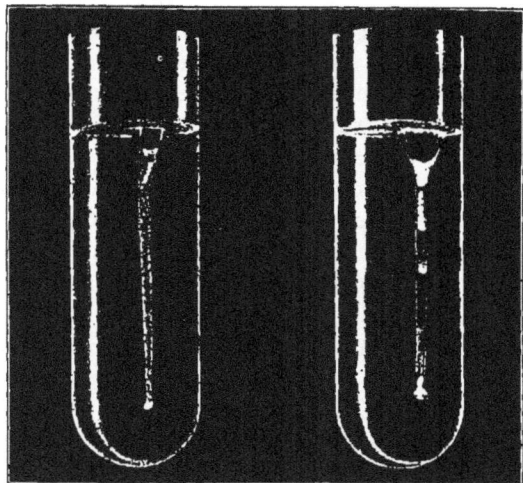

1 IG. 83.—Spirillum cholera Asiatica ; gelatin puncture-cultures aged forty-eight and sixty hours (Shakespeare).

growth takes place along the entire puncture, but develops best at the surface, where it is in contact with the atmosphere. An almost immediate liquefaction of the medium begins, and, keeping pace with the rapidity of the growth, is more marked at the surface than lower down. The result of this is the occurrence of a short, rather wide funnel at the top of the puncture. As the growth continues evaporation of the medium takes place slowly, so that the liquefied gelatin is lower than the solid surrounding portions, and appears to be surmounted by an air-bubble.

The luxuriant development of the spirilla in gelatin produces considerable solid material to sediment and fill up the lower third or lower half of the liquefied area. This solid material consists of masses of spirilla which have probably completed their life-cycle and become inactive. Under the microscope they exhibit the most varied involution-forms. The liquefaction reaches the sides of the tube in from five to seven days. Liquefaction of the medium is not complete for several weeks. According to Fränkel, in eight weeks the organisms in the liquefied culture all die, and cannot be transplanted. Kitasato, however, has found them living and active on agar-agar after ten to thirty days, and Koch was able to demonstrate their vitality after two years.

When planted upon the surface of agar-agar the spirilla produce a white, shining, translucent growth along the entire line of inoculation. It is in no way peculiar. The vitality of the organism is retained much better upon agar-agar than upon gelatin, and, according to Fränkel, the organism can be transplanted and grown when nine months old.

The growth upon blood-serum likewise is without distinct peculiarities, and causes gradual liquefaction of the medium.

Upon potato the spirilla grow well, even when the reaction of the potato is acid. In the incubator at a temperature of 37° C. a transparent, slightly brownish or yellowish-brown growth, somewhat resembling the growth of glanders, is produced. It contains large numbers of long spirals.

In bouillon and in peptone solution the cholera organisms grow well, especially upon the surface, where a folded, wrinkled mycoderma is formed. Below the mycoderma the culture fluid generally remains clear. If the glass be shaken and the mycoderma broken up, fragments of it sink to the bottom.

In milk the development is also luxuriant, but takes place in such a manner as not visibly to alter its appearance. The existence of cholera organisms in milk is, however, rather short-lived, for the occurrence of any acidity at once destroys them.

Wolffhügel and Riedel have shown that if the spirilla are planted in sterilized water they grow with great rapidity after a short time, and can be found alive after months have passed. Fränkel points out that this ability to grow and remain vital for long periods in sterilized water does not guarantee the same power in unsterilized water, for in the latter the simultaneous growth of other bacteria in a few days serves to extinguish the cholera germs.

One of the characteristics of the cholera spirillum is the metabolic production of indol. The detection of this substance is easy if the spirilla are grown in a transparent colorless solution. As the cholera organisms also produce nitrites, all that is necessary is to add a drop or two of chemically pure sulphuric acid to the culture-medium for the production of the well-known reddish color.

Several toxic products of the metabolism of the spirilla have been isolated. Brieger, Fränkel, Roux and Yersin have isolated toxalbumins; Villiers, a toxic alkaloid fatal to guinea-pigs; and Gamaléia, two substances about equally toxic.

The cholera spirilla can be found with great constancy in the intestinal evacuations of all cholera cases, and can often be found in the drinking-water, milk, and upon vegetables, etc. in cholera-infected districts. There can be little doubt that they find their way into the body through the food and drink. Many cases are reported in the literature upon cholera that show how the disease-germs enter the drinking-water, and are thus distributed; how they are sometimes thoughtlessly sprinkled over vegetables, offered for sale in the streets, with water from polluted gutters; how they enter milk with water used to dilute it; how they are carried about in clothing and upon foodstuffs; how they can be brought to articles of food upon the table by flies which have preyed upon cholera excrement; and how many other interesting infections are made possible. The literature upon these subjects is so vast that in a sketch of this kind it is scarcely possible to mention even the most instructive examples. One physician is reported to have been infected with cholera while experimenting with the spirilla in Koch's laboratory.

The evidence of the specificity of the cholera spirillum when collected shows that it is present in the choleraic dejections with great regularity, and that it is as constantly absent from the dejecta of healthy individuals and those suffering from other diseases; but these facts do not admit of satisfactory proof by experimentation upon animals. Animals are never affected by any disease similar to cholera during the epidemics, nor do foods mixed with cholera discharges or with pure cultures of the cholera spirillum affect them. This being true, we are prepared to receive the further information that subcutaneous injections of the spirilla are often without serious consequences, though cultures differ very much in this respect, some always causing a fatal septicemia in guinea-pigs, others being as constantly harmless.

Intraperitoneal injection of the virulent cultures produces a fatal peritonitis in guinea-pigs.

One reason that animals and certain men are immune to the disease seems to be found in the distinct acidity of the normal gastric juice, and the destruction of the spirilla by it. Supposing that this might be the case, Nicati and Rietsch, Von Ermengen and Koch, have suggested methods by which the micro-organisms can be introduced directly into the intestine. The first-named investigators ligated the common bile-duct of guinea-pigs, and then injected the spirilla into the duodenum with a hypodermic needle. The result was that the animals usually died, sometimes with choleraic symptoms ; but the excessively grave nature of the operation upon such a small and delicately constituted animal as a guinea-pig greatly lessens the value of the experiment. Koch's method is much more satisfactory. By injecting laudanum into the abdominal cavity of guinea-pigs the peristaltic movements are checked. The amount given for the purpose is very large, about 1 gram for each 200 grams of body-weight. It generally narcotizes the animals for a short time, but they recover without injury. After administering the opium the contents of the stomach are neutralized by introducing through a pharyngeal catheter 5 c.cm. of a 5 per cent. aqueous solution of sodium carbonate. With the gastric contents thus alkalinized and the peristalsis paralyzed a bouillon culture of the spirilla is introduced. The animal recovers from the manipulation, but shows an indisposition to eat, is soon observed to be weak in the posterior extremities, subsequently is paralyzed, and dies within forty-eight hours. The autopsy shows the intestine congested and filled with a watery fluid rich in spirilla—an appearance which Fränkel declares to be exactly that of cholera. In man, as well as in these artificially injected animals, the spirilla are never found in the blood or the tissues, but only in the intestine, where they frequently enter between the basement membrane and the epithelial cells, and aid in the detachment of the latter.

Issaëff and Kolle found that when virulent cholera spirilla are injected into the ear-veins of young rabbits the animals die on the following day with symptoms resembling the algid stage of human cholera. The autopsy in these cases showed local lesions of the small intestine very similar to those observed in cholera in man.

Guinea-pigs are also susceptible to intraperitoneal injections of the spirillum, and speedily succumb. The symptoms are—rapid fall of temperature, tenderness over the abdomen, and collapse. The autopsy shows an abundant fluid exudate containing the micro-organism, and injection and redness of the peritoneum and viscera.

Although in reading upon cholera at the present time we find very little skepticism in relation to Koch's "comma bacillus," we do find occasional doubters who believe with Von Pettenkoffer that the disease is mias-

21

matic. Pettenkoffer's theory is that the disease has
much to do with the ground-water and its drying zone.
He regards as the principal cause of the disease the development of germs in the subsoil moisture during the
warm months, and their impregnation of the atmosphere
as a miasm to be inhaled, instead of ingested with food
and drink. This idea of Pettenkoffer's, combined with
his other idea that individual predisposition must precede the inception of the disease, is scarcely compatible
with what has gone before, and cannot possibly be made
to explain the march of the disease from place to place
with caravans, or its distribution over extended areas
when fairs and religious gatherings among the Hindoos
break up, the people from an infected centre carrying
cholera with them to their homes.

While it is an organism that multiplies with great
rapidity under proper conditions, the cholera spirillum
is not possessed of much resisting power. Sternberg
found that it was killed by exposure to a temperature
of 52° C. for four minutes. Kitasato, however, found
that ten or fifteen minutes' exposure to a temperature
of 55° C. was not always fatal. In the moist condition the organism may retain its vitality for months,
but it is very quickly destroyed by desiccation, as was
found by Koch, who observed that when dried in a thin
film its power to grow was destroyed in a few hours.
Kitasato found that upon silk threads the vitality might
be retained longer. Abel and Claussen have shown that
it does not live longer than twenty to thirty days in fecal
matter, and often disappears in one to three days. The
organism is very susceptible to the influence of carbolic
acid, bichlorid of mercury, and other germicides.

The organism is also destroyed by acids. Hashimoto
found that it could not live longer than fifteen minutes
in vinegar containing 2.2–3.2 per cent. of acetic acid.

This low vital resistance of the microbe is very fortunate, for it enables us to establish safeguards for the prevention of the spread of the disease. Excreta, soiled

clothing, etc. are readily rendered harmless by the proper use of disinfectants. Water and foods are rendered innocuous by boiling or cooking. Vessels may be disinfected by thorough washings with jets of boiling water thrown upon them through hose. Baggage can be sterilized by superheated steam.

It often becomes a matter of importance to detect the presence of cholera in drinking-water, and, as the dilution in which the bacteria exist in such a liquid may be very great, much difficulty is experienced in finding them by ordinary methods. One of the most expeditious methods that have been recommended is that of Löffler, who adds 200 c.cm. of the water to be examined to 10 c.cm. of bouillon, allows the mixture to stand in an incubator for twelve to twenty-four hours, and then makes plate-cultures from the superficial layer of the liquid, where, if present, the development of the spirilla will be most rapid because of the presence of air. A similar method can be used to detect the spirilla when their presence is suspected in feces.

Gruber and Wiener, Haffkine, Pawlowsky, and Pfeiffer have all succeeded in immunizing animals against the toxic substances removed from cholera cultures or against living cultures properly injected. There seems, according to the researches of Pfeiffer, to be no doubt that in the blood of the protected animals a protective substance is present. In the peritoneal infection of guinea-pigs the spirilla grow vigorously in the peritoneal cavity, and can be found in immense numbers after twelve to twenty-four hours. If, however, together with the culture used for inoculation, a few drops of the protective serum be introduced, Pfeiffer found that instead of multiplying the organisms underwent a peculiar granular degeneration and disappeared, the unprotected animal dying, the protected animal remaining well.

Pfeiffer and Vogedes[1] have suggested the application of this "immunity-reaction" for the positive differentia-

[1] *Centralbl. für Bakt. und Parasitenk.*, March 21, 1896, Bd. xix., No. 11.

tion of cholera spirilla in cultures. A hanging-drop of a 1 : 50 mixture of powerful anti-cholera serum and a particle of cholera culture is made and examined under the microscope. The cholera spirilla at once become inactive, and are in a short time converted into little rolled-up masses. If the culture added be a spirillum other than the true spirillum of cholera, instead of destruction of the micro-organisms following exposure to the serum, they multiply and thrive in the mixture of serum and bouillon.

The specific immunity-reaction of the cholera serum has been carefully studied by Loburnheim,[1] and is specific against cholera alone. The protection is not due to the strongly bactericidal property of the serum, but to its stimulating effect upon the body-cells. If the serum be heated to 60°–70° C., and its bactericidal power thus destroyed, it is still capable of producing immunity.

The immunity produced by the injection of the spirilla into guinea-pigs continues in some cases as long as four and a half months, but the power of the serum to confer immunity is lost much sooner.

Of the numerous attempts which have from time to time been made, and are still being made, to produce immunity against cholera in man or to cure cholera when once established in the human organism, nothing very favorable can at the present time be said. Experiments in this field are not new: we find Dr. Ferrán administering hypodermic injections of pure virulent cultures of the cholera spirillum in Spain as early as 1885, in the hope of bringing about immunity. The more modern work of Haffkine seems to be followed by a distinct diminution of mortality in protected individuals. According to the work of this investigator, two vaccines are used, one of which, being mild, prepares the animal (or man) for a powerful vaccine, which, were it not preceded by the weaker form, would bring about extensive tissue-

[1] *Zeitschrift für Hygiene*, xx., p. 438.

necrosis and perhaps death. Protection certainly seems to follow the operation of these vaccines.

Haffkine's studies embrace more than 40,000 inoculations performed in India. From his latest paper (Dec., 1895) the following extract will show the results:

"1. In all those instances where cholera has made a large number of victims, that is to say, where it has spread sufficiently to make it probable that the whole population, inoculated and uninoculated, were equally exposed to the infection,—in all these places the results appeared favorable to inoculation.

"2. The treatment applied after an epidemic actually breaks out tends to reduce the mortality even during the time which is claimed for producing the full effect of the operation. In the Goya Garl, where weak doses of a relatively weak vaccine had been applied, this reduction was to half the number of deaths ; in the coolies of the Assam-Burmah survey-party, where, as far as I can gather from my preliminary information, strong doses have been applied, the number of deaths was reduced to one-seventh. This fact would justify the application of the method independently of the question as to the exact length of time during which the effect of this vaccination lasts.

"3. In Lucknow, where the experiment was made on small doses of weak vaccines, a difference in cases and deaths was still noticeable in favor of the inoculated fourteen to fifteen months after vaccination in an epidemic of exceptional virulence. This makes it probable that a protective effect could be obtained even for long periods of time if larger doses of a stronger vaccine were used.

"4. The best results seem to be obtained from application of middle doses of both anticholera vaccines, the second one being kept at the highest possible degree of virulence obtainable.

"5. The most prolonged observations on the effect of middle doses were made in Calcutta, where the mortality from the eleventh up to the four hundred and fifty-ninth day after vaccination was, among the inoculated, 17.24

times smaller, and the number of cases 19.27 times smaller than among the not inoculated."

Pawlowsky and others have found that the dog is susceptible to cholera, and have utilized the observation to prepare an antitoxic serum in considerable quantities. The dogs were first immunized with attenuated cultures, then with more and more virulent cultures, until a serum was obtained whose value was estimated at 1 : 130,000 upon experimental animals.

Freymuth and others have endeavored to secure favorable results from the injection of blood-serum from convalescent patients into the diseased. One recovery out of three cases treated is recorded—not a very glittering result.

In all these preliminaries the foreshadowing of a future therapeusis must be evident, but as yet nothing really satisfactory has been achieved.

Spirilla resembling the Cholera Spirillum.

The Finkler and Prior Spirillum.—Somewhat similar to the spirillum of cholera, and in some respects closely related to it, is the spirillum obtained from the feces of a case of cholera nostras by Finkler and Prior in 1884. It is a rather shorter, stouter organism, with a more pronounced curve, than the cholera spirillum, and rarely forms the long spirals which characterize the latter. The central portion is also somewhat thinner than the ends, which are a little pointed and give the organism a less uniform appearance than that of cholera (Fig. 84). Involution-forms are very common in cultures, and occur as spheres, spindles, clubs, etc. Like the cholera spirillum, each organism is provided with a single flagellum situated at its end, and is actively motile. Although at first thought to be a variety of the cholera germ, marked differences of growth were soon observed, and showed the organism to be a separate species.

The growth upon gelatin plates is quite rapid, and leads to such extensive liquefaction that four or five dilutions must frequently be made before the growth of a single colony can be observed. To the naked eye the colonies

FIG. 84.—Spirillum of Finkler and Prior, from an agar-agar culture; × 1000 (Itzerott and Niemann).

appear as small white points in the depths of the gelatin (Fig. 85). They, however, rapidly reach the surface,

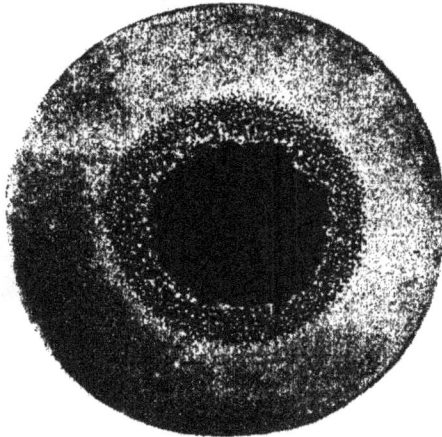

FIG. 85.—Spirillum of Finkler and Prior: colony twenty-four hours old, as seen upon a gelatin plate; × 100 (Fränkel and Pfeiffer).

begin liquefaction of the gelatin, and by the second

day appear about the size of lentils, and are situated in little depressions. Under the microscope they are of a yellowish-brown color, are finely granular, and are surrounded by a zone of sharply circumscribed liquefied gelatin. Careful examination with a high power of the microscope shows a rapid movement of the granules of the colony.

In gelatin punctures the growth takes place rapidly along the whole puncture, forming a stocking-shaped liquefaction filled with cloudy fluid which does not precipitate rapidly ; a rather smeary, whitish mycoderma is generally formed upon the surface. The much more extensive and more rapid liquefaction of the medium, the wider top to the funnel-shaped liquefaction at the surface,

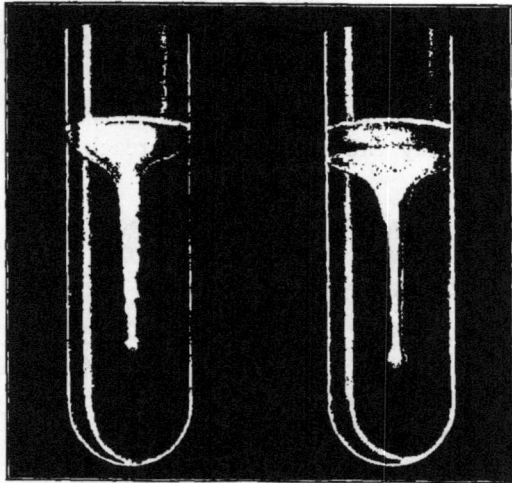

FIG. 86.—Spirillum of Finkler and Prior: gelatin puncture-cultures aged forty-eight and sixty hours (Shakespeare).

the absence of the air-bubble, and the clouded nature of the liquefied material, all serve to differentiate it from the cholera spirillum.

Upon agar-agar the growth is also very rapid, and in a short time the whole surface of the culture-medium is covered with a moist, thick, slimy coating, which may have a slightly yellowish tinge.

The cultures upon potato are also very different from those of cholera, for instead of a temperature of 37° C. being required for a rapid development, the Finkler and Prior spirilla grow rapidly at the room-temperature, and produce a grayish-yellow, slimy, shining layer, which may cover the whole of the culture-medium.

Blood-serum is rapidly liquefied by the growth of the organism.

Buchner has shown that in media containing some glucose an acid reaction is produced.

The spirillum does not grow well, if at all, in milk, and speedily dies in water.

The organism does not produce indol.

The spirillum can be stained well by the ordinary dyes, and seems, like the cholera spirillum, to have a special affinity for the aqueous solution of fuchsin.

In connection with this bacillus the question of pathogenesis is a very important one. At first it was suspected that it was, if not the spirillum of cholera itself, a very closely allied organism. Later it was regarded as the cause of cholera nostras. At present its exact pathological significance is a question. It was in one case secured by Knisl from the feces of a suicide, and has been found in carious teeth by Müller.

When injected into the stomach of guinea-pigs treated according the method of Koch, about 30 per cent. of the animals die, but the intestinal lesions produced are not the same as those produced by the cholera spirillum. The intestines in such cases are pale and filled with watery material having a strong putrefactive odor. This fluid teems with the spirilla.

It seems very unlikely, from the collected evidence, that the Finkler and Prior spirillum is associated with pathogenesis in the human species. As Fränkel points out, it is probably a frequent and harmless inhabitant of the human intestine.

The Spirillum of Denecke.—Another organism with a distinct resemblance to the cholera spirillum is one described by Denecke as occurring in old cheese (Fig. 87). Its form is much the same as that of the spirillum of cholera, the shorter individuals being of equal diameter throughout. The spirals which are produced are longer than those of the Finkler and Prior spirillum, and are more tightly coiled than those of the cholera spirillum.

Like its related species, this micro-organism is actively motile. It grows at the room-temperature, as well as at 37° C., in this respect, as in its reaction to stains, much resembling the other two.

Upon gelatin plates the growth of the colonies is much more rapid than that of the cholera spirillum, but slower than that of the Finkler and Prior spirillum. The col-

Fig. 87.—Spirillum Denecke, from an agar-agar culture; × 1000 (Itzerott and Niemann).

onies appear as small whitish, round points, which soon reach the surface of the gelatin and commence liquefaction. By the second day they are about the size of a pin's head, have a yellow color, and occupy the bottom of a conical depression. The appearance is much like that of a plate of cholera spirilla.

The microscope shows the colonies to be of irregular

shape and coarsely granular. The color is yellow, and is pale at the edges, gradually becoming intense toward the centre. The colonies are surrounded at first by distinct lines of circumscription, later by clear zones, which, according to the illumination, are pale or dark. From this description it will be seen that the colonies differ from those of cholera in the prompt liquefaction of the gelatin, their rapid growth, yellow color, irregular form, and distinct lines of circumscription.

In gelatin punctures the growth takes place all along the track of the wire, and forms a cloudy liquid which precipitates at the apex in the form of a coiled mass. Upon the surface a delicate imperfect yellowish myco-

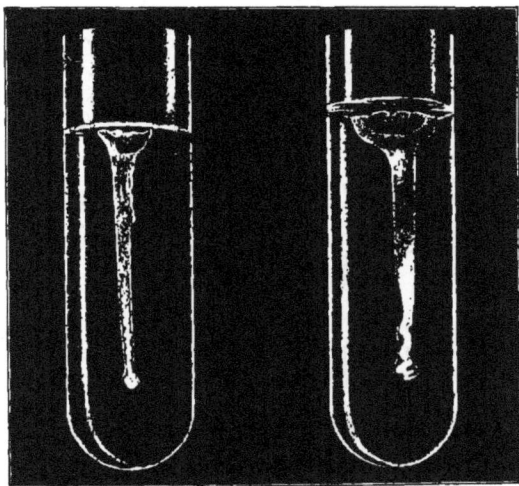

FIG. 88.—Spirillum Denecke: gelatin puncture-cultures aged forty-eight and sixty hours (Shakespeare).

derma forms. Liquefaction of the entire gelatin generally requires about two weeks.

Upon agar-agar this spirillum grows as a thin yellowish layer which does not seem inclined to spread widely.

The culture upon potato is luxuriant if grown in the incubating oven. It appears as a distinct yellowish moist film, and when examined microscopically is seen to contain long beautiful spirals.

The organism sometimes produces indol, but is irregular in its action in this respect.

The spirillum of Denecke is mentioned only because of its morphological relation to the cholera spirillum, not because of any pathogenesis which it possesses. It probably is not associated with any human disease. Experiments, however, have shown that when the spirilla are introduced into the intestines of guinea-pigs whose gastric contents are alkalinized and whose peristalsis is

FIG. 89.—Spirillum Metchnikoff, from an agar-agar culture; × 1000 (Itzerott and Niemann).

paralyzed with opium, about 20 per cent. of the animals die from intestinal disease.

The Spirillum of Gamaléia (Spirillum Metchnikoff). —Very closely related to the cholera spirillum in its morphology and vegetation and possibly, as has been suggested, a descendant of the same original stock, is the spirillum which Gamaléia cultivated from the intestines of chickens affected with a disease similar to chicken-cholera. This spirillum is a curved organism, a trifle shorter and thicker than the cholera spirillum, a little more curved, and with similar rounded ends (Fig. 89).

It forms long spirals in appropriate media, and is actively motile. Each spirillum is provided with a terminal flagellum. No spores have been positively demonstrated.

The organism, like the cholera vibrio, is very susceptible to the influence of acids, high temperatures, and drying, so that spores are probably not formed. It grows well both at the temperature of the room and at that of incubation.

The thermal death-point is 50° C., continued for five minutes.

The bacterium stains easily, the ends more deeply than the center. It is not stained by Gram's method.

Upon gelatin plates a remarkable similarity to the

FIG. 90.—Spirillum Metschnikoff; puncture-culture in gelatin forty-eight hours old (Fränkel and Pfeiffer).

colonies of the cholera spirillum is developed, yet there is a difference, and Pfeiffer points out that "it is comparatively easy to differentiate between a plate of pure cholera spirillum and a plate of pure Spirillum Metchnikoff, yet it is almost impossible to pick out a few colonies of the latter if mixed upon a plate with the former."

Fränkel regards this bacterium as a kind of intermediate species between the cholera and the Finkler-Prior spirilla.

The colonies upon gelatin plates appear in about twelve hours as small whitish points, and rapidly develop, so that by the end of the third day large saucer-shaped areas of liquefaction resembling colonies of the Finkler-Prior spirilla occur. The liquefaction of the gelatin is quite rapid, the resulting fluid being turbid. Generally there will be upon a plate of Vibrio Metchnikoff some colonies which closely resemble cholera by occupying small conical depressions in the gelatin. Under a high power of the microscope the contents of the colonies, which appear to be of a brownish color, are observed to be in rapid motion. The edges of the bacterial mass are fringed with radiating organisms (Fig. 90).

In gelatin tubes the culture is very much like that of cholera, but develops more slowly.

Upon the surface of agar-agar a yellowish-brown growth develops along the whole line of inoculation.

On potato at the room-temperature no growth occurs, but at the temperature of the incubator a luxuriant yellowish-brown growth takes place. Sometimes the color is quite dark, and chocolate-colored potato cultures are not uncommon.

In bouillon the growth which occurs at the temperature of the incubator is quite characteristic, and very different from that of the cholera spirillum. The entire medium becomes clouded, of a grayish-white color, and opaque. A folded and wrinkled mycoderma forms upon the surface.

When glucose is added to the bouillon no fermentation or gas-production results.

When grown in litmus milk the original blue color is changed to pink in a day, and at the end of another day the color is all destroyed and the milk coagulated. Ultimately the clots of casein sediment in irregular masses, and clear colorless whey is supernatant.

The addition of sulphuric acid to a culture grown in a

medium rich in peptone produces the same rose color observed in cholera cultivations.

The organism is pathogenic for animals, but not for man. Pfeiffer has shown that chickens, pigeons, and guinea-pigs are highly susceptible animals. The birds when inoculated under the skin generally die—pigeons always. W. Rindfleish has pointed out that this positive fatal outcome of the introduction of the spirillum into pigeons makes it a valuable diagnostic point for the differentiation of this spirillum from that of cholera. According to his researches, the simple subcutaneous injection of the most virulent cholera cultures is never fatal to pigeons. The birds only die when the injections are made into the muscles in such a manner that the muscular tissue is injured and becomes a *locus minoris resistentiæ*. When guinea-pigs are treated according to the method of Koch for the inoculation of cholera, the temperature of the animal rises for a short time, then abruptly falls to 33° C. or less. Death follows in twenty to twenty-four hours. A distinct inflammation of the intestine, with exudate and numerous spirilla, may be found. The spirilla can also be found in the heart's blood and in the organs of such guinea-pigs. When the bacilli are introduced by subcutaneous inoculation, the autopsy shows a bloody edema and a superficial necrosis of the tissues.

In the blood and all the organs of pigeons and young chickens the organisms can be found in such large numbers that Pfeiffer has suggested the term "vibrionensepticæmie" for the condition. In the intestines very few alterations are noticeable, and very few spirilla can be found.

Gamaléia has shown that pigeons and guinea-pigs can be made immune by inoculating them with cultures sterilized for a time at a temperature of 100° C. Mice and rabbits are immune except to very large doses.

Spirillum Berolinensis.—This organism (Fig. 91), which was discovered by Neisser in the summer of 1893, is of great interest in comparison with the spirillum of cholera and its related forms. Its morphology is in every particular exactly like that of the cholera spirillum, but its growth is a little more rapid. It grows upon the same culture-media and at the same temperature. The colonies are, however, quite different.

Upon the second day, when grown upon gelatin plates, the colonies of the Spirillum Berolinensis appear finely granular and paler than those of cholera. The borders are generally smooth and circular. As it becomes older the colony takes on a slightly brownish color, and may be nodulated or radiately lobulated. The gelatin is very slowly liquefied.

FIG. 91.—Spirillum Berolinensis, from an agar-agar culture; × 1000 (Itzerott and Niemann).

In puncture-cultures the development takes place along the entire puncture, and causes a gradual liquefaction of the gelatin.

Upon agar-agar the growth is generally similar to that of the cholera spirillum, but at times is copious, dry, and ragged, and suggests leather by its appearance.

When introduced intraperitoneally into guinea-pigs the animals die in from one to two days.

The indol reaction is exactly like that given by cultures of the cholera spirillum. The spirillum does not stain by Gram's method.

Spirillum Dunbar.—This organism (Fig. 92) was de-

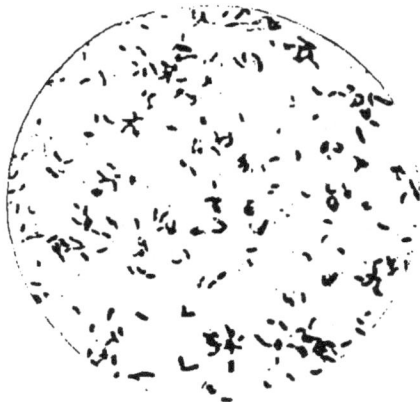

Fig. 92.—Spirillum Dunbar, from agar-agar; × 1000 (Itzerott and Niemann).

scribed in 1893 by Dunbar and Oergel, who secured it from the water of the Elbe River. It much resembles the cholera spirillum, but it never exhibits sigmoid forms. It stains poorly, the ends taking the color much better than the central portion.

Gelatin is liquefied by the growth of this organism more quickly than by the cholera spirillum. The colonies upon gelatin and the puncture-cultures in gelatin are identical with those of the cholera spirillum.

. On agar-agar a luxuriant whitish-yellow layer is produced.

In bouillon and peptone solution the addition of dilute sulphuric acid produces the red color of nitro-indol.

It is said that cultures grown at a temperature of 22° C. phosphoresce in the dark.

The spirillum seems to be pathogenic for guinea-pigs when introduced into the stomach according to Koch's method for cholera.

Spirillum Danubicus.—This organism (Fig. 93) also

22

much resembles cholera.   It was first isolated by Heider in 1892.   In appearance it is rather delicate and decidedly curved.   It is often united in sigmoid and semicircular forms, and exhibits long spirals in old cultures.   It is actively motile, each organism presenting a terminal flagellum.

The growth upon gelatin plates is rapid.   Small light-gray colonies, resembling those of cholera, but exhibiting a dentate margin, are observed.   The growth in gelatin punctures also much resembles cholera, and the agar-agar growth can scarcely be distinguished from it.

The potato growth has a distinct yellowish-brown color.

Milk is coagulated in three or four days.

FIG. 93.—Spirillum Danubicus, from an agar-agar culture; × 1000 (Itzerott and Niemann).

This spirillum does not produce indol.

Heider found the spirillum pathogenic for guinea-pigs.

Spirillum I. of Wernicke.—This organism is about twice as large as the cholera spirillum, liquefies gelatin more rapidly, produces indol, and is feebly pathogenic for guinea-pigs.

Spirillum II. of Wernicke.—This spirillum is smaller than the cholera spirillum, liquefies gelatin more slowly, produces indol, and is highly pathogenic for rabbits, guinea-pigs, pigeons, and mice.

Spirillum Bonhoffi.—This organism (Fig. 94) was found in water by Bonhoff. It has a decided resem-

FIG. 94.—Spirillum Bonhoffi, from a culture upon agar-agar; × 1000 (Itzerott and Niemann).

blance to the cholera spirillum, but is rather stouter and less curved. Curved forms—*i. e.* semicircles, sigmoids, and spirals—occur in old cultures especially.

These organisms are colored badly with ordinary stains, dahlia seeming to be the most appropriate color, and accomplishing the process better if warmed. The organism is motile, and has a long flagellum attached to one end.

The colonies develop slowly upon gelatin plates, first appearing in forty-eight hours as little grayish points. The margin of the colony is sharply circumscribed; the interior is broken up. The gelatin is *not* liquefied. In gelatin punctures there is no liquefaction observable.

Upon agar-agar the development at the temperature of the incubator, which is more rapid than that at the temperature of the room, results in the production of a bluish-gray layer.

The growth upon potato has a brownish color. The

growth in bouillon and in peptone solutions is accompanied by the production of indol.

The spirillum is pathogenic for mice, guinea-pigs, and canary birds.

Spirillum Weibeli.—This spirillum (Fig. 95) was found in 1892 by Weibel in spring-water which had a long time

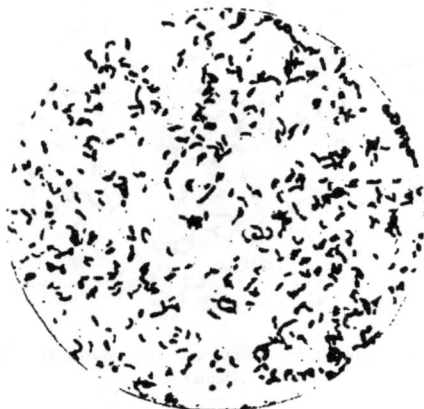

FIG. 95.—Spirillum Weibeli, from agar-agar; × 1000 (Itzerott and Niemann).

before been infected by cholera. It is short, rather thick, and distinctly bent, often forming S-shaped figures.

The colonies before liquefaction sets in are described as pale-brown, transparent, circular, and homogeneous. Liquefaction is much more rapid than in cholera, and causes the borders of the colonies to become irregular. In the centre of each colony a little depression is observed.

In gelatin puncture-cultures the growth is rapid, beginning first upon the surface, where a large flat, saucer-shaped liquefaction, extending to the sides of the tube, forms. Scarcely any growth takes place in the puncture, but the superficial liquefaction, separated by a horizontal line from the normal gelatin, descends slowly.

Upon agar-agar a grayish-white layer is formed.

No growth has been obtained upon potato.

In alkaline peptone solution a slow but luxuriant growth takes place.

Spirillum Milleri.—This spirillum (Fig. 96) was found in the mouth by Miller in 1885. It resembles the cholera

FIG. 96.—Spirillum Milleri, from an agar-agar culture; × 1000 (Itzerott and Niemann).

spirillum somewhat, but is much more like the spirillum of Finkler and Prior, with which many bacteriologists think it identical.

Upon gelatin the colonies are small, finely granular, have a narrow border-zone and a pale-brown color. The gelatin is rapidly liquefied.

Upon agar-agar a thick yellowish layer is produced.

The organism seems not to be pathogenic.

Spirillum Aquatilis.—Günther in 1892 found this organism (Fig. 97) in the water of the river Spree. It is similar to the cholera spirillum in shape, has a long terminal flagellum, and is motile.

The colonies which form upon gelatin are circular, have smooth borders, and look very much as if bored out with a tool. They have a brown color and are finely granular. In gelatin puncture-cultures the growth occurs almost exclusively at the surface.

The agar-agar cultures are similar to those of cholera.

Scarcely any development occurs in bouillon. By the

growth of the organism sulphuretted hydrogen gas is produced.

The spirillum does not grow at all upon potato.

Günther did not find the organism to be pathogenic.

Spirillum Terrigenus.—This species, also discovered by Günther, was secured from earth. It generally occurs in a slightly curved form, but sometimes is spiral. It is actively motile and has a terminal flagellum.

The colonies, which appear in twenty-four hours, are small, structureless, and transparent, and later take on a "fat-drop" appearance.

Upon agar-agar a thin white coating is formed. Milk is coagulated by the growth of the organism. No indol is produced.

The organism does not stain by Gram's method, and is said not to be pathogenic for guinea-pigs or for mice.

FIG. 97.—Spirillum aquatilis, from an agar-agar culture; × 1000 (Itzerott and Niemann).

Vibrio Schuylkiliensis.—This form, closely resembling the cholera spirillum, was found by Abbott[1] in sewage-polluted water from the Schuylkill River at Philadelphia. The colonies upon gelatin plates resemble very closely those of Spirillum Metschnikovi. In gelatin puncture-cultures the appearance is exactly like the true cholera spir-

[1] *Jour. of Exper. Med.*, vol. i., No. 3, July, 1896, p. 419.

illum. At times the growth may be a little more rapid. The growth on agar is very luxuriant, and gives off a pronounced odor of indol. Löffler's blood-serum is apparently not a perfectly adapted medium, but upon it the organisms grow, with resulting liquefaction. Upon potato at the point of inoculation there is a thin, glazed, more or less dirty yellow, shading to brownish deposit that is sometimes surrounded by a flat, dry, lusterless zone.

In litmus milk a slightly reddish tinge is found after twenty-four hours at body temperature. After forty-eight hours this is increased and the milk is coagulated. In peptone solutions indol is produced. No gas is produced in glucose-containing culture-media. The organism is a facultative anaërobic spirillum. The thermal death-point is 50° C. for five minutes.

The organism is pathogenic for pigeons, guinea-pigs, and mice. The pathogenesis is much like that of the Spirillum Metschnikovi. No Pfeiffer's phenomenon was observed with the use of the serum of immunized animals.

Immunity was produced in pigeons, and it was found that their serum was protective against both the Vibrio Schuylkiliensis and Spirillum Metschnikovi, the immunity thus produced being of about ten days' duration.

In a second paper by Abbott and Bergy[1] it was shown that the vibrios were found in river water during all four seasons of the year, and in all parts of the river within the city, both at low and at high tide. They were also found in the sewage emptying into the river. The spirilla were also found in the water of the Delaware River as frequently as in that from the Schuylkill.

One hundred and ten pure cultures of spirilla were isolated from the sources mentioned and subjected to routine tests. It was found that few or none of them were identical in all points. There seems, therefore, to be a family of river spirilla related to each other like the different colon bacilli are related.

[1] *Journal of Experimental Medicine*, vol. ii., No. 5, p. 535.

The opinion of the writers is that "the only trustworthy difference between many of these varieties and the true cholera spirillum is the specific reaction with serum from animals immune from cholera, or by Pfeiffer's method of intraperitoneal testing in such animals."

In discussing these spirilla of the Philadelphia waters Bergy[1] says:

"The most important point with regard to the occurrence of these organisms in the river water around Philadelphia, is the fact that similar organisms have been found in the surface-waters of the European cities in which there had recently been an epidemic of Asiatic cholera, notably at Hamburg and Altona. . . . The foremost bacteriologists of Europe have been inclined to the opinion that the organisms which they found in the surface-waters of the European cities were the remains of the true cholera organism, and that the deviations in the morphologic and biologic characters from those of the cholera organism were brought about by their prolonged existence in water. No such explanation of the occurrence of the organisms in Philadelphia waters can be given."

[1] *Jour. of the Amer. Med. Assoc.*, Oct. 23, 1897.

CHAPTER V.

PNEUMONIA.

THE term "pneumonia," while generally understood
to refer to the lobar disease particularly designated as
croupous pneumonia, is a vague one, really comprehending a variety of inflammatory conditions of the lung
quite dissimilar in character. This being true, no one
should be surprised to find that a single organism cannot
be described as "specific" for all. Indeed, pneumonia
must be considered as a group of diseases, and the various
microbes found associated with it must be described suc-
cessively in connection with the peculiar phase of the
disease in which they occur.

1. **Lobar or Croupous Pneumonia.**—The bacterium,
which can be demonstrated in at least 75 per cent. of the
cases of lobar pneumonia, which is now almost universally accepted as the cause of the disease, and about
whose specificity very few doubts can be raised, is the
pneumococcus of Fränkel and Weichselbaum.

Priority of discovery in the case of the pneumococcus
seems to be in favor of Sternberg, who as early as 1880 described an identical organism which he secured from his
saliva. Curiously enough, Pasteur seems to have cap-
tured the same organism, also from saliva, in the same
year. The researches of the observers whose names are
attached to the organism were not completed until five
years later. It is to Fränkel, Telamon, and particularly
to Weichselbaum, however, that we are indebted for the
discovery of the relation which the organism bears to
pneumonia.

The organism (Fig. 98) is variable in its morphology.
When grown in bouillon it is oval, has a pronounced dis-

position to occur in pairs, and not infrequently forms chains of five or six members, so that some have been disposed to look upon it as a streptococcus (Gamaléia). In the fibrinous exudate from croupous pneumonia, in the rusty sputum, and in the blood of rabbits and mice containing them the organisms are arranged in pairs, exhibit a distinct lanceolate shape, the pointed ends generally approximated, and are usually surrounded by a distinct halo or capsule of clear, colorless, homogeneous material, thought by some to be a swollen cell-wall, by

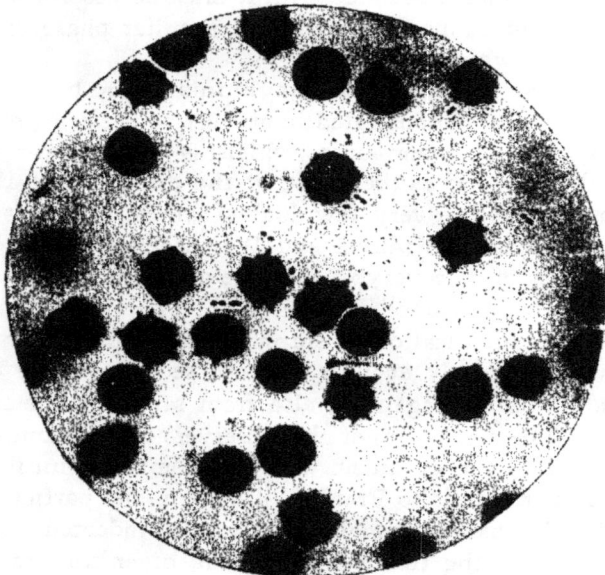

FIG. 98.—Diplococcus pneumoniæ, from the heart's blood of a rabbit; × 1000 (Fränkel and Pfeiffer).

others a mucus-like secretion given off by the cells. When grown ordinarily in culture-media, and especially upon solid media, the capsules are absent.

The organism is without motility, has no spores, and does not seem to be able to resist any unfavorable conditions when grown artificially. It stains well with the ordinary solutions of the anilin dyes, and gives most beautiful pictures in blood and tissues when stained by Gram's method. The capsule does not stain.

To demonstrate the capsule, the glacial acetic acid method may be used. The cover-glass is spread with a thin film of the material to be examined, which is dried and fixed as usual. Glacial acetic acid is dropped upon it for an instant, poured (not washed) off, and at once followed by anilin-water, gentian-violet, in which the staining continues several minutes. Finally, the preparation is washed in water, and may be examined at once in water or mounted in balsam after drying. The capsules are probably more distinct when the examination is made in water.

The pneumococcus is no stranger to us; it may sometimes be found in the saliva of healthy individuals, and the inoculation of human saliva into rabbits frequently causes a septicemia in which the bacillus is found abundantly in the blood and tissues. Because of its frequent presence in the saliva it was described by Flügge as the Bacillus septicus sputigenus.

When desired for purposes of study, it may be obtained by inoculating rabbits with pneumonic sputum and recovering the organisms from their heart's blood, or it may be secured from the rusty sputum of pneumonia by the method employed by Kitasato for securing tubercle bacilli from sputum. A single mouthful of fresh sputum is secured, washed in several changes of sterile water to free it from bacteria of the mouth and pharynx, carefully separated, and a central portion transferred to an appropriate culture-medium.

The organism grows upon all the culture-media except potato, but only between the temperature-extremes of 24° and 42° C.; the best development is at 37° C. The growth is always limited, probably because the formic acid produced serves to check it. The addition of an unusual amount of alkali to the culture-medium favors the growth.

The organisms readily lose their virulence in culture-media, and cease to be pathogenic after a few days.   In his experiments with antipneumococcic serum Washbourn found, however, that a pneumococcus isolated from pneumonia sputum and passed through one mouse and nine rabbits developed a permanent virulence when kept on agar-agar made carefully, so that it was not heated beyond 100° C., and alkalinized 4 c.cm. of normal caustic soda solution beyond the neutral point determined with rosalic acid, to each liter.   The agar-agar is first streaked with sterile rabbit's blood, then inoculated.   The cultures are

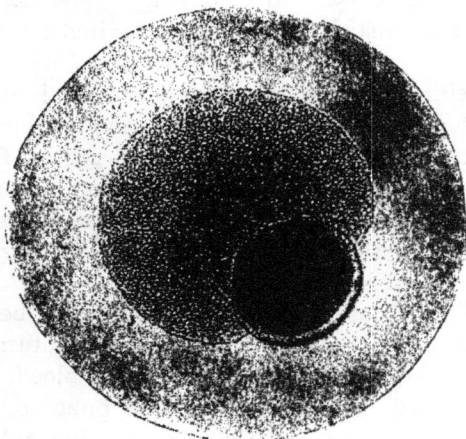

FIG. 99.—Diplococcus pneumoniæ : colony twenty-four hours old upon gelatin ;
× 100 (Fränkel and Pfeiffer).

kept at 37.5° C.   Not only is this true, but ordinarily they seem to be unable to accommodate themselves to a purely saprophytic life, and unless continually transplanted to new media die in a week or two, sometimes sooner.

Kinyoun recommended to the writer that virulence could be retained for a considerable time by keeping blood from an infected rabbit, in a hermetically sealed glass tube, on ice.   This plan seems to work admirably if the blood is not kept too long.

The colonies which develop at 24° C. upon 15 per cent. gelatin plates are described as small, round, circumscribed, finely granular white points which grow slowly, never attain any considerable size, and do not liquefy the gelatin (Fig. 99).

If, instead of gelatin, agar-agar be used and the plates kept at the temperature of the body, the colonies which develop upon the plates appear as transparent, delicate, drop-like accumulations, scarcely visible to the naked eye, but under the microscope distinctly granular, the central darker portion being frequently surrounded by a paler marginal zone.

In gelatin puncture-cultures, made with 15 instead of the usual 10 per cent. of gelatin, the growth takes place along the entire path of the wire in the form of little whitish granules distinctly separated from each other. The growth in gelatin is always very limited.

Upon agar-agar and blood-serum the growth consists of minute, transparent, semi-confluent, colorless, dew-drop-like colonies, which die before attaining a size which permits of their being seen without careful in-spection.

In bouillon the organisms grow well, clouding the medium very slightly.

Milk is quite well adapted as a culture-medium, its casein being coagulated.

No growth can be secured upon potato at any temperature or by any manipulation yet known.[1]

When it is desired to maintain or increase the virulence of a culture it must be very frequently passed through the body of a rabbit. The degree to which the virulence can be raised in this way is remarkable. C. W. Lincoln has succeeded in reducing the fatal dose for rabbits to $\frac{1}{100000000}$ of a c.cm.

If a small quantity of a pure culture of the virulent

---

[1] Ortmann asserts that the pneumococcus can be grown on potato at 37° C., but this is not generally confirmed. The usual acid reaction of the potato would indicate that it was a very unsuitable culture-medium.

organism is introduced into a mouse, rabbit, or guinea-pig, the animal dies in one or two days. Exactly the same result can be obtained by the introduction of a piece of the lung-tissue from croupous pneumonia, by the introduction of some of the rusty sputum, and generally by the introduction of saliva.

The post-mortem shows that an inflammatory change has taken place at the point of inoculation, with a fibrinous exudate resembling somewhat that in diphtheria. At times, and especially in dogs, there may be a little pus formed. The other appearances are those of a general disturbance. The spleen is much enlarged, is firm and red brown. The blood in all the organs contains large numbers of the bacteria, most of which exhibit a distinct lanceolate form and have their capsules very distinct. The disease is a pure septicemia unassociated with pronounced tissue-changes.

In cases of the kind described the lungs show no pneumonic changes. Likewise, if the hypodermic needle used for injection be plunged through the breast-wall into the pulmonary tissue, no pneumonia results. Monti, however, claims to have found that a true characteristic pneumonia results from the injection of cultures into the trachea of susceptible animals. This observation lacks confirmation.

Not all animals are susceptible. Guinea-pigs, mice, and rabbits are highly sensitive to the operations of the organism; dogs are comparatively immune.

From this brief review of the peculiarities of the pneumococcus it must be obvious that its reputation in pneumonia depends more upon the regularity with which it is found in that disease than upon its capacity to produce a similar affection in the lower animals.

As in numerous other diseases, we are unable to furnish an absolute proof of specificity according to the postulates of Koch.

The disease is peculiar in that recovery from it is followed either by no immunity or by one of such brief duration as to allow of frequent relapses; and it is well known that many cases show a subsequent predisposition to fresh attacks of the disease. This brevity of immunity lessens the probability that in the future we shall discover an antitoxin that shall be powerful in its influence upon the course and termination of the disease.

The experiments of G. and F. Klemperer, a few years ago, showed that the serum of immunized rabbits protected animals inoculated with the pneumococcus. The principle failed, however, when applied to human medicine. The treatment of pneumonia by the injection of blood-serum from convalescents has also been abandoned as useless and dangerous.

Washbourn has recently prepared an *antipneumococcic serum* which is efficacious in protecting rabbits against ten times the fatal dose of live pneumococci. In general, the lines upon which he operated were those of Behring, Marmorek's work with the streptococcus furnishing most of the details. A pony was subjected to immunization for a period of five months, allowed to rest three or four months until the live pneumococci introduced were all destroyed, and then bled. Two cases of human pneumonia seem to have received some benefit from the injection of large doses of this serum.

The pneumococcus causes other lesions than croupous pneumonia; thus, Foa, Bordoni-Uffreduzzi, and others have found it in cerebrospinal meningitis; Fränkel, in pleuritis; Weichselbaum, in peritonitis; Banti, in pericarditis; numerous observers have found it in acute abscesses; Gabbi has isolated it from a case of suppurative tonsillitis; Axenfeld has observed an epidemic of conjunctivitis caused by it; and Zaufal, Levy, and Schräder and Netter have been able to demonstrate its presence in the pus of otitis media. It has also been reported as occurring in the joints in arthritis following pneumonia.

The pneumococcus is often present in the mouths of healthy persons. The conditions under which it enters the lung to produce pneumonia are not known.

In the opinion of most authorities, something more than the simple entrance of the bacterium into the lung is required for the production of the disease, but what that something is, is still a matter of doubt. It would seem to be some systemic depravity, and in support of this view we may point out that pneumonia is very frequent, and almost universally fatal, among drunkards. Whether, however, any vital depression or systemic depravity will predispose to the disease, or whether it depends for its origin upon the presence of a certain leucomaïne, time and further study will be required to tell.

Bacillus Pneumoniæ of Friedländer (Fig. 100).—An un-

FIG. 100.—Bacillus pneumoniæ of Friedländer, from the expectoration of a pneumonia patient; × 1000 (Fränkel and Pfeiffer).

fortunate accident has applied the name "pneumococcus" to an organism very different from the one just described. It was discovered by Friedländer in 1883 in the exudate from the lung in croupous pneumonia, and, being thought by its discoverer to be the cause of the disease, very naturally was called the pneumococcus, or, more correctly, the *pneumobacillus*. The grounds upon which the pathogeny of the organism was supposed to depend were very insufficient, and the bacillus of Friedländer—or, as Flügge prefers to call it, the Bacillus pneumoniæ—has ceased to be regarded as specific, and is now looked upon as an accidental organism whose presence in the lung is, in most cases, unimportant.

As the two organisms are similar in more respects than their names, Friedländer's bacillus requires at least a brief description.

It is distinctly a bacillus, but sometimes, when occurring in pairs, has a close resemblance to the pneumococcus of Fränkel and Weichselbaum. Very frequently it forms chains of four or more elements. It is also commonly surrounded by a transparent capsule. It is nonmotile, has no spores and no flagella. It stains well with the ordinary anilin dyes, but does not retain the color when stained by Gram's method.

Fränkel points out that Friedländer's error in supposing this bacillus to be the chief parasite in pneumonia depended upon the fact that his studies were made by the plate method. If some of the pneumonic exudate be mixed with gelatin and poured upon plates, the bacilli grow into colonies at the end of twenty-four hours, and appear as small white spheres which spread upon the gelatin to form white masses of a considerable size. Under the microscope these colonies are rather irregular in outline and somewhat granular.

The bacillus grows at as low a temperature as 16° C., and, according to Sternberg, has a thermal death-point of 56° C.

When a colony is transferred to a gelatin puncture-culture, quite a massive growth occurs. Upon the surface a somewhat elevated, rounded white mass is formed, and in the track of the wire innumerable little colonies spring up and become confluent, so that a "nail-growth" results. No liquefaction occurs. When old the cultures sometimes become brown in color.

Upon the surface of agar-agar at ordinary temperatures quite a luxuriant white or brownish-yellow, smeary, cir-

23

cumscribed growth occurs. The growth upon blood-serum is the same.

Upon potato the growth is abundant, quickly covering the entire surface with a thick yellowish-white layer, which sometimes contains bubbles of gas. Gas is also sometimes developed in gelatin cultures.

A most superficial comparison will suffice to show the great difference in vegetation between these two so-called pneumococci.

Friedländer had considerable difficulty in causing any pathogenic changes by the injection of his bacillus into animals. Rabbits and guinea-pigs were immune, and the only actual pathogenic results which Friedländer obtained were in mice, into whose lungs and pleura he injected the cultures. The remarks of Fränkel upon such mouse-operations, which do not add much weight to experiments, have already been quoted.

In the *status præsens* of bacteriologic knowledge the bacillus of Friedländer is regarded as an organism of very feeble pathogenic powers, generally a harmless saprophyte, but which may at times aid in producing inflammatory changes when in the tissues of the human body.

2. **Catarrhal Pneumonia.**—This form of pulmonary inflammation occurs in local areas, generally situated about the distribution of a bronchiole. It cannot be said to have a specific micro-organism, as almost any irritant foreign materials accidentally inhaled can cause it. The majority of the cases, however—and especially those which are distinctly peribronchial—are caused by the presence of the staphylococcus and streptococcus of suppuration. Friedländer's bacillus may also aid in producing local inflammations.

3. **Tubercular Pneumonia.**—At times the process of pulmonary tuberculosis is so rapid, and associated with the production of so much semi-liquid, semi-necrotic material, that the auto-infection of the lung is greatly favored ; the tubercle-bacilli are distributed to the entire lung or to large parts of it, and a distinct inflammation

occurs. Such a pneumonia may be caused by the tubercle bacillus alone, but more often it is aided by accompanying staphylococci, streptococci, tetragenococci, pneumococci, pneumobacilli, and other organisms apt to be present in a lung in which tuberculosis is in progress and ulceration and cavity-formation are advanced.

4. **Mixed Pneumonias.**—It frequently happens that pneumonia occurs in the course of, or shortly after the convalescence from, influenza. In these cases a mixed infection is present, and there is no difficulty in determining that both the influenza bacillus and the pneumococcus are present. Again, sometimes the pneumococci and staphylococci operate simultaneously, and produce a purulent pneumonia with abscesses as the conspicuous feature. As almost any combination of the described bacteria is possible in the lungs, and as these combinations will all produce varying inflammatory conditions, it must be left for the student to imagine what the particular characters of each may be.

Among these mixed pneumonias may be mentioned those called by Klemperer and Levy "complicating pneumonias," occurring in the course of typhoid, etc.

C. THE SEPTIC DISEASES.

CHAPTER I.

ANTHRAX.

THE disease of cattle known as anthrax or "splenic fever" is of infrequent occurrence in this country and in England. In France, Germany, Hungary, Russia, Persia, and the East Indian countries it is a dreaded and common malady which robs herdsmen of many of their valuable stock. Siberia perhaps suffers most, the disease being so exceedingly common and malignant as to deserve the name "Siberian pest." Certain local areas, such as the Tyrol and Auvergne, in which it seems to be constantly present, serve as distributing foci from which the disease spreads rapidly in summer, afflicting many animals, and ceasing its depredations only with the advent of winter. It seems to be distinctly a disease of the summer season.

The animals most frequently affected are cows and sheep. Among our laboratory animals white mice, guinea-pigs, and rabbits are highly susceptible; dogs, cats, most birds, and amphibians are almost perfectly immune. White rats are infected with difficulty. Man is only slightly susceptible, the manifestation of the disease as seen in the human species being different from the same disease in the lower animals in that it is usually a local affection—malignant carbuncle—and only at times gives rise to a general infection.

Anthrax was one of the first of the specific diseases proven to be caused by a definite micro-organism. As early as 1849, Pollender discovered small rod-shaped bodies in the blood of animals suffering from anthrax, but the exact relation which they bore to the disease was not pointed out until 1863, when Davaine, by a series of interesting experiments, proved to most unbiased minds their etiological significance. The further confirmation of Davaine's conclusions and actual proof of the matter rested with Pasteur and Koch, who, observing that the bacilli bore spores, cultivated them successfully outside the body, and then produced the disease by the inoculation of pure cultures.

The anthrax bacilli (Fig. 101) are large rods with a

FIG. 101.—Bacillus anthracis: colony three days old upon a gelatin plate; adhesive preparation; × 1000 (Fränkel and Pfeiffer).

rectangular form, caused by the very slight rounding of the corners. They measure 5–20 μ in length and are from 1 μ to 1.25 μ in breadth. The pronounced tendency is toward the formation of long threads, in which, however, the individuals can generally be made out; at times isolated rods occur. In the threads the bacilli seem enlarged a little at the ends, and give somewhat the appearance of a bamboo cane. The formation of spores is prolific: each spore has a distinct oval shape, is transparent, and does not alter the contour of the bacillus in which it occurs. Spores are generally formed in the presence of oxygen upon the surfaces of the culture-media. When a spore is placed under favorable conditions for its development and is carefully watched, it may be observed to increase in length a trifle, then to undergo a rupture at one end, from which the new bacillus projects. The spores of anthrax (Fig. 102), being large and easily ob-

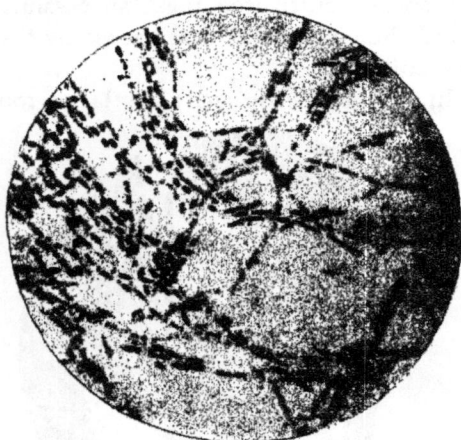

FIG. 102.—Bacillus anthracis, stained to show the spores; × 1000 (Fränkel and Pfeiffer).

tainable, are excellent subjects for the study of sporulation, for the action of germicides and antiseptics, and for demonstration by stains. When dried upon threads of silk they will retain their vitality for several years, and are highly resistant to heat and disinfectants.

Spores of anthrax are killed by five minutes' exposure to a temperature of 100° C., and are killed in five minutes in a 5 per cent. solution of carbolic acid, or, at least, are deprived of their vegetative property in relation to culture-media. It is said by some that spores subjected to 5 per cent. carbolic acid can germinate when introduced into susceptible animals. Spores are also killed by simple wetting with 1 : 100,000 bichlorid-of-mercury solution.

The bacilli are not motile and are not provided with flagella. They stain well with ordinary solutions of the anilin dyes, and can be beautifully demonstrated in the tissues by Gram's method and by Weigert's fibrin method. Picro-carmin, followed by Gram's method, gives a beautiful, clear picture. The spores can be stained with carbol-fuchsin, the bacilli decolorized with a very weak acid and then counter-stained with a watery solution of methyl blue.

. Upon the surface of gelatin plate-cultures the bacillus forms beautiful and highly characteristic colonies (Fig. 103). To the naked eye they appear first as minute

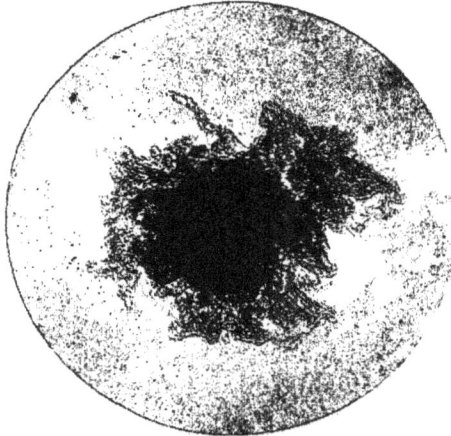

FIG. 103.—Bacillus anthracis: colony upon a gelatin plate; × 100 (Fränkel and Pfeiffer).

round whitish dots occurring upon the surface, and causing liquefaction of the gelatin as they increase in size. Under the microscope they can be seen in the gelatin as egg-shaped, slightly brownish granular bodies, not attaining their full development except upon the surface, where they spread out into flat, irregular, transparent growths bearing a partial resemblance to tufts of curled wool. From a tangled centre large numbers of curls extend, each made up of parallel threads of bacilli. As soon as the colony attains any considerable size liquefaction begins. These colonies make beautiful adhesive preparations. If a perfectly clean cover-glass be passed once through a flame and laid carefully upon the gelatin, the colonies can generally be picked up entire when the glass is removed. Such a specimen can be dried, fixed, and stained in the same manner as an ordinary cover-glass preparation.

In gelatin puncture-cultures the growth is even more characteristic than are the colonies. The bacilli begin to grow along the entire track of the wire, most luxuriantly at the surface, where oxygen is plentiful. As the growth progresses fine filaments like bristles, extend from the puncture into the neighboring gelatin giving the growth somewhat the appearance of an evergreen tree inverted (Fig. 104).

FIG. 104.—Bacillus anthracis : gelatin puncture-culture seven days old (Günther).

The more superficial of these threads reach about half-way to the sides of the tube, while the deeper ones are shorter and shorter, until near the apex branches cease. When the projections are pretty well developed a distinct surface-growth will be discerned, and if the tube be tilted, one can observe that the gelatin beneath it has liquefied. As the growth becomes older the liquefaction increases, until ultimately the entire gelatin is fluid and the growth is precipitated.

Upon agar-agar the characteristics are few. The growth takes place all along the line of inoculation as a slightly translucent, slightly wrinkled layer with irregular edges, from which sufficient bacillary threads project to give it a ciliated appearance to the naked eye. When the culture is old the agar-agar turns a distinct brown. Spore-formation is luxuriant upon agar-agar.

On potato the growth is white, creamy, sometimes rather dry in appearance. Sporulation is marked.

Blood-serum cultures lack peculiarities; the culture-medium is slowly liquefied.

The bacillus only grows between the extremes of 20° and 45° C., best at 37° C. The exposure of the organism to the temperature of 42–43° C. for twenty-four hours is sufficient to destroy its virulence.

The culture-media should always be faintly alkaline, as anthrax bacilli will not grow in the presence of free acid.

The micro-organism under consideration is a parasitic microbe, yet is one which, because of its spores, can, in a latent form, exist without the animal organism until appropriate conditions for its natural development are presented.

Ordinarily, the infection takes place either through the *respiratory tract* or through the *alimentary canal.*

Buchner has shown that when animals are allowed to inhale anthrax spores they die of typical anthrax. The spores establish themselves in the alveoli of the lung, penetrate the epithelium, enter the vascular system, and soon give rise to typical lesions. Strange to say, the appearance caused by the inhalation of the bacilli in their perfect form is entirely different, for a rapid multiplication occurs without sporulation, and causes a violent irritative pneumonia with serous or sero-fibrinous exudate in which large numbers of the bacilli occur. In these cases there may be no general infection.

When the bacilli are taken into the stomach in food they meet with a rapid death because of the acidity of the gastric juice. Should spores, however, be ingested, they are able to endure the gastric juice, to pass into the intestine, and, as soon as proper conditions of alkalinity are encountered, to develop into bacilli. They develop rather rapidly, surround the villi with thick networks of bacillary threads, separate the epithelial cells, enter the lymphatics, and thus find the appropriate environment for the production of a general infection.

Sometimes the bacillus enters the body through a wound, cut, scratch, or fly-bite. This is especially the case with men who come in contact with diseased cattle. As has already been pointed out, a malignant pustule is apt to follow, and may cause death. Men whose occupations bring them in contact with skins and hair from animals dead of anthrax are not only liable to wound-infection, but are sometimes the subjects of a pulmonary form of the disease—"wool-sorter's disease"—caused by inspiration of the spores attached to the wool.

The disease as we see it in the laboratory is accompanied by few but marked lesions. The ordinary method of inoculation is to cut away a little of the hair from the abdomen of a guinea-pig or rabbit or the root of a mouse's tail, make a little subcutaneous pocket with a snip of a pair of sterile scissors, and introduce the spores or bacilli from a pure culture upon a rather heavy platinum wire, the end of which is flattened, pointed, and perforated. An animal inoculated in this way generally dies, according to the species, in from twenty-four hours to three days. The symptoms are weakness, fever, loss of appetite, and sometimes a bloody discharge from nose and bowels. There is much subcutaneous edema. At the autopsy very little change is observed at the seat of inoculation. The subcutaneous tissue beneath it for a considerable distance around is occupied by a peculiar colorless gelatinous edema which contains the bacilli. The abdominal cavity shows injection and congestion of its viscera. The spleen is considerably enlarged, is dark in color, and of mushy consistence. The liver is somewhat enlarged. When the thorax is opened, the lungs may be slightly congested, but otherwise no changes are to be found.

When the various organs, which present no appreciable changes to the naked eye, are subjected to a microscopic examination, the appropriate staining methods bring out a most remarkable and beautiful change. The capillary system is almost universally occupied by bacilli, which extend throughout its meshworks in long threads. Most beautiful bundles of these bacillary threads can, at times, be found in the glomeruli of the kidney and in the minute capillaries of the intestinal villi. In the larger vessels, where the blood-stream is rapid, the bacteria are relatively few, so that the burden of bacillary obstruction is borne by the minute vessels. The condition is thus one of pure septicemia, and bacilli can be secured in pure cultures from the blood and tissues.

The susceptibility of the anthrax bacillus to the influence of heat, cold, antiseptics, etc. not only permitted Buchner, Behring, and others to produce biological curiosities in the form of bacilli unable to bear spores and robbed of their pathogenic powers, but also suggested to Pasteur the important practical measure of protective vaccination. Pasteur found that the inoculation of non-virulent bacilli into cows and sheep, and their reinoculation with slightly virulent bacilli, gave them the ability to withstand the action of highly virulent organisms. Löffler, Koch, and Gaffky, however, found that these immunized animals were not absolutely protected from intestinal anthrax.

The methods of diminishing the virulence of the anthrax bacilli are numerous. Toussaint, who was certainly the first to produce immunity in animals by injecting them with sterile cultures of the bacillus, found that the addition of 1 per cent. of carbolic acid to blood of animals dead of anthrax destroyed the virulence of the bacilli; Chamberland and Roux found it removed when 0.1–0.2 per cent. of bichromate of potassium was added to the culture-medium; Chauveau used atmospheric pressure to the extent of six to eight atmospheres and found the virulence diminished; Arloing found that direct sunlight operated similarly; Lubarsch found that the inoculation of the bacilli into immune animals, such as the frog, and their subsequent recovery from its blood, diminishes the virulence markedly.

Protection can be afforded in still other ways. The

simultaneous inoculation of bacteria not at all related to anthrax will sometimes recover the animal, as Hüppe found. Hankin found in the cultures chemical substances, especially an albuminose, which exerted a protective influence. Chamberland has shown that protective inoculation by Pasteur's method has diminished the death-rate from 10 per cent. for sheep and 5 per cent. for cattle to about 0.94 per cent. for sheep and 0.34 per cent. for cattle, so that the utility of the method is scarcely questionable. In 1890, Ogata and Jasuhara showed that in the convalescents from anthrax among their experimental animals an antitoxic substance was present in the blood in such quantities that 1 : 800 parts per body-weight of dog's serum containing the antitoxin would protect a mouse. Similar results have been attained by Marchoux.

Experiments of interest have been performed to show that the natural immunity enjoyed by many animals can be destroyed. Behring found that if the alkalinity of the blood of rats was diminished, they could become affected with anthrax, and numerous observers have shown that when anthrax bacilli and unrelated organisms, such as the erysipelas cocci, Bacillus prodigiosus, and Bacillus pyocyaneus, are simultaneously introduced into immune animals, the immunity is destroyed and the animals succumb to the disease. Frogs have been made to succumb to the disease by exposure to a temperature of 37° C. after inoculation. Pasteur destroyed the immunity of fowls by a cold bath after inoculation.

In the natural order of events anthrax in cattle is probably the result of the inhalation or ingestion of the spores of the bacilli from the pasture. At one time much discussion arose concerning the infection of the pasture. It was argued that, the bacilli being enclosed in the tissues of the diseased animals, the infection of the pasture must be due to the distribution of the germs from the buried cadaver to all parts of the field, either through the activity of earth-worms, which ate of the earth surrounding the corpse and then deposited the spores in their excrement at remote areas (Pasteur), or to currents of moisture in the soil. Koch seems, however, to have demonstrated the fallacy of the theories by showing that the conditions under which the bacilli find themselves in buried cadavers are exactly opposed to those favorable to fructification or sporulation, and that in all probability the majority of bacteria suffer the same fate as the animal cells, and disintegrate, especially if the animal be buried at a depth of two or three meters.

Fränkel points out particularly that no infection of the soil by the dead animal could be worse than the pollution of its surface by the bloody stools and urine, rich in bacilli, discharged upon it by the animal before death, and that in all probability it is the live, and not the dead, animals that are to be blamed as sources of infection.

As every animal affected with anthrax is a source of danger to the community in which it lives, to the men who handle it as well as the animals who browse beside it, such animals, as soon as the diagnosis is made, should be killed, and, together with the hair and skin, be burned. When this is impracticable, Fränkel recommends that they be buried to a depth of at least 1½–2 meters, so that the sporulation of the bacilli is impossible. The dejecta should also be carefully disinfected with 5 per cent. carbolic-acid solution.

Of course, animals can be infected through wounds. This mode of infection is, however, more common among men, who suffer from the local disease manifested as the malignant carbuncle, than among animals.

Occasionally bacilli are encountered presenting all the morphological and cultural characteristics of the anthrax bacillus, but devoid of any disease-producing power—Bacillus anthracoides, etc. Exactly what relation they may bear to the anthrax bacillus is uncertain. They may be entirely different organisms, or they may be individuals whose pathogeny has been lost through unfavorable environment.

CHAPTER II.

TYPHOID FEVER.

THE bacillus of typhoid fever (Fig. 105) was discovered by Eberth and Koch in 1880, and was first secured in

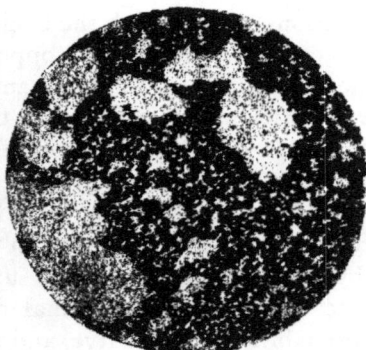

FIG. 105.—Bacillus typhi, from a twenty-four-hours-old agar-agar culture; × 650 (Heim).

pure culture from the spleen and affected lymphatic glands by Gaffky four years later.

The organism is a small, short bacillus about 1–3 μ (2–4 μ Chantemesse, Widal) in length and 0.5–0.8 μ broad (Sternberg). The ends are rounded, and it is rather exceptional for the bacilli to be united in chains, though this arrangement is common in potato cultures. The size and morphology vary distinctly with the nature of the culture-medium and the age of the culture. Thoinot and Masselin in describing these morphological peculiarities mention that when grown in bouillon it is a very slender bacillus; in milk it is a large bacillus; upon agar-agar and potato it is very thick and short; and in old gelatin cultures it forms very long filaments.

366

The organisms are actively motile, the motility probably being caused by the numerous flagella with which
the bacilli are provided. The flagella stain well by
Löffler's method, and, as they are numerous (ten to
twenty) and readily demonstrable, the typhoid bacillus is
the favorite subject for their study. The movements of

FIG. 106.—Bacillus typhi, from an agar-agar culture six hours old, showing the
flagella stained by Löffler's method; × 1000 (Fränkel and Pfeiffer).

the short bacilli are oscillating, those of the longer individuals serpentine.

The organism stains quite well by the ordinary methods, but loses the color entirely when stained by Gram's
method. Its peculiarity of staining is the readiness with
which the bacillus gives up its color in the presence of
solvents, so that it is particularly difficult to stain it in
tissue.

When sections are to be stained the best method is to
allow the tissue to remain in Löffler's alkaline methylene
blue for from fifteen minutes to twenty-four hours, then
wash in water, dehydrate rapidly in alcohol, clear up in
xylol, and mount in Canada balsam. Ziehl's method
also gives good results. The sections are stained for fifteen minutes in a solution of distilled water 100, fuch-

sin 1, and phenol 5. After staining they are washed in distilled water containing 1 per cent. of acetic acid, dehydrated in alcohol, cleared, and mounted. In such preparations the bacilli may be found in little groups, which are easily discovered, under a low power of the microscope, as reddish specks, and readily resolved into bacilli with the high power of the oil-immersion lens.

In bacilli stained by this alkaline methylene-blue solution dark-colored dots may sometimes be observed near the ends of the rods. These dots were at first regarded as spores, but are now denominated polar granules, and are thought to be of no importance.

The typhoid bacillus is both saprophytic and parasitic. It finds abundant conditions in nature for its growth and development, and, enjoying strong resisting powers, can accommodate itself to environment much better than the majority of pathogenic bacteria, and can be found in water, air, soiled clothing, dust, sewage, milk, etc. contaminated directly or indirectly by the intestinal discharges of diseased persons.

The bacillus is also occasionally present upon green vegetables sprinkled with water containing it, and epidemics are reported in which the infection was traced to oysters from a certain place where the water was infected through sewage. Newsholme[1] found that in 56 cases of typhoid fever about one-third was attributable to the eating of raw shell-fish. In such cases the evidence accumulated serves to show that the shell-fish were from sewage-polluted beds. The bacillus probably enters milk occasionally in water used to dilute it.

The resistant powers of the organisms have already been described as great. They can grow well at the room-temperature. The thermal death-point is given by Sternberg as 60° C. The bacilli can, according to Klemperer and Levy, remain vital for three months in distilled water, though in ordinary water the commoner and more

[1] *Brit. Med. Jour.*, Jan., 1895.

vigorous saprophytes outgrow them and cause their disappearance in a few days. When buried in the upper layers of the soil the bacilli retain their vitality for nearly six months. Robertson [1] found that when planted in soil and occasionally fed by pouring bouillon upon the surface, the typhoid bacillus maintained its vitality for twelve months. He suggests that it may do the same in connection with leaky drains.

Cold has no effect upon typhoid bacilli, for freezing and thawing several times are without injury to them. They have been found to remain alive upon linen for from sixty to seventy-two days, and upon buckskin for from eighty to eighty-five days. Sternberg has succeeded in keeping hermetically sealed bouillon cultures alive for more than a year. In the experience of the author, unless transplanted rather frequently, cultures upon agaragar are apt to die out. In the presence of chemical agents the bacillus is also able to retain its vitality, 0.1 to 0.2 per cent. of carbolic acid added to the culture-media being without effect upon its growth. At one time the tolerance to carbolic acid was thought to be characteristic, but it is now known to be shared by other bacteria. The bacilli seem to be killed in a short time by thorough drying.

The bacillus is best secured in pure culture, either from an enlarged lymphatic gland or from the splenic pulp of a case of typhoid. To secure the bacillus in this way the autopsy should be made as soon after death as possible, lest the Bacillus coli invade the tissue.

Cultures of the typhoid bacillus may be obtained, but with difficulty, from the alvine discharges of typhoid patients. In examining this material, however, it must be remembered that the bacilli are certain to be present only in the second and third weeks.

As numerous saprophytic bacteria are present in the feces, the resistance which the typhoid bacillus exhibits to carbolic acid can be made use of in obtaining the pure

[1] *Brit. Med. Jour.*, Jan. 8, 1898.

culture. To each of several tubes of melted gelatin 0.05 per cent. of carbolic acid is added. This addition is most easily calculated by supposing the average amount of gelatin contained in a tube to be 10 c.cm. To the average tube $\frac{1}{10}$ c.cm. of a 5 per cent. solution of carbolic acid is added, and gives very nearly the desired quantity. A minute portion of the feces is broken up with a platinum loop and stirred in the tube of melted gelatin; a drop from this dilution is transferred to the second tube, a drop from it to a third, and then the contents of each tube are poured upon a sterile plate or into a Petri dish,

FIG. 107.—Bacillus typhi abdominalis: superficial colony two days old, as seen upon the surface of a gelatin plate; × 20 (Heim).

or rolled, according to Esmarch's plan, in the manner already described. The carbolic acid present in these cases prevents the great mass of saprophytes from developing, but allows the perfect development of the typhoid bacillus (Fig. 107) and its near congener, the Bacillus coli communis (Fig. 110).

The colonies that develop upon such gelatin plate-cultures are seen under the microscope to be brownish-yellow in color, spindle-shaped, and sharply circumscribed. When superficial they are larger and form a bluish iridescent layer with notched edges. These colonies are often described as resembling grape-vine leaves.

The center of the superficial colonies is the only portion which shows the yellowish-brown color. The margins of the colony appear somewhat reticulated. The gelatin is not liquefied.

Unfortunately, the appearances of the colonies of the Bacillus typhi and the Bacillus coli communis are identical, and make it next to impossible to select a single colony of either with any certainty. The only solution of the problem is to transfer a large number of colonies to some culture-medium in which a characteristic of one or the other species is manifested, and then study the growth; or to grow the colonies upon some special medium in which differences, such as rapidity of growth or acid-production, etc. cause the colonies of the different species to assume characteristic appearances.

A method recently suggested by Elsner[1] has materially aided the separation of these allied bacteria by using a culture-medium upon which the two bacilli develop differently.

The Elsner medium can be made by allowing 1 kgm. of grated potatoes (the small red German potato is best) to macerate in 1 liter of water over night. The juice is carefully pressed out, and filtered cold to get rid of as much starch as possible. The filtrate is now boiled and filtered again. The next step is a neutralization, in which Elsner used litmus as an indicator, and added 2.5–3 c.cm. of a $\frac{1}{10}$ normal solution of sodium hydrate to each 10 c.cm. of the juice. Abbott prefers to use phenolphthalein as an indicator. The final reaction should be slightly acid. Ten per cent. of gelatin (no peptone or sodium chlorid) is now dissolved in the solution, which is boiled for the purpose, and must then be again neutralized to the same point as before. After filtration, the medium receives the addition of 1 per cent. of potassium iodid. It is filled into tubes and sterilized.

When water or feces suspected of containing the typhoid bacillus are mixed in this medium and poured

[1] *Zeitschrift für Hygiene*, xxii., Heft 1, 1895; Dec. 6, 1896.

upon plates, no bacteria develop well except the colon bacillus and the typhoid bacillus.

These two bacteria, however, differ very markedly in their appearance upon the medium, for the colon bacillus appears as usual in twenty-four hours, while at that time, if present, the typhoid bacillus will have produced no colonies discoverable by the microscope.

It is only after forty-eight hours, long after the colon colonies have attained considerable size and are conspicuous, that the little colonies of the typhoid bacillus appear as small, round, shining, dew-like points, which are finely granular and in marked contrast to their coarsely granular predecessors. Unfortunately, many of the small colonies that develop in Elsner's medium subsequently prove to be those of the colon bacillus.

Kashida[1] prefers to make the differential diagnosis by observing the marked acid production of the Bacillus coli upon a medium consisting of bouillon containing 1½ per cent. of agar, 2 per cent. of milk-sugar, 1.0 per cent. of urea, and 30.0 per cent. of tincture of litmus. The culture-medium should be blue. When liquefied and inoculated with the colon bacillus, poured into Petri dishes, and stood for sixteen to eighteen hours in the incubator, the blue color passes off and the culture-medium becomes red. If a glass rod dipped in hydrochloric acid be held over the dish, vapor of ammonium chlorid is given off. The typhoid bacillus produces no acid in this medium, and there is consequently no change in its color.

For the differentiation of the typhoid bacillus from the allied bacillary forms, Hiss[2] recommends the use of two special media. The first consists of 5 grams of agar-agar, 80 grams of gelatin, 5 grams of Liebig's beef-extract, 5 grams of sodium chlorid, and 10 grams of glucose to the liter. The agar is dissolved in the 1000 c.cm. of water, to which have been added the beef-extract and sodium chlorid. When the agar is completely melted the gelatin is added

[1] *Centralbl. f. Bakt. u. Paristenk.*, Bd. xxi., Nos. 20 and 21, June 24, 1897.
[2] *Journal of Experimental Medicine*, Nov., 1897, vol. ii., No. 6.

and thoroughly dissolved by a few minutes' boiling. The medium is then titrated to determine its reaction, phenolphthalein being used as the indicator, and enough HCl or NaOH added to bring it to the desired reaction—*i. e.* a reaction indicating 1.5 per cent. of normal acid. To the clear medium add one or two eggs, well beaten in 25 c. cm. of water; boil for forty-five minutes, and filter through a thin filter of absorbent cotton. Add the glucose after cleaning.

The medium is used in tubes, in which it is planted by the ordinary puncture. The typhoid bacillus alone, of many of the allied forms studied, has the power of *clouding this medium* uniformly without showing streaks or gas-bubbles.

The second medium is used for *plating*. It contains 10 grams of agar, 25 grams of gelatin, 5 grams of beef-extract, 5 grams of sodium chlorid, and 10 grams of glucose. The method of preparation is the same as for the tube-medium, care always being taken to add the gelatin after the agar is thoroughly melted, so as not to alter this ingredient by prolonged exposure to high temperature. This preparation should never contain less than 2 per cent. of normal acid. Of all the organisms with which Hiss experimented, the Bacillus typhosus alone displayed the power of producing *thread-forming colonies* upon this medium.

The colonies of the typhoid bacillus when deep in the medium appear small, generally spherical, with a rough, irregular outline, and by transmitted light are of a vitreous greenish or yellowish-green color. The most characteristic feature consists of well-defined filamentous outgrowths, ranging from a single thread to a complete fringe about the colony. The young colonies are, at times, composed solely of threads. The fringing threads generally grow out nearly at right angles to the periphery of the colony.

The colonies of the colon bacillus are, on the average, larger than those of the typhoid bacillus; they are spherical or of a whetstone form, and by transmitted light are darker, more opaque, and less refractive than the typhoid colonies. By reflected light, to the unaided eye they are pale yellow. The surface-colonies are large, round, irregularly spreading, and are brown or yellowish-brown in color. Hiss claims that by the use of these reagents the typhoid bacillus can be readily detected in typhoid stools.

When transferred to gelatin puncture-cultures the bacilli develop along the entire track of the wire, with the formation of minute confluent spherical colonies. A small thin whitish layer develops upon the surface near the center. The gelatin is not liquefied, but sometimes is slightly clouded in the neighborhood of the growth. The growth upon the surface of obliquely solidified gelatin, agar-agar, or blood-serum is not very luxuriant. It forms a thin, moist, translucent, non-characteristic band with smooth edges.

Upon potato a growth formerly regarded as characteristic takes place. When the potato is inoculated and stood in the incubating-oven, no growth can be detected at the end of the second day, unless the observer be skilled and the examination thorough. If, however, the medium be touched with a platinum wire, it is discovered that its entire surface is covered with a rather thick, invisible layer of a sticky vegetation which the microscope shows to be made up of bacilli. No other bacillus gives the same kind of growth upon potato. Unfortunately, it is not constant, for occasionally there will be encountered a typhoid bacillus which will show a distinct yellowish or brownish color. The typical growth seems to take place only when the reaction of the potato is acid.

In bouillon the only change produced by the growth of the bacillus is a diffuse cloudiness.

In milk a slight and slow acidity is produced. The growth in milk is not accompanied by coagulation.

The chief hindrance to the ready isolation of the typhoid bacillus is the closely-allied Bacillus coli communis. This organism, being habitually present in the intestine, exists there in typhoid fever, and adds no little complication to the bacteriological diagnosis by responding in exactly the same manner as the typhoid bacillus to the action of carbolic acid, by having colonies almost exactly like those of typhoid, by growing in exactly the same manner upon gelatin, agar-agar, and blood-serum, by clouding bouillon in the same way, by being of almost exactly the same shape and size, by having flagella, by being motile, and, in fact, by so many pronounced similarities as almost to warrant the assertion of some that it and the typhoid bacillus are identical.

Not the least significant fact about the colon bacillus is that it is also pathogenic and capable of exciting acute inflammatory processes which are not infrequent, and which sometimes serve to increase the seriousness of typhoid fever.

At the present time we are in more or less of a quandary about this extraordinary resemblance, but base our differentiation of the species upon certain constant, slight, but distinct differences.

The typhoid bacillus does not produce indol.

The open lymphatics and vessels of the intestinal ulcers of typhoid favor the absorption of the bacteria in the digestive tract, and the colon bacillus enters the blood no longer to be a saprophyte, but now to be a virulent pus-producer, and in many cases of typhoid we find suppurations and other milder inflammations due to this microbe. This is also a stumbling-block, for the typhoid bacillus when distributed through the blood may act in exactly the same manner.

The typhoid bacillus may enter the body, at times, through dust (Klemperer and Levy), but no doubt, in the great majority of cases, enters the digestive tract at once through the mouth. It may possibly enter through the rectum at times, as illustrated by the mention which Eichhorst makes of the infection of soldiers in military barracks through the wearing of drawers previously worn by comrades who had suffered from typhoid.

When ingested the resisting power of the bacillus permits it to pass uninjured through the acid secretions of the stomach and to enter the intestine, where the chief local disturbances are set up.

The bacilli enter the solitary glands and Peyer's patches, and multiply slowly during the one to three weeks of the incubation of the disease. The immediate result of their residence in these lymphatic structures is increase in the number of cells, and ultimately the necrosis and slough-

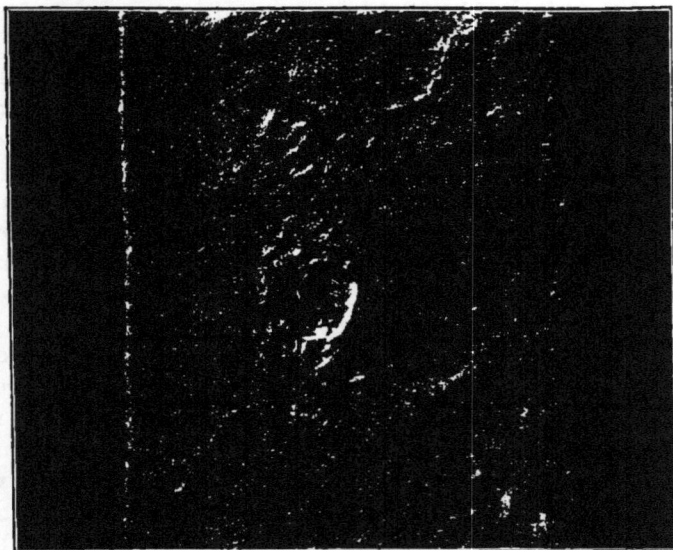

FIG. 108.—Intestinal perforation in typhoid fever. Observe the threads of tissue obstructing the opening. (Museum of the Pennsylvania Hospital.) (Keen, *Surgical Complications and Sequels of Typhoid Fever.*)

ing which cause the typical post-mortem lesion (Fig. 108). From the intestinal lymphatics the bacilli pass, in all probability, to the mesenteric glands, which become enlarged and softened, and finally extend to the spleen and liver, and sometimes to the kidneys. The growth of the bacilli in the kidneys causes the albuminuria of the disease. Sometimes under these conditions the bacilli may be found in the urine. P. Horton Smith[1] found the bacilli in the urine in three out of seven cases which he investigated. They did not occur before the third week, and remained in one case twenty-two days after cessation of the fever. Sometimes they were present in immense numbers. Their occurrence, no doubt, depends upon their growth in the kidney and descent with the urine. It is of importance from a sanitary point of view to remember that the urine as well as the feces is infectious. Occasionally the bacilli succeed in entering the general circulation, and, finding a lodgement at some remote part of the body, set up local inflammatory processes sometimes terminating in suppuration.

Weichselbaum has seen general peritonitis from rupture of the spleen in typhoid fever with escape of the bacilli. Ostitis, periostitis, and osteomyelitis are very common results of the lodgement of the bacilli in bony tissue, and Ohlmacher has found the bacilli in suppurations of the membranes of the brain. The bacilli are also encountered in other local suppurations occurring in or following typhoid fever. Flexner and Harris[2] have seen a case in which the distribution of the bacilli was sufficiently widespread to constitute a real septicemia, the bacillus being isolated from various organs of the body, and shown to be the true bacillus of Eberth by all the specific laboratory tests, but in which there were no intestinal lesions.

The bacilli can be found in the intestinal lesions, in the mesenteric glands, in the spleen, in the liver, in the kidneys, and in any local lesions which may be present. Their scattered distribution and their occurrence in minute clumps have already been alluded to. They should always be sought for at first with a low power of the microscope.

Ordinarily no bacilli can be found in the blood, but it has been shown that the blood in the roseolæ some-

---

[1] *Brit. Med. Jour.*, Feb. 13, 1897.
[2] *Bull. of the Johns Hopkins Hospital*, Dec., 1897.

times contains them, so that the eruption may be regarded as one of the local irritative manifestations of the bacillus.

The amount of local disturbance, in proportion to the constitutional disturbance, is, in the majority of cases, slight, and almost always partakes of a necrotic character, which suggests that in typhoid we have to do with a toxic bacterium whose disease-producing capacity resides in the elaboration of a toxic substance. This, indeed, is true, for Brieger and Fränkel have separated from bouillon cultures a toxalbumin which they thought to be the specific poison. Klemperer and Levy also point out further clinical proof in certain exceptional cases dying with the typical picture of typhoid, yet without characteristic post-mortem lesions, the only confirmation of the diagnosis being the discovery of the bacilli in the spleen.

Pfeiffer and Kolle found that the toxic substance resided only in the bodies of the bacilli, and could not, like the toxins of diphtheria and tetanus, be dissolved in the culture-medium. This was an obstacle to their immunization-experiments as well as those of Löffler and Abel, later to be described, for the only method of immunizing animals to large quantities of the bacilli was to make massive agar-agar cultures, scrape the bacilli from the surface, and distribute them through nutrient bouillon.

When injected into guinea-pigs the typhotoxin of Brieger is productive of increased secretion of saliva, increased rapidity of respiration, diarrhea, and mydriasis, and usually causes a fatal termination in from twenty-four to forty-eight hours.

As the discovery of the bacilli in the spleen, and especially the securing of a pure culture of the bacilli from the spleen, are sometimes attended with considerable difficulty because of the dissemination of the colonies throughout the organ, E. Fränkel recommends that as soon as the organ is removed from the body it be wrapped in cloths wet with a solution of bichlorid of mercury and kept for three days in a warm room, in order that a considerable and massive development of the bacilli may take place.

Typhoid fever is a disease which is communicable to animals with difficulty. They are not affected by bacilli in fecal matter or in pure culture mixed with the food, and are not diseased by the injection into them of blood from typhoid patients. Gaffky failed completely to produce any symptoms suggestive of typhoid fever in rabbits, guinea-pigs, white rats, mice, pigeons, chickens, and calves, and found that Java apes could feed daily upon food polluted with typhoid germs for a considerable time, yet without symptoms. The introduction of pure cultures into the abdominal cavity of most animals is without effect. Fränkel and Simon found that when pure cultures were injected into mice, rabbits, and guinea-pigs the animals died.

Germano and Maurea found that mice succumbed in from one to three days after intraperitoneal injection of 1–2 c. cm. of a twenty-four-hour-old bouillon culture. Subcutaneous injections in rabbits and dogs caused abscesses.

Lösener found the introduction of 3 mgr. of an agar-agar culture into the abdominal cavity of guinea-pigs to be fatal.

When animals are treated in the manner described in the chapter upon Cholera—*i. e.* the gastric contents rendered alkaline, a large quantity of laudanum injected into the peritoneal cavity, and the bacilli introduced through an esophageal catheter—Klemperer, Levy, and others found that there was produced an intestinal condition which very much resembled typhoid as it occurs in man. The virulence of the bacillus can be very greatly increased by rapid passage from guinea-pig to guinea-pig.

In the experiments of Chantemesse and Widal the symptoms following the injection of virulent culture into guinea-pigs were briefly as follows: "Very shortly after the inoculation there is a rise of temperature, which continues from one to four hours, and is succeeded by a depression of the temperature, which continues to the fatal issue. Meteorism and great tenderness of the abdomen are observed. At the autopsy a sero-fibrinous or sero-purulent peritonitis is observed—sometimes hemorrhagic. There is also generally a pleurisy, either serous or hemorrhagic. All the abdominal viscera are congested. The intestine is congested—contains an abundant mucous secretion. The Peyer patches are enlarged. The spleen is enlarged, blackish, and often hemorrhagic. In cases which are prolonged the liver is discolored. The kidneys are congested, the adrenals filled with blood.

"In such cases the bacillus can be found upon the inflamed serous membranes, in the inflammatory exudates, in the spleen in large numbers, in the adrenals, the liver, the kidneys, and sometimes in the lungs. The blood is also infected, but to a rather less degree.

"In cases described as chronic, the bacillus disappears completely in from five to twenty-four hours, and produces but one lesion, a small abscess at the point of inoculation.

"Sanarelli has observed that if some of the poisonous products of the colon bacillus or the Proteus vulgaris be injected into the abdominal cavity of an animal recovering from a chronic case, it speedily succumbs to typical typhoid fever."

Petruschky[1] found that mice that recovered from subcutaneous injections of typhoid cultures frequently suffered from a more or less widespread necrosis of the skin at the point of injection.

I experienced great difficulty in immunizing a horse to the disease, because every injection of virulent living organisms was followed by a necrosis equalling in size the distended area of subcutaneous tissue.

Large quantities of filtered cultures produce symptoms similar to those resulting from inoculation with the bacilli. The toxic product of the bacilli is, however, practically insoluble, and, according to the experiments of Löffler and Abel and those of Pfeiffer and

[1] *Zeitschrift für Hygiene*, Bd. xii., 1892, p. 261.

Kolle, cannot be separated from the bodies of the bacilli producing it.

Animals can easily be immunized to this bacillus, and then, according to Chantemesse and Widal, develop in their blood an antitoxic substance capable of protecting other animals. Stern [1] has also found that in the blood of human convalescents a substance exists which has a protective effect upon guinea-pigs. His observation is in accordance with a previous one by Chantemesse and Widal, and has recently been abundantly confirmed.

The immunization of dogs and goats by the introduction of increasing doses of virulent cultures has been achieved by Pfeiffer and Kolle [2] and by Löffler and Abel. [3] From these animals serums were secured not exactly antitoxic, but anti-infectious or anti-microbic in operation, and possessed of marked specific germicidal action upon the typhoid bacilli when simultaneously introduced into the peritoneal cavity of guinea-pigs.

The action of the typhoid serum is specific, and exerts exactly the same action upon the typhoid bacilli as the cholera serum exerts upon the cholera spirilla, killing and dissolving them (Pfeiffer's phenomenon).

So far, no serum has been produced that is efficacious in human medicine.

The specific reaction of the serum can be used to differentiate cultures of the colon and typhoid bacilli, the typhoid bacilli alone exhibiting the specific effect of the typhoid serum.

Christophers [4] found that the serum from typhoid patients occasionally caused agglutinations in cultures of the colon bacillus, but concludes that this does not lessen the specificity of the reaction, as there may be two combined specific actions of these serums. Experiments on rabbits established that typhoid and colon serums could be produced, each specific in its agglutin-

---

[1] *Zeitschrift für Hygiene*, xvi., 1894, p. 458.   [2] *Ibid.*, 1896.

[3] *Centralbl. f. Bakt. u. Parasitenk.*, Bd. xix., No. 23, p. 51, Jan. 23, 1896.

[4] *Brit. Med. Jour.*, Jan. 8, 1898.

ating power upon bouillon cultures of its respective organism.

Löffler and Abel also prepared a colon serum which exerted a like specific action upon the colon bacillus, but was without effect upon the typhoid bacillus.

The serum of immunized animals has been found to destroy the motility of the typhoid bacilli in a few moments, and to cause them to group together. Widal found that the serum of convalescents and of individuals suffering from the acute disease possessed the same power, and suggested that this specific action might prove a valuable adjunct in diagnosis.

Wyatt Johnston[1] and McTaggert worked upon the subject, and found that a drop of blood from a typhoid patient, dried upon paper and kept for some time, when moistened and brought in contact with a culture of the bacilli was still potent to bring about a characteristic effect. When such a preparation in the "hanging drop" is watched under the microscope the typhoid bacilli are found to be paralyzed in from one minute to half an hour, and subsequently to collect in masses—agglutinations. This reaction may occasionally be brought about by normal blood if insufficiently diluted, but is characteristic enough to be very useful in the diagnosis of obscure cases. In a later paper Johnston states that to obtain a satisfactory reaction an attenuated typhoid bacillus is more useful than a highly virulent one.

My own experiments have satisfied me of the value of the test, both for making a diagnosis of the disease and for confirming the species of the bacillus in doubtful cases.

It is now the opinion of all observers that cessation of motion and agglutination of the bacteria, resulting from the contact of typhoid bacilli and typhoid serum, are inconclusive for diagnostic purposes unless the reaction follows the combination of a *suitable culture* and a *definite quantity of serum.*

[1] *Montreal Med. Journal*, March, 1897.

The thorough investigations of Wyatt Johnston and his associates in Montreal have shown that reliable reactions can only be secured when the cultures employed are of an ordinarily virulent typhoid bacillus, and are grown in an alkaline medium for about twenty-four hours.

I prefer fresh agar-agar cultures, distributed throughout sterile clean water, rather than bouillon cultures, because of the larger number of bacteria in the former, the consequently greater number of agglutinations formed, and the readiness with which they are found upon microscopic examination. It is necessary, however, to make a microscopic examination of the diluted culture before adding the serum or blood, in order to be sure that there are no natural clumps of bacteria present to simulate the specific agglutinations. This is of great importance. The natural clumps of bacilli are more apt to occur in cultures grown upon fresh, moist agar-agar than upon that kept for a short time until the surface has become partially dried. The chief difficulty experienced in making the test seems, at present, to reside in the preparation of the blood in accurate dilution—*i. e.* securing it in measured amounts.

The original method of Widal, to collect about 5 c.cm. of blood in a test-tube by the introduction of a hypodermic needle into a vein, is a rather more serious and disturbing operation than most patients care to undergo for purposes of diagnosis.

Blood dried upon paper, as suggested by Johnston, or upon glass, while extremely convenient for transportation, is not susceptible of accurate dilution for quantitative estimation.

Cabot has successfully made dilutions with a medicinedropper, by using one drop of blood and as many drops of culture, dropped from the same instrument, as were necessary for the desired dilution. This method seems to be very practical, but can only be employed at the bedside, or where it is not necessary to keep or transport the blood.

In the absence of a satisfactory method of securing definite small quantities of blood for immediate or subsequent use, I was led to make some experiments with capillary tubes to determine their possible value for the purpose.

It is a well-known physical phenomenon that in clean capillary tubes fluids are attracted to a height varying according to the diameter of the tube and the density of the fluid. In tubes of equal diameter the height of the column is invariably the same.

Such tubes can be made by heating a piece of ordinary glass tubing, such as is to be found in every laboratory, in a Bunsen flame for a few minutes until it becomes red and soft, removing the glass from the flame, and then pulling upon the ends steadily and slowly until the tube is drawn out to the desired diameter. The errors to be avoided in making the tubes will be—heating too much and making the glass too soft, drawing out the tube while still in the flame, and drawing too rapidly. The result of these erroneous methods will be that the tubes are much shorter and finer than is desired. A few moments' practice will show just how the manipulation should be done to secure the best results.

The fact was, however, established that tubes of about the same diameter showed almost no variation in the quantity of liquid contained. So little was the difference in the length of the column and the weight of the contained blood in tubes recognized by the eye to have uniform caliber that I have no hesitation in recommending an application of the capillary tube for securing small measured quantities of blood for the specific typhoid tests and similar experiments.

The application of the method is simple and consists in:

1. Accurately weighing the amount of blood that enters a capillary tube of a size arbitrarily selected as a standard.

2. The manufacture of a large number of tubes of the same size.

3. The dilution of the known quantity of blood contained in the tube with a measured quantity of the bouillon, or diluted agar-agar, culture of the bacillus.

The standard tube that I adopted had a diameter about equal to the E string of a violin. A larger or smaller tube would have done quite as well. In such a tube the column of blood rises about an inch and weighs about 0.018 gram. As personal equation in judging size is a marked source of error, the experimenter must work out his own standard tube and not adopt that which has just been given. It is important to know the length of the column that has a certain weight, because, as each tube is not separately measured and graduated, the two chief means of avoiding error will be (1) to have the tubes as nearly as possible of equal diameter, and (2) to prove them to be so by observing that the columns of fluid they contain when used are of the same length, rejecting one after another all the tubes which seem to the eye to have the proper caliber, but in which the column is obviously longer or shorter than that of the original tube.

Keeping the standard tube before him as a guide, and using a Bunsen flame—which is better than a blowpipe, because it does not heat the glass so rapidly and make it so soft—the experimenter prepares one hundred or more capillary tubes as nearly as possible of the same size as that of the standardized tube. All the irregular sizes are rejected, and the suitable sizes cut into portions about three inches long. These pieces, which should number several hundred (it is economy to make a large number at a time), are now carefully sorted, being compared with the standard tube at both ends, and thrown away if too large or too small at either. It is best to sort the tubes twice on different days, or have several different persons go over them all. Of course, some tubes of quite different caliber will, in spite of all precautions, remain in the bundle, but this is no serious matter, because at the last moment the height of the

25

column to which the blood rises can be taken as a proof of actual variation. It may be true that no two of the tubes have exactly—absolutely—the same contents, but when the given precautions are taken the variation will be so small as to make no significant error in the results obtained.

The use of the tubes is extremely simple. The ordinary puncture is made in the lobule of the ear or the finger-tip of the patient, and one end of one of the tubes touched to the surface of the oozing drop and held there until the blood ceases to rise in the tube. So little blood is required that a number of tubes may be filled with the blood from a single puncture if desired. The blood in the tube coagulates in a few minutes, and can be allowed to dry, or be drawn to the central portion of the tube and sealed in by fusing the ends in a flame if it be desired to keep it moist.

When the agglutination reaction is to be made the blood should not be blown out of the tube, as the total quantity contained is small and a large relative quantity will remain in the tube. A better method is to crush the tube in a small crucible or other diminutive vessel and dissolve its contents directly in the culture.

The proper proportionate amount of culture is measured with a finely graduated pipette (graduated to thousandths of a cubic centimeter), the calculation according to the standard tube of the writer's experiments being: dilution 1 : 10 = 0.153 c.cm. of the culture; dilution 1 : 100 = 1.53 c.cm. of the culture; dilution 1 : 1000 = 15.3 c.cm. of the culture.

The now recognized specific reaction is supposed to take place in dilutions of 1 : 50, which would require 0.71 + c.cm. of the bouillon or diluted agar culture.

The culture is measured into the little crucible, the blood-containing portion of the capillary tube broken off, dropped in, and subsequently crushed to minute fragments and stirred about with a clean, rounded, glass rod, and a drop of the mixture placed as a "hanging drop" upon the stage of a microscope and examined for the agglutinations.

As recent extended observations have shown that occasionally the blood of healthy men and animals has the power of producing the agglutinations, the consensus of opinion now seems to be in favor of the view that a certain dilution of the blood is required for a satisfactory diagnosis, and that all reactions with concentrations greater than one part of blood in fifty of culture may be questionable, while less concentrated dilutions are almost positively diagnostic. A time-limit must be placed upon the experiment. For the weak dilution not more than two hours should be required for a perfect reaction, and for the stronger solution correspondingly less time should be required.

A curious fact that should not be overlooked is that the agglutinating substance is not constantly present in the blood, but sometimes alternates, being present for several days and then absent for a day or two.

The agglutinating power of the blood occurs early in the course of typhoid, and in typical cases seems to be present in the first week of actual illness.

A point that should not be forgotten is that the agglutination of the bacilli seems to be a phenomenon quite independent of any immunity possessed by the individual, and therefore is not an "immunity-reaction." Just what the agglutinating substance is, has not yet been determined.

The agglutinations are occasionally caused by the serum and dried blood from other diseases than typhoid, but in a collection of 4000 cases it was shown that the errors from this source were only about 5 per cent.

Malvoz [1] has experimented with a number of chemicals, and has found that formaldehyd, corrosive sublimate, peroxid of hydrogen, strong alcohol, and anilin colors (such as chrysoidin, vesuvin, and safranin) have the power to produce the typical agglutinations even in very dilute solutions.

[1] *Ann. de l'Inst. Pasteur*, xl., 7, 1897.

Wright and Semple assert that dead cultures of the typhoid bacillus may be used for the test, as bacilli killed by a temperature of 60° C. agglutinate perfectly. They have the advantage of being easily kept.

Rumpf,[1] and Kraus and Buswell[2] report a number of cases of typhoid which were favorably influenced by the introduction hypodermically of small quantities of sterilized cultures of Bacillus pyocyaneus. These experiments are still too new to deserve extended mention.

Following the lines of experimentation suggested by Haffkine's researches upon preventive vaccination against cholera Asiatica, Pfeiffer and Kolle, and Wright and Semple have used the subcutaneous injection of sterilized cultures as a prophylactic measure. One c. cm. of a bouillon culture sterilized. by heat is thought to be sufficient. Wright and Semple report 18 cases in which it was used, and by experiment showed the blood to be changed similarly to that of typhoid patients and convalescents. This change consisted in the destruction of motility and agglutination of the bacilli, as seen in Widal's reaction. It is hoped that we can gauge the duration of the immunity thus acquired by the frequent use of Widal's test.

One of the most important and practical points for the physician to grasp in relation to the subject of typhoid fever is the highly virulent character of the discharges from the bowels. In every case the greatest care should be taken for a proper disinfection of the feces, a rigid attention to all the details of cleanliness in the sick-room, and the careful sterilization of all articles which are soiled by the patient. If country practitioners were as careful in this particular as they should be, the disease would be much less frequent in regions remote from the filth and squalor of large cities with their unmanageable slums, and the distribution of the bacilli to villages and towns, by watercourses polluted in their infancy, might be checked.

[1] *Deutsche med. Wochenschrift,* 1893, No. 41.
[2] *Wien. klin. Wochenschrift,* July 12, 1894.

CHAPTER III.

BACILLUS COLI COMMUNIS.

THE Bacillus coli was first isolated from human feces in 1885 by Emmerich, who thought that it was the specific cause of Asiatic cholera. Many investigators have since studied its peculiarities, until at the present time it is one of the best-known bacteria.

It is habitually present in the fecal matter of most animals except the horse, and in water and soil contaminated

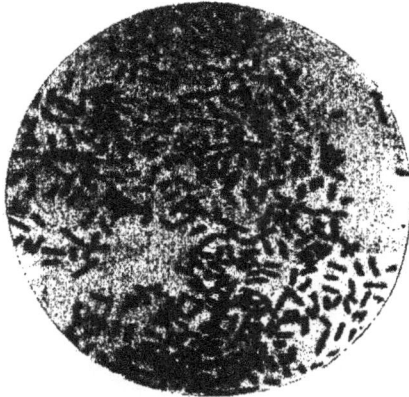

FIG. 109.—Bacillus coli communis, from an agar-agar culture; × 1000 (Itzerott and Niemann).

with it. With water or dust it gains entrance into the mouth, where it can frequently be found, and occurs accidentally in foods and drinks. During life the organism sometimes enters wounds externally from the surface of the body or internally from the intestine, and is a cause of suppuration—or at least occurs in the pus. The Bacillus pyogenes fœtidus of Passet is almost certainly identical with it.

The bacillus is rather variable culturally, and is somewhat polymorphic. Probably both size and form depend to a certain extent upon the culture-medium on which it grows. On the average, it measures 1–3 × 0.4–0.7μ. It usually occurs in the form of short rods, but very short coccus-like elements and quite elongate forms are often found in the same culture. The individual bacilli are frequently isolated or in pairs. Chains are the exception. They are provided with flagella, which are very variable in number, generally from four to a dozen, though there may be more. It forms no spores.

The bacillus stains well with the ordinary aqueous solutions of the anilin dyes, but does not retain the stain after immersion in Gram's solution.

The bacillus is motile, though in this particular it is subject to irregularity, the organisms from some cultures

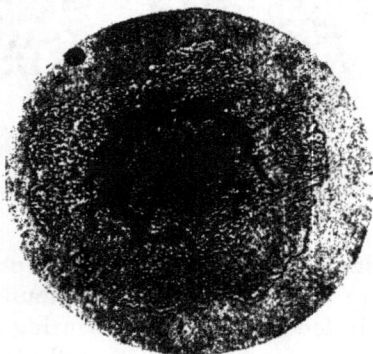

FIG. 110.—Bacillus coli communis: superficial colony two days old upon a gelatin plate; × 21 (Heim).

always swimming actively, even when the culture is some days old, others being exceedingly sluggish even when young and actively growing, and a few cultures seem to consist of bacilli that do not move at all. Fresh cultures which, when grown at incubation temperature, consist of entirely non-motile bacteria are probably *Bacillus coli immobilis*, not Bacillus coli communis.

The bacillus is readily cultivated upon the ordinary

media. Upon gelatin plates the colonies develop in twenty-four hours. Those situated below the surface appear round, yellow-brown, and homogeneous. As they grow older they increase in size and become opaque. The superficial colonies are larger and spread out upon the surface. Their edges are dentate and resemble grape-leaves, often showing radiating ridges suggestive of the veins of a leaf. They may have a slightly concentric appearance. The colonies rapidly increase in size and become more and more opaque. The gelatin is not liquefied.

In gelatin punctures the culture, developing rapidly upon the surface, and also in the needle's track, causes the formation of a nail-like growth. The head of the nail may reach the walls of the test-tube. Not infrequently gas is formed in ordinary gelatin, and when 1 per cent. of glucose is dissolved in the medium the gas-production is often so copious and rapid as to form large bubbles, which by their distention subsequently break it up into irregular pieces. Sometimes the gelatin becomes slightly clouded as the bacilli grow.

Upon agar-agar along the line of the inoculation a grayish-white, translucent, smeary growth takes place. It is devoid of any characteristics. The entire surface of the culture-medium is never covered, the growth remaining confined to the inoculation-line, except where the moisture of the condensation-fluid allows it to spread out at the bottom. Kruse says that in old cultures crystals may form. I have never seen them.

Bouillon is soon evenly clouded by the development of the bacteria. Sometimes a delicate pellicle forms upon the surface. There is rarely much sediment in the culture.

Würtz found that the bacillus produced ammonia in culture-media free from sugar, and thus caused an intense alkaline reaction in the culture-media. The cultures usually give off an odor that varies somewhat, but is, as a rule, unpleasant.

Indol is formed in both bouillon and pepton solutions. Phenol is not produced. Litmus added to the culture-media is ultimately decolorized by the bacilli.

The presence of indol is probably best determined by Salkowski's method. To the culture 1 c.cm. of a 0.02 per cent. aqueous solution of potassium nitrate and a few drops of concentrated sulphuric acid are added. If a rose color develops, indol is present.

Nitrates are reduced to nitrites by the growth of the bacillus.

Upon potato the growth is luxuriant. The bacillus forms a yellowish-brown, glistening layer spreading from the line of inoculation over about one-half to two-thirds of the potato. The color shown by the potato-cultures varies considerably, sometimes being very pale, sometimes quite brown. It cannot, therefore, be taken as a characteristic of much importance. Sometimes the potato becomes greenish in color. Sometimes the growth on potato is almost invisible.

In milk there are rapid coagulation and acidulation, with the evolution of much gas.

The bacillus seems to require very little nutriment. It grows in Uschinsky's asparagin solution, and is frequently found living in river and well waters.

It is quite resistant to antiseptics and germicides, and grows in culture-media containing from 0.1–0.2 per cent. of carbolic acid. It lives for months upon artificial media.

The bacillus begins to penetrate the intestinal tissues almost immediately after death, and is the most frequent contaminating micro-organism met with in cultures made at autopsy. Exactly how it penetrates the tissues is not known. It may spread by direct continuity of tissue, or *via* the blood-vessels.

While under normal conditions a saprophytic bacterium, the colon bacillus is far from harmless. It not infrequently is found in the pus of abscesses remote from the intestine, and is almost always found in suppurations connected with the intestines, as, for example, appendicitis.

It is a question whether the colon bacillus is always virulent, or whether it becomes virulent under abnormal conditions. Klencki[1] found that it was very virulent in the ileum, and less so in the colon and jejunum, especially in dogs. He also found that the virulence was greatly increased in a strangulated portion of intestine. Other observers, as Dreyfuss, found that the colon bacillus as it occurs in normal feces is non-pathogenic. Most experimenters, however, believe that pathological conditions, such as disease of the intestine, ligation of the intestine, etc., cause increased virulence.

Adelaide Ward Peckham, in an elaborate study of the "Influence of Environment on the Colon Bacillus,"[2] concludes that while the conditions of nutrition and development in the intestine seem to be most favorable, the colon bacillus is ordinarily not virulent, because "its first force is spent upon the process of fermentation, and as long as opportunities exist for the exercise of this function the affinities of this organism appear to be strongest in this direction.

"Moreover, the contents of the intestine remain acid until they reach the neighborhood of the colon, and by that time the tryptic peptons have been formed and absorbed to a great extent.

"During the process of inflammation in the digestive tract a very different condition may exist. The peptic and tryptic enzymes may be partially suppressed. Fermentation of carbohydrates and proteid foods then begins in the stomach, and continues after the mass of food is passed on into the intestine. The colon bacillus cannot, therefore, spend its force upon fermentation of sugars, because they are already broken up and an alkaline fermentation of the proteids is in progress. It also cannot form peptons from the original proteids, for it does not

[1] *Ann. de l'Inst. Pasteur*, 1895, No. 9.
[2] *Journal of Experimental Medicine*, Sept., 1897, vol. ii., No. 4, p. 549.

possess this property, and unless trypsin is present it must be dependent upon the proteolytic activity of other bacteria for a suitable form of proteid food. Perhaps these bacteria form an albuminate molecule, which like leucin and tyrosin cannot be broken up into indol, and thus there might be caused an important modification of the metabolism of the colon bacillus, which might have either an immediate or remote influence upon its acquisition of disease-producing properties, for our own experiments indicate that the power to form indol, and the actual forming of it, are to some extent an indication of the possession of pathogenesis.''

To the laboratory animals the colon bacillus is pathogenic in varying degree. Intraperitoneal injections into mice cause their death in from one to eight days if the culture is virulent. Guinea-pigs and rabbits also succumb to intraperitoneal and intravenous injection. Subcutaneous injections are of less effect, and in rabbits seem to produce abscesses only.

When the bacilli are injected into the abdominal cavity a sero-fibrinous or purulent peritonitis occurs, the bacilli being very numerous in the abdominal fluids.

The pathogeny of the colon bacillus is due to irritating, chemotactic substances in its protoplasm. The experiments of Pfeiffer and Kolle and Löffler and Abel have proved very conclusively that the poisonous principle is in, and cannot by any means be separated from the bodies of the bacteria.

Frequent transplantation lessens the virulence, passage through animals increases it.

Numerous observers have found that cultures of the bacillus obtained from cholera, cholera nostras, and other intestinal diseases are much more pathogenic than those obtained from normal feces or from pus.

Cumston,[1] from a careful study of thirteen cases of summer infantile diarrheas, comes to the following conclusions:

[1] *International Medical Magazine*, Feb., 1897.

The bacterium coli seems to be the pathogenic agent of the greater number of summer infantile diarrheas.

This organism is the more often associated with the streptococcus pyogenes.

The virulence, more considerable than in the intestine of a healthy child, is almost always in direct relation to the condition of the child at the time the culture is taken, and does not appear to be proportional to the ulterior gravity of the case.

The mobility of the Bacterium coli is in general proportional to its virulence. The jumping movement, nevertheless, does not correspond to an exalted virulence in comparison with the cases in which the mobility was very considerable, without presenting these jumping movements.

The virulence of the Bacterium coli found in the blood and other organs is identical with that of the Bacterium coli taken from the intestine of the same individual.

Lesage,[1] in studying the enteritis of infants, found that in 40 out of 50 cases depending upon the Bacillus coli the blood of the patient agglutinated the cultures obtained, not only from his own stools, but from those of all the other cases. From this uniformity of action Lesage very properly suggests that the colon bacilli in these cases are all of the same species.

The agglutinating reaction occurs only in the early stages and acute forms of the disease.

It is not difficult to immunize an animal against the colon bacillus. Löffler and Abel immunized dogs by progressively increased subcutaneous dosage of live bacteria, grown in solid culture and distributed through water. The injections at first produced hard swellings. The blood of the immunized animals possessed an active bactericidal influence upon the colon bacteria. It was not in the correct sense antitoxic.

In intestinal diseases, such as typhoid, cholera, and

[1] *Semaine Médicale*, Oct. 20, 1897.

dysentery, the bacillus not only seems to acquire an unusual degree of virulence, but because of the existing
denudation of mucous surfaces, etc., finds it easy to enter
the general system, with the result of secondary remote
suppurative lesions in which it is the essential factor.
When absorbed from the intestine it frequently enters
the kidney and is excreted with the urine, causing, inci-
dentally, local inflammatory areas in the kidney, and
occasionally cystitis.   A case of urethritis is reported to
have been caused by it.

In infants cholera infantum may not infrequently be
caused by the colon bacillus, though probably in this
disease other bacteria play a very important *role.*

The bile-ducts are very often invaded by the bacillus,
which may cause inflammation, obstruction, suppuration,
or calculous formation.

The bacillus has also been met in puerperal fever,
Winckel's disease of the new-born, endocarditis, meningitis, liver-abscess, bronchopneumonia, pleuritis, chronic
tonsillitis, and urethritis.

For the determination of the colon bacillus the important points are the motility, the indol reaction, the
milk-coagulation, and the active gas-production.   As,
however, all of these features are shared by other bacteria to a greater or less degree, the only positive differential point upon which very great reliance can be placed
is the immunity-reaction of the serum of an immunized
animal, which not only protects susceptible animals from
the effects of inoculation, but produces with fresh cultures of the bacillus exactly the same reaction as that
observed in connection with the blood and serum of
typhoid patients, and convalescents and immunized animals.   This reaction has been considered at length in
speaking of typhoid fever.

For the few who are convinced that the colon and
typhoid bacilli are identical, the fact that the typhoid
serum is specific for the typhoid bacillus, and the colon
serum for the colon bacillus, with rare exceptions,

should be important evidence of their separate individuality.

The author has no doubt that the Bacillus coli communis is not a single species of bacteria, but is a name applied to a group whose individual differences are thus far too similar to enable us to differentiate them. This opinion seems to be shared by other bacteriologists, some of whom have attempted to separate the bacillus into groups, types, or families.

In order to establish a *type species* of the Bacillus coli communis, Smith[1] says:

"I would suggest that those forms be regarded as true to this species which grow on gelatin in the form of delicate, bluish, or more opaque, whitish expansions with irregular margin, which are actively motile when examined in the hanging drop from young surface-colonies taken from gelatin plates which coagulate milk within a few days; grow upon potato, either as a rich-pale or brownish-yellow deposit, or merely as a glistening, barely recognizable layer, and which give a distinct indol reaction. Their behavior in the fermentation-tube must conform to the following scheme:

"*Variety α:*

"One per cent. dextrose-bouillon (at 37° C.). Total gas approximately $\frac{1}{2}$; HCO_2 approximately $\frac{3}{4}$; reaction strongly acid.

"One per cent. lactose-bouillon: as in dextrose-bouillon (with slight variations).

"One per cent. saccharose-bouillon; gas-production slower than the preceding, lasting from seven to fourteen days. Total gas about $\frac{2}{3}$; HCO_2 nearly $\frac{3}{4}$. The final reaction in the bulb may be slightly acid or alkaline, according to the rate of gas-production.

"*Variety β:*

"The same in all respects, excepting as to its behavior in saccharose-bouillon; neither gas nor acids are formed in it."

[1] *American Journal of the Medical Sciences,* 1895, 110, p. 287.

Characteristics for Differentiation.

TYPHOID BACILLUS.	COLON BACILLUS.
Bacilli usually slender.	Bacilli inclined to be a little thicker.
Flagella numerous (10–20), long, and wavy.	Flagella fewer (8–10).
Growth not very rapid, not particularly luxuriant.	Growth rapid and luxuriant. This character is by no means constant.
Upon Elsner's culture-medium develops slowly, the colonies remaining small.	Upon Elsner's medium develops more rapidly, the colonies being larger. (Sometimes the colonies are small and remain so.)
Upon fresh acid potato the so-called " invisible growth" formerly thought to be differential.	Upon potato a brownish-yellow, distinct pellicle.
Acid-production in whey not exceeding 3 per cent. Sometimes slight in ordinary media, and sometimes succeeded by alkaline production.	Acid-production well marked.
Grows in media containing sugars without producing any gases.	Gas-production well marked.
Produces no indol.	Indol-production marked.
Growth in milk unaccompanied by coagulation.	Milk coagulated.
In Maassen's asparagin-glycerin solution the bacillus does not grow.	Grows in Maassen's solution.
Gives the Widal reaction with the serum of typhoid blood.	Does not react with typhoid blood.

CHAPTER IV.

YELLOW FEVER.

THE bacteriology of yellow fever has been studied by Domingos Freire, Carmona y Valle, Sternberg, Havelburg, and most recently by Sanarelli.

Sternberg, whose work is extensive and important, says: "Facts relating to the endemic and epidemic prevalence of yellow fever, considered in connection with the present state of knowledge concerning the etiology of other infectious diseases, justify the belief that yellow fever is due to a living organism capable of development under favorable local and meteorological conditions external to the human body, and of establishing new centers of infection when transported to distant localities."

Sternberg, at the Tenth International Medical Congress (Berlin, 1890), reported the study of 42 yellow fever autopsies in which aërobic and anaërobic cultures were made from the blood, liver, kidney, urine, stomach, and intestines, but the specific infectious agent was not found, and the most approved bacteriological methods failed to demonstrate the constant presence of any particular micro-organism in the blood and tissues of yellow fever cadavers. The micro-organism most frequently encountered was the Bacillus coli communis.

A few scattered bacilli were found in the liver and other organs at the moment of death, but when a portion of liver was preserved in an antiseptic wrapper and kept for twenty-four to forty-eight hours the large number of bacteria that developed were of many varieties, the most common being the Bacillus coli communis and the Bacillus cadaveris.

The blood, urine, and crushed liver-tissue obtained

from a recent autopsy are not pathogenic in moderate amounts for rabbits or guinea-pigs. Liver-tissue preserved at 28° F. in an antiseptic wrapper is very pathogenic for guinea-pigs when injected subcutaneously, but Sternberg found that this pathogenesis was not true of yellow fever livers only, as it developed also in control-autopsies.

Extended research of the alimentary canal in yellow fever showed the intestine to contain a great number of bacteria, but no pure or nearly pure culture of any single species, as in cholera. Few liquefying bacteria were found, and the most abundant bacterium was, as in health, the Bacterium coli communis.

The most important micro-organism met with was Bacillus *x* (Sternberg), which was isolated by the culture-method from a considerable number of cases, and may have been present in all. It was not present in any of the control-experiments. It was very pathogenic for rabbits when injected into the abdominal cavity. Sternberg says: "It is possible that this bacillus is concerned in the etiology of yellow fever, but no satisfactory evidence that this is the case has been obtained by experiments upon the lower animals, and it has not been found in such numbers as to warrant the inference that it is the veritable infectious agent."

The latest researches upon yellow fever are those of Sanarelli.[1] In studying the cadavers of yellow fever Sanarelli found them either entirely sterile or universally invaded by certain microbic species, such as the Streptococcus pyogenes, the colon bacillus, the protei, etc., which cannot be the cause of the disease. In the second case he examined he was fortunate enough to find what he is satisfied is the specific microbe, the *Bacillus icteroides*. In 11 autopsies he never found the organism alone, but always associated with the ordinary bacteria mentioned above. The Bacillus icteroides must be sought for in the blood and tissues, and not in the gastro-intes-

[1] *Brit. Med. Journ.*, July 3, 1897.

tinal cavity. In the latter it is never found. The isolation of the specific microbe was only possible in 58 per cent. of the cases, and in some rare instances may be accomplished during life.

The bacillus, at first sight, presents nothing morphologically characteristic. It is a small bacillus with rounded ends, generally united in pairs in the culture and in small groups in the tissues. It is 2–4 μ in length, and, as a rule, two or three times longer than broad (Fig. 111). It is pleomorphous, and has flagella. By employing suitable methods it can be found in the organs

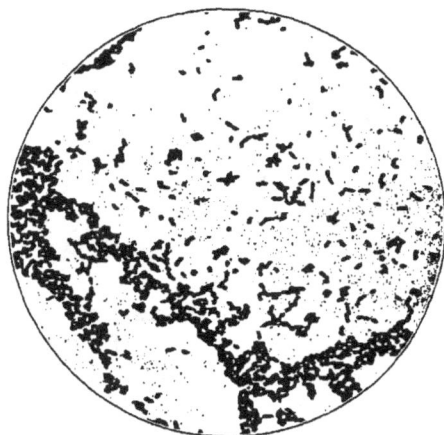

FIG. 111.—Bacillus icteroides (Sanarelli).

of yellow fever cadavers, usually united in little groups, always situated in the small capillaries of the liver, kidney, etc. The best method of demonstration is to keep a fragment of liver, obtained from a body soon after death, in the incubator at 37° C. for twelve hours and allow the bacteria to multiply in the fresh tissue before examination.

The bacillus can be cultivated upon the ordinary media. Upon gelatin plates it forms rounded, transparent, granular colonies, which during the first three or four days present somewhat the appearance of leukocytes.

The granular appearance becomes continuously more marked, and usually a central or peripheric nucleus, completely opaque, is seen. In time the entire colony becomes opaque, but does not liquefy gelatin.

Stroke-cultures on obliquely solidified gelatin exhibit brilliant, opaque, little drops similar to drops of milk.

In bouillon it develops slowly, without either pellicle or flocculi.

The culture upon agar-agar is said to be characteristic.

If grown at 37° C., the peculiar appearances of the colonies do not develop; but if the culture is kept at 20°–22° C., the colonies appear rounded, whitish, opaque, and prominent, like drops of milk. This appearance of the colonies shows well if the cultures are kept for the first twelve to sixteen hours at 37° C., and afterward at room-temperature, when the colonies will show a flat central nucleus, transparent and bluish, surrounded by a prominent and opaque zone, the whole resembling a drop of sealing-wax. Sanarelli refers to this appearance as constituting the diagnostic feature of Bacillus icteroides. It can be obtained in twenty-four hours.

The growth upon potato corresponds to the classic description of that of the bacillus of typhoid fever.

The bacillus is a facultative anaërobe. It cannot be colored by Gram's stain. It slowly ferments lactose, more actively ferments glucose and saccharose, but is not capable of coagulating milk. It strongly resists drying, dies in water at 60° C., and is killed in seven hours by the solar rays. It can live for considerable time in sea-water.

The bacterium is pathogenic for the majority of the domestic animals. All mammals seem more or less sensitive to the pathogenic action of the bacillus; birds are often immune. Guinea-pigs are invariably killed by either intraperitoneal or subcutaneous injection of 0.1 c.cm. White mice are killed in five days; guinea-pigs in eight to twelve days; rabbits in four to five days. The morbid changes present include splenic tumor, hypertrophy of the thymus, and adenitis. In the rabbit there are, in addition, nephritis, enteritis, albuminuria, hemoglobinuria, and hemorrhages into the body-cavities.

The dog is the most susceptible animal. When it is injected intravenously the disease-process that results is almost immediately manifested with such violent symptoms and such complex lesions as to recall the clinical and anatomical picture of yellow fever in the human being. The most prominent symptom in experimental yellow fever in the dog is vomiting, which begins directly after the penetration of the virus into the blood and continues for a long time. Hemorrhages appear after the vomiting, the urine is scanty and albuminous, or there is suppression, which shortly precedes death. Once grave jaundice was observed.

At the necropsy the lesions met are highly interesting, and are almost identical with those observed in man. Most conspicuous is the profound steatosis of the liver. The liver-cells, even when examined fresh, appear completely degenerated into fat, this appearance corresponding to that found in fatal cases of yellow fever. The same result may be obtained by injecting the liver directly or through the abdominal wall. The kidneys are the seat of acute parenchymatous nephritis, sometimes with marked fatty degeneration. The whole digestive tract is the seat of hemorrhagic gastro-enteritis comparable in intensity only to poisoning by cyanid of potassium.

Experiments upon monkeys were also of interest, inasmuch as they demonstrated the possibility of obtaining fatty degeneration more extensive than is observed in man. In one case the liver was transformed into a mass of fatty substance similar to wax.

Goats and sheep are also very sensitive to the icteroid virus, and the lesions described also occur in them.

The death of a yellow fever victim is the result of one of three causes:

1. It may be due to the specific infection principally, when the Bacillus icteroides is found in the cadaver in a certain quantity and in a state of relative purity.

2. It may be due to the septicemias established during the course of the disease, the cadaver then presenting an almost pure culture of the other microbes.

3. It may be due in large measure to renal insufficiency, when the cadaver is found nearly sterile.

The black vomit is due to the action of gastric acidity upon the blood which has extravasated in the stomach in consequence of the toxic products of the Bacillus icteroides.

The Bacillus icteroides produces a toxin the result of whose action corresponds to the essential symptoms of the disease. Animals immune to the infection, or only partially susceptible to it, are not much affected by the toxin. Susceptible animals, such as dogs, are profoundly affected. Ten to fifteen minutes after injecting the toxin the animals experience a general rigor; abundant lachrymation begins, followed by continued vomiting, first of food, then of mucus. In a short time the animals lie helpless and extended. Hematuria frequently occurs. If the dose be moderate, the dog recovers quickly from the violent attack; but if the quantity of toxin be very large or repeated on successive days, it finally succumbs, presenting the anatomical lesions already described as due to infection.

The proofs of the specificity of the Bacillus icteroides are not limited to the animal experiments quoted. Sanarelli also adduces five experimental inoculations upon men. These inoculations were not made with the bacteria—*i. e.* were not infection experiments—but were made with the filtered sterile toxin, whose action could be more easily controlled. "The injection of the filtered cultures in relatively small doses reproduced in man typical yellow fever, accompanied by all its imposing anatomical and symptomatological retinue. The fever, congestions, hemorrhages, vomiting, steatosis of the liver, cephalalgia, collapse—in short, all that complex of symptomatic and anatomical elements which in their combination constitute the indivisible basis of the diagnosis of yellow fever. This fact is not only striking evidence in favor of the specific nature of the Bacillus icteroides, but it places the etiological and pathologic conception of yellow fever on an altogether new basis."

The discovery of the Bacillus icteroides, and especially of its toxin, entirely changes our view of the pathology of the disease. Instead of being a disease of the gastro-intestinal tract, as one would conclude from the symptoms, "all the symptomatic phenomena, all the functional alterations, all the anatomical lesions of yellow fever, are only the consequence of an eminently steatogenous, emetic, and hemolytic action of the toxic substances manufactured by the Bacillus icteroides."

The mode by which the Bacillus icteroides enters the body to produce the disease has not been made out. The digestive and respiratory tracts are the most likely routes.

Sanarelli points out that when it happens that a mould develops near the Bacillus icteroides, the products of material exchange of this hyphomycete or the transformation effected by it, are sufficient to nourish the bacillus and enable it to live and multiply, whereas it would be otherwise condemned to a more or less early death.

There seems to be no particular mould possessed of this power, as of six experimented upon all were capable of it. Sanarelli is of the opinion that in the holds of ships and in damp places generally the presence of moulds favors the development of the Bacillus icteroides.

About the same time that Sanarelli published his researches, Havelburg announced[1] the discovery of an entirely different bacillus. Without entering into a long description of Havelburg's bacillus, which seems to be far less established in its specificity, the following are the chief characteristic and differential points:

The bacillus is found in the stomach and intestine and in the "black vomit." It is almost the sole organ-

[1] *Ann. de l'Inst. Pasteur*, 1897.

ism found in the blood in the stomach. There seems to be very little toxin in the blood of patients with yellow fever, 30–40 c. cm. of blood being necessary to kill a guinea-pig when injected subcutaneously. The injection of 1–2 c. cm. of blood from the stomach, however, caused death of the guinea-pig. In its body an almost pure culture of the bacillus of Havelburg was found. This experiment was repeated twenty-one times without a failure.

The micro-organism is an exceedingly small straight bacillus 1 μ in length and 0.3–0.5 μ in breadth, and may be single or in pairs, never occurring as filaments. The stained specimens are more deeply colored at the ends than at the center, so that the bacillus somewhat resembles the bacillus of fowl-cholera and looks somewhat like a diplococcus. It has no flagella, is not motile, and does not seem to produce spores.

Upon gelatin plates the colonies appear in twenty-four hours as small, round, white points, and increase in size during the next twenty-four hours. The older colonies are yellowish, finely granular discs, with delicately serrated borders. The gelatin is not liquefied.

In gelatine puncture-cultures a "nail-growth" is produced, consisting of a delicate line of colonies along the puncture and a broad surface-growth.

The growth on agar-agar is not characteristic, as is that of Sanarelli's bacillus. Bouillon becomes clouded by the development of the organism. Rapid fermentation and gas-production occur in media containing sugar. A grayish growth occurs on potato. Milk is curdled in twenty-four hours. The bacillus produces large quantities of indol and sets free H_2S. Development in acid media is rapid. The organism is a facultative anaërobic. Guinea-pigs and mice are very susceptible: white rats far less so. Dogs suffer only from local abscesses at the point of injection. The bacillus rapidly alternates in virulence. No toxin seems to be produced by it.

Havelburg is of the opinion that "yellow fever is a disease of which the specific toxic agent enters the stomach and intestines, where it develops. It is only exceptionally and in small numbers that it makes its way from these positions to other organs. He thinks the toxic substances formed in the stomach and intestine are probably the result of the breaking down of the bodies of the bacilli by the digestive juices, and that to the absorption of these the various tissue-changes and fatal terminations are to be referred.

In a lengthy and interesting review and comparison of Sanarelli's and his own work, Sternberg[1] concludes that the Bacillus icteroides of Sanarelli is identical with the Bacillus *x*, which he had discovered in yellow fever cadavers as early as 1888, and felt disposed to describe as the specific cause of the disease, except for a few facts, such as finding it in only one-half of the cases, etc. Sternberg seems inclined to believe in Sanarelli's work, and asserts his intention to further investigate Bacillus *x*. Bacillus *x* was, however, isolated from the alimentary canal, in which Sanarelli's bacillus is said not to exist, and was isolated from the liver of a case of tuberculosis, which takes away considerable of the evidence of its specificity.

In a later paper[2] Sanarelli discusses the validity of Sternberg's claim to priority of discovery, and points out a sufficient number of differences in the original descriptions of the organisms to establish conclusively the individuality of the Bacillus icteroides.

It would seem, from a careful consideration of the recent literature, that Havelburg had very little ground for considering his bacillus specific, and that it is not possible for Sternberg to establish the identity of the Bacillus *x* with the Bacillus icteroides, while at the same time Sanarelli's descriptions and arguments are convincingly in favor of the accuracy of his own work and the specificity of his bacillus.

---

[1] *Centralbl. für Bakt. und Parasitenk.*, Sept. 6, 1897, Bd. xxii., Nos. 6 and 7.

[2] *Ibid.*, Bd. xxii., Nos. 22 and 23, p. 668.

Sanarelli's labors have not ceased with his careful study of the Bacillus icteroides, but have been carried into the important field of serum-therapy. By careful manipulation he has succeeded in immunizing the horse and ox to large doses of the bacillus, injecting into a vein so as to prevent the intense local reaction, and has found that the serum of these animals has the power to protect guinea-pigs from lethal doses of the bacillus. He hopes that the serum will also be efficacious in the treatment of yellow fever in the human being.

CHAPTER V.

CHICKEN-CHOLERA.

THE barnyards of Europe, and sometimes of America, are occasionally visited by an epidemic disease which affects pigeons, turkeys, chickens, ducks, and geese, and causes almost as much destruction among them as the occasional epidemics of cholera and small-pox produce among men. Rabbit-warrens are also at times seriously affected by the epidemic. When fowls are ill with the disease, they fall into a condition of weakness and apathy which causes them to remain quiet, seemingly almost paralyzed, and ruffle up the feathers. The eyes are closed shortly after the illness begins, and the birds gradually fall into a stupor from which they do not awaken. The disease leads to a fatal termination in twenty-four to forty-eight hours. During its course there is profuse diarrhea, the very frequent fluid, slimy, grayish-white discharges containing numerous micro-organisms.

The bacilli which are responsible for this disease were first observed by Perroncito in 1878, and afterward thoroughly studied by Pasteur. They are short, broad bacilli with rounded ends, sometimes united to each other, with the production of moderately long chains (Fig. 112). Pasteur at first regarded them as cocci, because when stained with a penetrating anilin dye the poles stain intensely, but a narrow space between them remains almost uncolored. This peculiarity is very marked, and sharp observation is required to observe the outline of the intermediate substance. The bacillus does not form spores, and does not stain by Gram's method. When examined in the living condition it is found to be non-motile.[1]

The bacillus readily succumbs to the action of heat and dryness. The cultures upon gelatin plates after about two days appear as irregular, small, white points. The deep colonies reach the surface slowly, and do not attain any considerable size. The gelatin is not lique-fied. The microscope shows the colonies to be irregularly

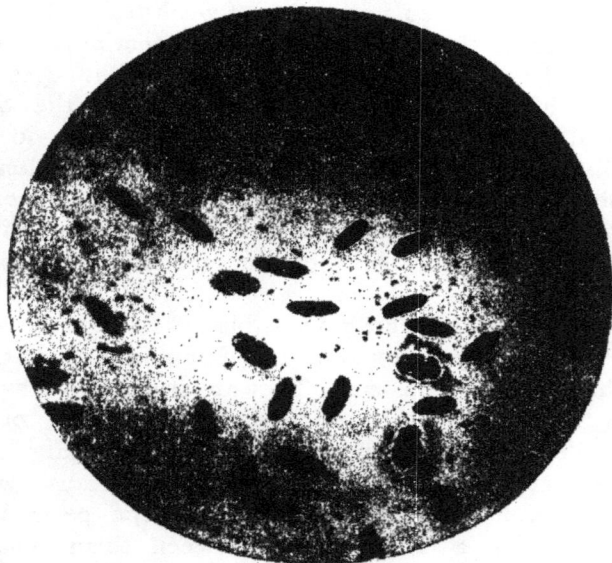

FIG. 112.—Bacillus of chicken-cholera, from the heart's blood of a pigeon; × 1000 (Fränkel and Pfeiffer).

rounded disks with distinct smooth borders. The color is yellowish-brown, and the contents are granular. Some-times there is a distinct concentric arrangement.

In gelatin puncture-cultures a delicate white line occurs along the entire path of the wire. When viewed through a lens this line is seen to consist of aggregated mi-nute colonies. Upon the surface the development is

[1] Most authorities state that the bacillus is not motile, but Thoinot and Mas-selin assert that it is so. *Precis de Microbie*, 2d ed., 1893.

much more marked, so that the growth resembles a nail with a pretty good-sized flat head. If, instead of a puncture, the inoculation be made upon the surface of obliquely solidified gelatin, a much more pronounced growth takes place, and along the line of inoculation a dry, granular coating is formed. This growth is quite similar to that upon agar-agar and blood-serum, which growths are white, shining, rather luxuriant, and devoid of characteristics. No growth occurs in the absence of oxygen.

Upon potato no growth occurs except at the incubation temperature. It is a very insignificant, yellowish-gray, translucent film.

The introduction of cultures of this bacillus into the tissues of chickens, geese, pigeons, sparrows, mice, and rabbits is sufficient to produce fatal septicemia. Feeding chickens, pigeons, and rabbits with material infected with the bacillus is also sufficient to produce the disease with pronounced intestinal lesions. Guinea-pigs usually seem immune, though they succumb to very large doses, especially when given intraperitoneally.

The autopsy shows that when the bacilli are introduced subcutaneously a true septicemia results, with the addition of a hemorrhagic exudate and gelatinous infiltration at the seat of inoculation. The liver and spleen are enlarged; circumscribed, hemorrhagic, and infiltrated areas occur in the lungs ; the intestine shows an intense inflammation with red and swollen mucosa, and occasional ulcers following small hemorrhagic spots. Pericarditis is of frequent occurrence. The bacilli are found in all the organs. If, on the other hand, the disease has been produced by feeding, the bacilli are chiefly to be found in the intestine. Pasteur found that when pigeons were inoculated into the pectoral muscles, if death did not come on rapidly, portions of the muscle (*sequestræ*) underwent degeneration and appeared anemic, indurated, and of a yellowish color.

The bacillus of chicken-cholera is one whose peculiarities can be made use of for protective vaccination.

Pasteur discovered that when cultures are allowed to remain undisturbed for several months, their virulence is greatly lessened, and new cultures planted from these are also attenuated. When chickens are inoculated with such cultures, no other change occurs than a local inflammatory reaction by which the birds are protected against virulent bacilli. From this observation Pasteur worked out a system of protective vaccination in which fowls can first be inoculated with very weak, then with stronger, and finally with highly virulent cultures, with a resulting protection and immunity. Unfortunately, the method is too complicated to be very practical. Use has, however, been made of the ability of this bacillus to kill rabbits, and in Australia, where they are pests, they are being exterminated by the use of bouillon culture. It is estimated that two gallons of bouillon culture will destroy 20,000 rabbits irrespective of infection by contagion.

The bacillus of chicken-cholera seems not only to be specific for that disease, but seems able, when properly introduced into various other animals, to produce several different diseases. Indeed, no little confusion has arisen in bacteriology by the description of what is now pretty generally accepted to be this very bacillus under the various names of bacillus of rabbit-septicemia (Koch), Bacillus cuniculicida (Flügge), bacillus of swine-plague (Löffler and Schütz), bacillus of "Wildseuche" (Hüppe), bacillus of "Büffelseuche" (Oriste-Armanni), etc.

CHAPTER VI.

HOG-CHOLERA.

THE bacillus of hog-cholera (Bacillus suipestifer) was first found by Salmon and Smith,[1] but was for a long time confused with the bacillus of "swine-plague," which it closely resembles and with which it frequently occurs. It is a member of the group of which the Bacillus coli communis may be taken as a type. Since the careful studies of Smith,[1] however, the claims of the discoverers that the bacillus of hog-cholera is a separate and specific organism can hardly be doubted.

Hog-cholera, or "pig typhoid," as the English call it, is a common epidemic disease of swine, which at times kills 90 per cent. of the infected animals, and thus causes immense loss to breeders. Salmon estimates that the annual losses from this disease in the United States range from $10,000,000 to $25,000,000.

The disease is particularly fatal to young pigs. The symptoms are not very characteristic, and the animals often die suddenly without having appeared particularly ill, or after seeming ill but a few hours. The symptoms consist of fever (106°–107° F.), unwillingness to move, and more or less loss of appetite. The animals may appear stupid and dull, and have a tendency to hide in the bedding and remain covered by it. The bowels may be normal or constipated at the beginning of the attack, but later there is generally a liquid and fetid diarrhea, abundant, exhausting, and persisting to the end. The eyes are congested and watery, the secretion drying and

---

[1] *Reports of the Bureau of Animal Industry*, 1885–91.

[2] *Centralbl. für Bakt. und Parasitenk.*, Bd. ix., Nos. 8, 9, and 10, March 2, 1897.

gluing the lids together. The breathing is rapid, and there may be cough. Occasionally there is an eruption with crusts or scabs of various sizes on the skin, which is often congested. The animal becomes weak, stands with arched back and drawn abdomen, and walks with a weak, tottering gait.

The course of this disease varies from one or two days to two or three weeks.

At *post-mortem* examination petechiæ, ecchymoses, and extravasations of blood into the tissues are found to be common and form one of the principal changes in the acute

FIG. 113.—Bacillus of hog-cholera, showing flagella.

form of the disease. The spleen is enlarged to two or four times its normal size, and is soft and engorged with blood.

The extravasations of blood are common in the lymphatic glands, beneath the serous membranes of the thorax and abdomen, and particularly along the intestines; on the surface of the lungs and kidneys and in their substance. The contents of the intestine are sometimes covered with clotted blood. In the subacute form of the disease the principal changes are found in the large intestine, and consist of ulcers which appear as circular, slightly projecting masses varying in color from yellowish to black. Occasionally these ulcers are slightly depressed in outline. When cut across they are found to consist of a firm, solid growth extending nearly through the intestinal wall. They are most frequent in the cecum, upper half of the colon, and on the ileocecal valve. In the chronic form of the disease the spleen is rarely enlarged.

"In hog-cholera the first effect of the disease is believed to be upon the intestines, with secondary invasion of the lungs."

The most characteristic lesions of the disease are the petechiæ and ecchymoses, the ulcerations of the large intestine (Fig. 113), and the collapse and occasional bronchopneumonic changes in the lung.

The kidneys are nearly always affected, the urine containing albumin and tube-casts.

The specific bacillus of hog-cholera was secured by Smith from the spleens of more than 500 hogs. It occurs in all the organs and has also been cultivated from the urine.

The organisms appear as short rods with rounded ends, 1.2–1.5 μ long and 0.6–0.7 μ in breadth. They are very actively motile. No spore-production has ever been observed. In general the bacillus resembles in appearance that of typhoid fever. It stains readily by the ordinary methods, but not by Gram's method.

The bacilli possess numerous long flagella, easily demonstrable by the usual methods of staining (Fig. 114).

No trouble is experienced in cultivating the bacilli, which grow well in all the media.

Upon gelatin plates the colonies become visible in twenty-four to forty-eight hours; the deeper ones spherical with sharply defined borders. The surface is brownish by reflected light, and is without markings. They are rarely larger than 0.5 mm. in diameter and are homogeneous throughout. The superficial colonies have little tendency to spread upon the gelatin. Their borders may be circular and rounded, or irregular. They are said rarely to reach a greater diameter than 2 mm. The gelatin is not liquefied. There is nothing distinctly characteristic about the appearance of the colonies.

Upon agar-agar the superficial colonies attain a diam-

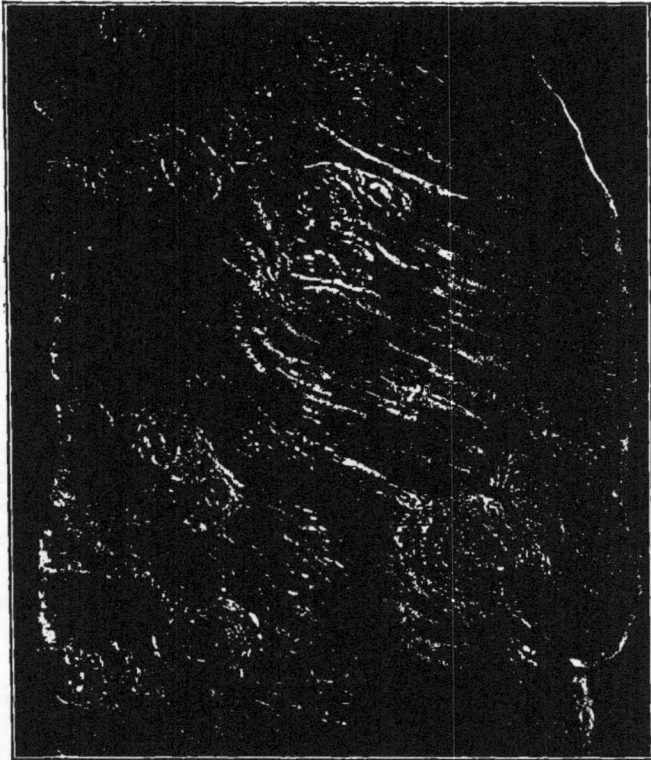

FIG. 114.—Ulceration of the intestine in a typical case of swine-fever (Crookshank).

eter of 4 mm. and have a gray translucent appearance with polished surface. They are round and slightly arched.

In gelatin punctures the growth takes the form of a nail with a flat head. There is nothing characteristic about it. The growth *in* the puncture shows it to be an optional anaërobe.

Linear cultures upon agar-agar present a translucent, rather circumscribed, grayish, smeary layer.

Upon potato a yellowish coating is formed, especially when the culture is kept in the thermostat.

Bouillon made with or without pepton is clouded in twenty-four hours. When the culture is allowed to stand for a couple of weeks without being disturbed a thin surface-growth can be observed.

Milk is an excellent culture-medium, but is not visibly changed by the growth of these bacteria. Its reaction remains alkaline.

The hog-cholera bacillus is a copious gas-producer, capable of breaking up sugars into CO_2, H, and an acid, which, formed late, eventually checks its further development. No indol and no phenol are formed in the culture-media.

The bacillus is hardy. Smith found it vital after being kept dry for four months. It ordinarily dies sooner, however. The thermal death-point is 54° C., maintained for sixty minutes.

The bacillus is markedly pathogenic for animals. Small quantities introduced subcutaneously into rabbits or mice kill them in from seven to twelve days. The animal appears quite well for three or four days, then begins to sit quietly in the cage and eat but little, or refuses to eat at all, until death takes place.

In Smith's experiments one-four-millionth of a cubic centimeter of a bouillon culture injected subcutaneously into a rabbit was sufficient to cause its death. Before death the temperature abruptly rises 2°–3° C., and remains high until death. Larger quantities may kill in five days. Injected intravenously in small doses the bacillus may cause death in forty-eight hours.

When the animal is subjected to a *postmortem* examination the spleen is found enlarged, firm, and dark red in color. The liver is found to contain small yellowish-white necrotic areas which sometimes occur in one, sometimes in several acini, and not infrequently surround the

27

interlobular veins. The kidneys are acutely inflamed and the urine is albuminous. The heart-muscle is spotted, gray, and fatty. In the intestinal tract the picture of the disease will be found to vary according to its duration.

The contents of the small intestine are yellowish, watery, and mucous; Peyer's glands are enlarged. In the neighborhood of the pylorus, ecchymoses and extensive extravasations of blood are common. The bacilli are found in all of the organs.

The house mouse is very susceptible to the disease; guinea-pigs much less so, $\frac{1}{10}$ c.cm. of a virulent culture often being required to kill them. Pigeons are still more refractory, and Smith found that $\frac{3}{4}$ c.cm. of a bouillon culture injected into the breast-muscles was required to kill them.

In spite of the fact that hog-cholera is a disease of swine, and that it is from dead swine that the bacilli are obtained, these animals are not very easily affected artificially. They show no symptoms when injected subcutaneously, but almost invariably die after intravenous injection of 1–2 c.cm. of a virulent culture.

Smith found that feeding with 200–300 c.cm. of a bouillon culture after a day's fasting, or with small quantities administered daily, would also cause death, with a widespread diphtheritic inflammation of the stomach and colon. Feeding with the organs of dead hogs produces the same lesions as the administration of the culture.

As early as 1886 Salmon and Smith found it possible to produce, in both very and partly susceptible animals, immunity to hog-cholera by gradually accustoming them to increasing doses of the bacteria. DeSchweinitz isolated from cultures of the bacteria two toxic substances, a ptomäin (sucholo-toxin) and an albumose (sucholo-albumin), together with cadaverin and methylamin. With these substances he seems to have been able to produce immunity. Selander and Metschnikoff found that immunity could be produced more quickly by the use of blood of infected rabbits exposed to 58° C. This blood was found to be exceedingly toxic.

DeSchweinitz[1] found that the introduction of progressingly increased amounts of cultures into cows caused the development in them of an antitoxic substance capable of protecting guinea-pigs from the disease.

Working in my laboratory, Pitfield[2] has found that after a single injection of a sterilized bouillon culture of the bacillus into the horse, the serum, which has originally slight agglutinative reactive power, is so changed as to show a decided reaction. If the horse be immunized to large doses of such sterile cultures, the serum reaction becomes so marked that with a dilution of 1 : 10,000 a typical reaction occurs in sixty minutes.

According to this experiment, in doubtful cases the use of this reaction should greatly facilitate the differentiation of the bacillus of hog-cholera from similar bacilli.

[1] *Centralbl. f. Bakt. u. Parasitenk.*, xx., p. 573.
[2] *Microscopical Bulletin*, 1897, p. 35.

CHAPTER VII.

SWINE-PLAGUE.

THE bacillus of swine-plague, or the Bacillus suisepti-
cus of Löffler and Schütz, and Salmon and Smith, so
closely resembles that of chicken-cholera that it is easily
confounded with it, and, indeed, at one time, they were
thought to be identical. The species has, however, sufficient well-marked characters to make its differentiation
clear (Fig. 115).

Swine-plague is a rather common and exceedingly

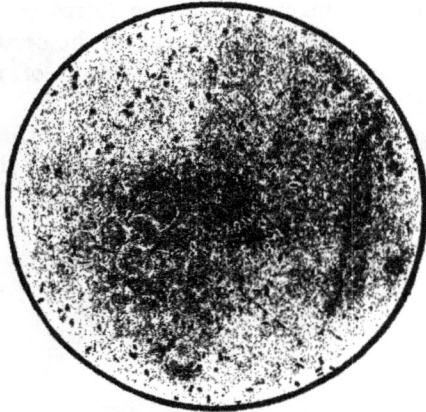

FIG. 115.—Bacillus of swine-plague (from photograph by E. A. de Schweinitz).

fatal disease. It occurs alone or in combination with
hog-cholera (*q. v.*), and because of the lack of sufficiently well-characterized symptoms—sick hogs appearing more or less alike—it is often mistaken for that
disease. The confusion resulting from the mixed cases
makes it impossible to determine exactly how fatal swine-
plague may be in uncomplicated cases.

The symptoms of swine-plague, while closely resembling those of hog-cholera, may differ from them in the existence of cough, swine-plague being prone to affect the lungs and oppress the breathing, which becomes frequent, labored, and painful, and associated with frequent cough, while hog-cholera chiefly presents intestinal symptoms.

The course of the disease is usually rapid, a fatal result often occurring in one or two days.

At autopsy the lungs are often found inflamed, and contain numerous small, pale, necrotic areas, and sometimes large cheesy masses one or two inches in diameter. Inflammations of the serous membranes affecting the pleura, pericardium, and peritoneum, and associated with fibrinous inflammatory deposits on the surfaces, are common. There may be congestion of the mucous membrane of the intestines, particularly of the large intestine, or the disease in this region may be an intense croupous inflammation with the formation of a fibrinous exudative deposit on the surface.

A hemorrhagic form of the disease is said to be common in Europe, but, according to Salmon, is rare in the United States.

The bacillus of swine-plague much resembles that of hog-cholera, and not a little that of chicken-cholera. It is a short organism, rather more slender than its congeners, not possessed of flagella, and is incapable of movement and produces no spores. Its vitality is low, and it is easily destroyed. Salmon says that it soon dies in water or by drying, and that the temperature for its growth must be more constant and every condition of life more favorable than for the hog-cholera germ. This germ is said to be widely distributed in nature, and is probably present in every herd of swine, though not pathogenic except when its virulence has been increased or the resistance of the animals diminished by some unusual conditions.

In its growth the bacillus of swine-plague is an optional anaërobic organism.

In general, its appearance in culture-media is very similar to that of the bacillus of hog-cholera. Kruse, however,[1] points out that when the bacillus grows in bouillon the liquid remains clear on account of the formation of a flocculent, stringy sediment. Upon ordinary acid potato the bacillus does not grow, but if the reaction of the medium be alkaline a grayish-yellow patch is formed. In its growth in milk slight acidity is produced, but the milk is not coagulated and the litmus color added to it is not decolorized.

The bacillus stains by the ordinary methods, sometimes only at the poles, then resembling very closely the bacillus of chicken-cholera. It is not colored by Gram's method.

The pathogenesis, while similar to that of the hog-cholera bacillus, presents some marked differences, especially in regard to the seat of the local manifestations, to which attention has already been called, and in the duration of the disease, which is much shorter. There is also considerable resemblance to the bacillus of chicken-cholera in pathogenesis, but the local reaction following injection of the culture partakes of the nature of a hemorrhagic edema, which is not present in chicken-cholera, and the cases often exhibit fatty metamorphosis of the liver.

Rabbits, mice, and small birds are all very susceptible to the disease, generally dying of septicemia in twenty-four hours; guinea-pigs are less susceptible, except the very young animals, which die without exception. Chickens are more immune, but usually succumb to large doses. Hogs die after subcutaneous injection of the bacilli, and suffer from marked edema at the point of injection, and septicemia. If injected into the lung, a pleuropneumonia with multiple necrotic areas in the lung follows. In these cases the spleen is not much swollen, there is slight gastro-intestinal catarrh, and the bacilli are present everywhere in the blood.

Animals cannot be infected by feeding.

[1] Flügge's *Mikroörganismen*, p. 419, 1896.

CHAPTER VIII.

TYPHUS MURIUM.

THE Bacillus typhi murium (Fig. 116), which created havoc among the mice in his laboratory, causing most of them to die, was discovered by Löffler in 1889. It is a short organism, somewhat resembling the bacillus of chicken-cholera. It is rather variable in its dimensions, and often grows into long, flexible filaments. No

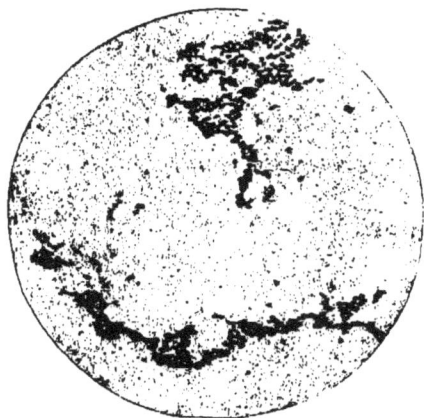

FIG. 116.—Bacillus typhi murium, from agar-agar; × 1000 (Itzerott and Niemann).

sporulation has been observed. It is a motile organism, with numerous flagella, like those of the typhoid-fever bacillus. It stains well with the ordinary dyes, but rather better with Löffler's alkaline methylene blue.

Upon gelatin plates the deep colonies are at first round, slightly granular, transparent, and grayish. Later they become yellowish-brown and granular. Superficial colonies are similar to those of the typhoid bacillus. In gelatin punctures there is no liquefaction. The growth takes place upon the surface principally, where a grayish-white mass slowly forms.

Upon agar-agar a grayish-white development devoid of peculiarities occurs.

Upon potato a rather thin whitish growth may be observed after a few days.

The bacillus grows well in milk, with the production of an acid reaction, but without coagulation.

The organism is pathogenic for mice of all kinds, which succumb in from one to two days when inoculated subcutaneously, and in eight to ten or twelve days when fed upon material containing the bacillus. The bacilli multiply rapidly in the blood- and lymph-channels, and cause death from a general septicemia.

Löffler expressed the opinion that this bacillus might be of use in ridding infested premises of mice, and the results of its use for this purpose have been highly satisfactory. He has succeeded in ridding a field so infested as to be useless for agricultural purposes by saturating some bread with bouillon cultures of the bacillus and distributing it near the holes inhabited by the mice. The bacilli that were eaten by the mice not only killed them, but also infected others which ate the dead bodies of the first victims, and so the extermination progressed until scarcely a mouse remained in the field. In discussing the practical applicability of the employment of cultures of this bacillus for the destruction of field-mice, Brunner[1] calls attention to certain conditions that are requisite for a satisfactory result. (1) It is necessary, first of all, to attack rather extensive areas of the invaded territory, and not to attempt to destroy the mice of a small field into which an indefinite number of fresh animals may immediately come from the surrounding fields. The country-people, who are the sufferers, should combine their efforts so as to extend the benefits widely. (2) The preparation of the cultures is a matter of im-

[1] *Centralbl. f. Bakt. u Parasitenk.*, Jan. 19, 1898, Bd. xxiii., No. 2.

portance. Agar-agar cultures are best, as being most readily transportable. They are broken up in water and well stirred, and the liquid poured upon a large number of small pieces of broken bread. These are next to be distributed with reasonable care. Instead of being carelessly scattered over the ground, they should be dropped into the fresh mouse-holes, and pushed sufficiently far in to escape the effects of sunlight upon the bacilli. Attention should be paid to holes in walls, under railway tracks, etc. and other places where mice live in greater freedom from disturbance than in the fields. (3) The attempted eradication of the mice should be begun at a time of year when the natural food is not plenty. By observing these precautions the mice can be eradicated with certainty, usually in a period of time not exceeding eight to twelve days. For this purpose, in the course of two years, no less than 250,000 cultures were distributed from the Bacteriological Laboratory of the Tierarznei Institut in Vienna. The bacilli are not pathogenic for the animals, such as the fox, weasel, ferret, etc. that feed upon the mice, do not affect man in any way, and so seem to occupy a useful place in agriculture by destroying the little but almost invincible enemies of the grain.

CHAPTER IX.

MOUSE-SEPTICEMIA.

In 1878, during his investigations upon the infectious traumatic diseases, Koch observed that when a minute amount of putrid blood or of meat-infusion was injected into mice the animals died of a septicemia caused by the multiplication in their blood of a minute bacillus to which he gave the name "Bacillus der Mäusesepticämie" (Fig. 117).

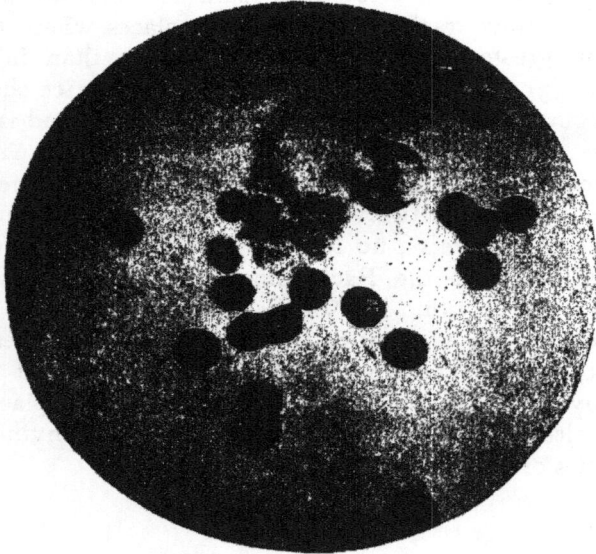

FIG. 117.—Bacillus of mouse-septicemia, from the blood of a mouse; × 1000 (Fränkel and Pfeiffer).

In 1885 the bacillus was again brought into prominence by Löffler and Schütz, who found a very similar, perhaps idéntical, organiṣm in the erysipelatous disease which attacks the swine of many parts of Europe.

There seem to be certain slight morphological and developmental differences between these two organisms, but Baumgarten, Günther, Sternberg, and others have regarded them as insufficient for the formation of separate species, and have boldly described the organisms as identical, while Lorenz has shown that immunity produced in the rabbit by one bacillus protects against the other. The described differences are, indeed, so very small that I think it well to follow in the path of the observers mentioned, pointing out in the description such points of difference as may arise.

The bacilli are extremely minute, measuring about $1.0 \times 0.2 \mu$ (Sternberg). Flügge, Fränkel, and Eisenberg find the Bacillus erysipelas suis somewhat shorter and stouter than that of mouse-septicemia: there seems to be a division of opinion upon this point.

Sporulation has been described by some observers, but nothing definite seems to be known upon this point.

Motility is ascribed by some (Schottelius and Fränkel) to the Bacillus erysipelas suis, and is denied to the bacillus of mouse-septicemia by others. The truth seems to be that the motility of both organisms is a matter of doubt.

No flagella have been demonstrated upon the bacillus. It grows quite well both at the room-temperature and at the temperature of incubation. It can grow well with or without oxygen, but perhaps flourishes a little better without than with it. It is killed by a temperature of 52° C. in fifteen minutes.

The colonies upon gelatin plates can first be seen on the second or third day, then appearing as transparent grayish specks with irregular borders, from which many branched processes extend

FIG. 118.—Colony of the bacillus of mouse-septicemia; × 80 (Flügge).

(Fig. 118). Fränkel describes them as resembling in shape the familiar branched cells occupying the lacunæ

of bone. When further developed the colonies flow together and give the plate a cloudy gray appearance. The gelatin is not liquefied, but is gradually softened and its evaporation thus aided.

In gelatin puncture-cultures the growth is quite characteristic, and the tendency of the bacillus to grow anërobically is well shown (Fig. 119). The develop-

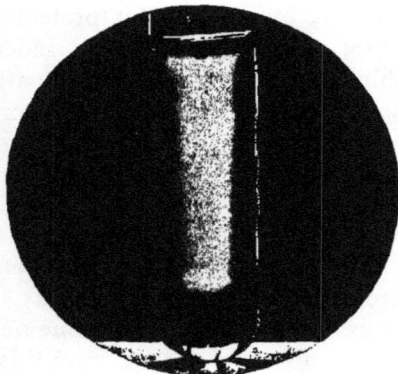

FIG. 119.—Bacillus of mouse-septicemia: gelatin puncture-culture three and a half days old (Günther).

ment takes place all along the line of puncture, but is more marked below than at the surface. The growth takes place in a peculiar form, resembling superimposed disks, each disk separate from its neighbors and consisting of an area of clouded grayish gelatin reaching almost to the walls of the tube. This growth develops slowly, and causes a softening rather than an actual liquefaction of the gelatin.

Upon agar-agar and blood-serum a very delicate, transparent grayish line develops along the path of the needle. It does not grow upon potato.

The bacillus grows at the room temperature, but much better at the temperature of the incubator.

The disease affects quite a variety of animals, notably hogs, rabbits, mice, white rats, pigeons, and sparrows.

The guinea-pig, which is generally the victim of laboratory experiments, is not susceptible to it. Field- and wood-mice, cattle, horses asses, dogs, cats, chickens, and geese are immune.

When mice are inoculated with a pure culture they soon become ill, lose their appetite, mope in a corner, and are not readily disturbed. As the disease becomes worse they assume a sitting posture with the back much bent; the eyelids are glued together by adhesive pus; and when death comes to their relief, in the course of forty to sixty hours after inoculation, they remain sitting in the same characteristic position.

When the ears of rabbits are inoculated with the bacillus from cases of erysipelas suis, a violent inflammatory edema and distinct redness occurs, much resembling erysipelas. This lesion gradually spreads, involves the head, then the body of the animal, and ultimately causes death.

When swine are affected, they are dull and weak, and have a kind of paralytic weakness of the hind quarters. The temperature is elevated ; red patches appear upon the skin and swell and become tender. Death follows in two or three days. Sixty per cent. of the diseased animals die.

In all animals the anatomical changes are much alike. The disease proves to be a septicemia, and the bacilli can be found in all the organs, especially the lungs and spleen. They are few in number in the streaming blood.

As the organisms stain well by Gram's method, this stain is of great value for their discovery in the tissues, and can be highly recommended.

Most of the bacilli occupy the capillary blood-vessels ; many of them are enclosed in leucocytes. The organs in such cases do not appear distinctly abnormal, except the spleen, which is considerably enlarged. The mesenteric and other lymphatics are also enlarged, and the gastric and intestinal mucous membranes are usually inflamed and mottled. The bacilli also occupy the intestinal contents, and Kitt, who discovered them in this position, points out that the infection of swine probably takes place by the entrance, along with the food, of the fecal matter of diseased animals into the alimentary apparatus of others.

Pasteur, Chamberland, Roux, and others have worked upon a protective vaccination based upon the attenuation of the virulence of the organism by passing it through rabbits. Two vaccinations are said to be necessary to produce immunity. The vaccinated animals, however, may be a source of infection to others, and should always be isolated. Klemperer in 1892 found that the blood-serum of immunized rabbits would save infected mice into which it was injected.

Lorenz in 1894 found an antitoxic substance in the blood of rabbits immunized to the disease. The effect of its injection into other animals is, however, only a temporary immunity. Later[1] he found it possible to protect hogs against the disease by injecting them first with a serum obtained from a hog immunized in the ordinary manner described by Pasteur, afterward with a feeble culture of the bacillus, and finally with virulent cultures. The strength of the serum should be determined by injecting varying quantities of it into mice infected with definite amounts of a culture of known virulence. The immunity produced by Lorenz lasted for a year.

[1] *Centralbl. f. Bakt. u. Parasitenk.*, Jan., 1896, p. 168.

CHAPTER X.

RELAPSING FEVER.

As long ago as 1873, Obermeier discovered that a flexible spiral organism, about 0.1 μ in diameter and from 20–40 μ in length, could be observed in the blood of patients suffering from relapsing fever.

Although many of the best bacteriologists of our day have occupied themselves with the study of this spirillum, we really have, at present, very little more knowledge than that given us by Obermeier.

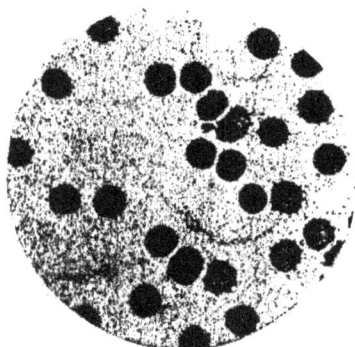

FIG. 120.—Spirochæta febris recurrentis; × 650 (Heim).

The spirilla (Fig. 120) are generally very numerous, are long, slender, and flexible (spirochæta), and possess a vigorous movement by flagella. The ends are rather pointed.

The spirillum stains well by ordinary methods, but not by Gram's method. It seems to be a strict parasite, and has never been cultivated artificially.

Of the pathogenesis of the organism there can be no doubt, as it is invariably present in relapsing fever and undergoes a peculiar cycle of changes according to the stage of the disease. During the pyrexia the organisms are found in the blood in active movement, swimming both by rotation on the long axis and by undulation. As soon as the crisis comes on they are found to be without motion, most of them enclosed in leucocytes and seemingly dead. The recurrence of the paroxysm has suggested to many that spores are formed in the spirillum, but no one has been successful in proving that this is the case. Koch, Carter, and Soudakewitch have all succeeded in giving the disease to monkeys, and Münch and Moczutkowsky have gone further and have produced it in men by introducing into them blood from diseased patients.

Soudakewitch finds that the removal of the spleen causes the disease to terminate fatally in monkeys.

CHAPTER XI.

BUBONIC PLAGUE.

THE bacillus of bubonic plague (Fig. 121) seems to have met an independent discovery at the hands of

FIG. 121.—Bacillus of bubonic plague (Yersin).

Yersin and Kitasato in the summer of 1894, during the activity of the plague then raging at Hong-Kong. There seems to be but little doubt that the micro-organisms described by the two observers are identical.

In a recent study of the plague, Ogata[1] states that while Kitasato found his bacillus in the blood of cadavers, Yersin seldom found his bacillus in the blood, but always in the enlarged lymphatic glands. Kitasato's bacillus retains the color when stained by Gram's method; Yersin's does not. Kitasato's bacillus is motile; Yersin's, non-motile. The colonies of Kitasato's bacillus when grown upon agar are round, irregular, grayish-white with

[1] *Centralbl. f. Bakt. u. Parasitenk.*, Bd. xxi., Nos. 20 and 21, June 24, 1897.

a bluish tint, and resemble glass-wool when slightly
magnified; Yersin's bacillus forms white, transparent colonies with iridescent edges.   Ogata, in the investigation

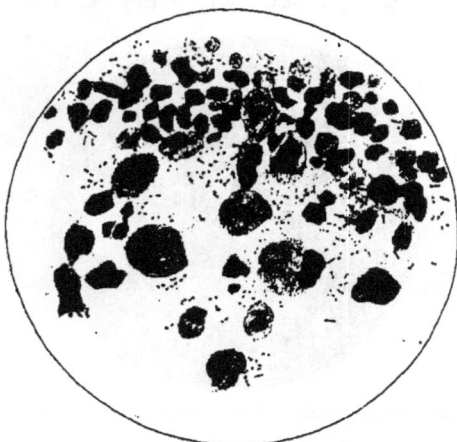

FIG. 122.—Bacilli of plague and phagocytes; × 800.· From human lymphatic
gland (Aoyama).

of the cases that came into his hands found a bacillus
that resembles that of Yersin, but not that of Kitasato.

The bubonic plague is an extremely fatal infectious
disease, whose ravages in the hospital in which Yersin
made his observations carried off 95 per cent. of the
cases.   It affects both men and animals, and is characterized by sudden onset, high fever, prostration, delirium,
and the occurrence of lymphatic swellings—buboes—
affecting chiefly the inguinal glands, though not infrequently the axillary, and sometimes the cervical, glands.
Death comes on in severe cases in forty-eight hours.   If
the case is of longer duration, the prognosis is said to be
better.   Autopsy in fatal cases reveals the characteristic
enlargement of the lymphatic glands, whose contents are
soft and sometimes purulent.

Wyssokowitz and Zabolotny[1] describe two forms of
the disease:

[1] *Ann. de l'Inst. Pasteur*, Aug. 25, 1897, xi., 8, p. 665.

1. Plague with buboes. ·

2. Plague without buboes, but with a primary specific pneumonia in which the bacilli occur in immense numbers in the affected pulmonary tissue, but sparingly in the blood and kidney.

The studies of Kitasato and Yersin show that in blood drawn from the finger-tips and in the softened contents of the glands a small bacillus is demonstrable. The organisms are small, stain much more distinctly at the ends than in the middle, so that they resemble diplococci, and in fresh specimens seem to be surrounded by a capsule. Kitasato compares the organism to the well-known bacillus of ckicken-cholera. It is feebly motile (according to Abel, entirely non-motile), and does not seem to form spores. Nothing is said in the original descriptions about the presence of flagella, though it is probable from the studies of Gordon [1] that some, at least, of the bacilli may be possessed of them. It does not stain by Gram's method.

When cultures are made from the softened contents of the buboes the bacillus may be obtained almost or quite pure, and is found to develop upon artificial culture-media. In bouillon a diffuse cloudiness results from the growth, as observed by Kitasato, though in Yersin's observations the culture more nearly resembled erysipelas cocci, and contained zoöglea attached to the sides and at the bottom of the tube of nearly clear fluid.

According to Haffkine,[2] when an inoculated bouillon culture is allowed to stand, perfectly at rest, on a solid shelf or table a characteristic appearance results. In from twenty-four to forty-eight hours, the liquid remaining limpid, flakes appear underneath the surface, forming little islands of growth, which in the next twenty-four to forty-eight hours grow down into a long stalactite-like jungle, the liquid always remaining clear. In four to

---

[1] *Centralbl. f. Bakt. u. Parasitenk.*, Sept. 6, 1897, Bd. xxii., Nos. 6 and 7, p. 170.

[2] *Brit. Med. Jour.*, June 12, 1897, p. 1461.

six days the islands are still more compact and solidified. If the vessel be disturbed, the islands fall like snow and are deposited at the bottom.

Upon gelatin plates at 22° C. the colonies may be observed in twenty-four hours by the naked eye. They are pure white or yellowish-white, spherical in the deep gelatin, flat upon the surface, and are about the size of a pin's head. The gelatin is not liquefied. The borders of the colonies are, upon microscopic examination, found to be sharply defined and to become more granular as their age increases. The superficial colonies occasionally are surrounded by a fine, semi-transparent zone.

In gelatin puncture-cultures the development is scant. The medium is not liquefied (?); the growth takes place in the form of a fine duct, little points being seen on the surface and in the line of puncture.

Upon agar-agar—glycerin agar-agar is best—the bacilli grow freely, the colonies being whitish in color, with a bluish tint by reflected light. Under the microscope they appear moist, with rounded, uneven edges. The small colonies are said to resemble little tufts of glass-wool; the larger ones have large round centers. Microscopic examination of the bacilli grown upon agar-agar reveals the presence of long chains resembling streptococci.

Klein[1] states that the colonies develop quite readily upon gelatin made from beef-bouillon (not infusion), appearing in twenty-four hours, at 20° C., as small, gray, irregularly rounded dots. Magnification shows the colonies to be serrated at the edges and made up of short, oval, sometimes double bacilli. Some colonies contrast markedly with their neighbors in that they are large, round, or oval, and consist of longer or shorter, straight or looped threads of bacilli. The appearance was much like that of the young colonies of the Proteus vulgaris. At first Klein regarded these as contaminations, but later he was led to believe that their occurrence was character-

[1] *Centralbl. f. Bakt. u. Parasitenk.*, xxi., Nos. 24 and 25, July 10, 1897.

istic of the plague bacillus. The peculiarities of these colonies cannot be recognized after forty-eight hours.

Involution-forms on partly desiccated agar-agar not containing glycerin are said by Haffkine to be characteristic. The microbes swell up and form large, round, oval, pea- or spindle-shaped or biscuit-like bodies, which may attain twenty times the normal size and in growing gradually lose the ability to take up the stain. Such involution-forms are not seen in liquid culture.

Hankin and Leumann [1] recommend for the differential diagnosis of the plague bacillus the addition of 2.5–3.5 per cent. of salt to the agar-agar. When transplanted from ordinary agar-agar to the salt agar-agar the involution-forms which are so characteristic of the plague bacillus form with exceptional rapidity.

Upon blood-serum the growth at the temperature of the incubator is luxuriant. It forms a moist layer of a yellowish-gray color, and is unaccompanied by liquefaction of the serum.

Upon potato no growth occurs at ordinary temperatures. When the potato is stood away for a few days in the incubator a scanty, dry, whitish layer develops.

Abel found the best culture-medium to be 2 per cent. alkaline pepton solution with 1 or 2 per cent. of gelatin, as recommended by Yersin and Wilson.

The bacillus develops under conditions of aërobiosis and anaërobiosis. In glucose-containing media it does not form gas. No indol is formed. Ordinarily the culture-medium is acidified by the development of an acid that persists for three weeks or more.

By frequent passage through animals of the same species the bacillus increases very much in virulence. Curiously enough, however, the observations of Knorr, substantiated by Yersin, Calmette and Borrel, show that the bacillus made virulent by frequent passage through mice is not increased in virulence for rabbits. [2]

---

[1] *Centralbl. f. Bakt. u. Parasitenk.*, Oct., 1897, Bd. xxii., Nos. 16 and 17, p. 438.　　　　[2] *Ann. de l'Inst. Pasteur*, July, 1895.

Kitasato found that mice, rats, guinea-pigs, and rabbits are all susceptible; pigeons are immune. Julian Hawthorne, in his paper in the *Cosmopolitan*, speaks of having seen cats and dogs dying of the disease, but no mention is made of these animals in the scientific papers I have read. When blood, lymphatic pulp, or pure cultures are inoculated into them, the animals become ill in from one to two days, according to their size. Their eyes become watery, they begin to show disinclination to take food or to make any bodily effort, the temperature rises to 41.5° C., they remain quietly in a corner of the cage, and die with convulsive symptoms in from two to five days.

Devell[1] has found that frogs are susceptible to the disease.

Wyssokowitz and Zabolotmy[2] found monkeys to be highly susceptible to plague, especially when inoculated subcutaneously. When so small an inoculation was made as a puncture with a pin dipped in a culture of the bacillus, the puncture being made in the palm of the hand or sole of the foot, the monkeys always died in from three to seven days. In these cases the local edema observed by Yersin did not occur. They point out the interest attaching to infection through so insignificant a wound and without local lesions.

According to Yersin, an infiltration or watery edema can be observed in a few hours about the point of inoculation. The autopsy shows the infiltration to be made up of a yellowish gelatinous exudation. The spleen and liver are enlarged, the former often presenting an appearance much like an eruption of miliary tubercles. Sometimes there is universal swelling of the lymphatic glands. Bacilli are found in the blood and in all the internal organs. Very often there are eruptions during life, and upon the inner abdominal walls there are petechiæ and occasional hemorrhages. The intestine is hyperemic, the

[1] *Centralbl. f. Bakt. u. Parasitenk.*, Oct. 12, 1897.
[2] *Ann. de l' Inst. Pasteur*, Aug. 25, 1897, xi., 8, p. 665.

adrenals congested. There are often sero-sanguinolent effusions into the serous cavities.

Klein [1] states that the intraperitoneal injection of the bacillus into guinea-pigs is of diagnostic value, producing in twenty-four to forty-eight hours a thick cloudy peritoneal exudate rich in leukocytes and containing characteristic chains of the plague bacillus.

Animals fed upon cultures or upon the flesh of other animals dead of the disease became ill and died with typical symptoms. When Klein inoculated animals with the dust of dwelling-houses in which the disease had occurred, some died of tetanus, one from plague. Many rats and mice in which examination showed the characteristic bacilli died spontaneously in Hong-Kong.

Yersin showed that flies also die of the disease. Macerating and crushing a fly in bouillon, he not only succeeded in obtaining the bacillus from the medium, but infected an animal with it.

Nuttall,[2] in reviewing Yersin's fly-experiment, found the statement true, and showed that flies fed with the cadavers of plague-infected mice died in a variable length of time. Large numbers of plague bacilli were found in their intestines. He also found that bed-bugs allowed to prey upon infected animals took up large numbers of the plague bacilli and retained them for a number of days. These bugs did not, however, infect healthy animals when allowed, subsequently, to feed upon them. Nuttall is not, however, satisfied that the number of his experiments upon this point was great enough to be conclusive.

Ogata found that the plague bacillus existed in the bodies of fleas found upon diseased rats. One of these he crushed between sterile object-glasses and introduced into the subcutaneous tissues of a mouse, which died in three days with typical lesions of the plague, a control-animal remaining well. Some guinea-pigs taken

---

[1] *Centralbl. f. Bakt. u. Parasitenk.*, xxi., No. 24, July 10, 1897, p. 849.

[2] *Ibid.*, Aug. 13, 1897.

for experimental purposes into a plague district, and kept carefully isolated, died spontaneously of the disease, presumably because of insect infection.

Yersin found that when cultivated for any length of time upon culture-media, especially agar-agar, the virulence was rapidly lost and the bacillus eventually died. On the other hand, when constantly inoculated from animal to animal the virulence of the bacillus is much increased.

The bacillus probably attenuates readily. Kitasato found that it did not seem able to withstand desiccation longer than four days; and Yersin found that although it could be secured from the soil beneath an infected house at a depth of 4–5 c.cm., the virulence of such bacilli was lost.

Kitasato found that the bacillus was killed by two hours' exposure to 0.5 per cent. carbolic acid, and also by exposure to a temperature of 80° C. Ogata found that the bacillus was instantly killed by 5 per cent. carbolic acid, and in fifteen minutes by 0.5 per cent. carbolic acid. In 0.1 per cent. sublimate solution it is killed in five minutes.

It seems possible to make a diagnosis of the disease in doubtful cases by examining the blood, but it is admitted that a good deal of bacteriologic practice is necessary for the purpose.

Abel finds that the blood may yield fallacious results because of the rather variable appearance of the bacilli, which are sometimes long and easily mistaken for other bacteria. He deems the best tests to be the inoculation of broth-cultures and subsequent inoculation into animals, which he advises should have been previously vaccinated against the streptococcus. Plague bacilli persist in the urine a week after convalescence.

Wilson, of the Hoagland Laboratory, found the thermal death-point of the organism was one or two degrees higher than that of the majority of pathogenic bacteria of the non-sporulating variety, and that, unlike cholera, the influence of sunlight and desiccation cannot be relied upon to limit its viability.

Kitasato's experiments first showed that it is possible to bring about immunity to the disease, and Yersin, working in India, and Fitzpatrick, in New York, have successfully immunized large animals (horses, sheep, goats). The serum of these immunized animals contains an antitoxin capable not only of preventing the disease, but also of curing it in mice and guinea-pigs and probably in man.

Haffkine in his experiments followed the line of preventive inoculation as employed against cholera. Bouillon cultures were used in which floating drops of butter were employed to make the islands of plague bacilli float. The cultures were grown for a month or so, successive crops of the island-stalactite growth as it formed having been precipitated by agitating the tube. In this manner there was obtained an "intense extra-cellular toxin" containing large numbers of the bacilli. The culture was killed by exposure to a temperature of 70° C. for one hour, and the mixture used in doses of about 3 c.cm. as a preventive inoculation. In the Byculla Gaol, where Haffkine's experiments numbered over one hundred, a decided prophylactic effect was observed in twelve to fourteen hours in men already advanced in the stage of incubation.

Wyssokowitz and Zabolotmy, whose studies have already been quoted, used 96 monkeys in the study of the value of the "plague-serums," and found that when the treatment is begun within two days from the time of inoculation the animals can be saved, even though symptoms of the disease are marked. After the second day the treatment cannot be relied upon. The dose necessary was 20 c.cm. of a serum having a potency of 1 : 10. If too little serum was given, the course of the disease was slowed, the animal improved for a time and then suffered a relapse, and died in from thirteen to seventeen days. The serum also produced immunity, but of only ten to fourteen days' duration. Immunity lasting three weeks was conferred by inoculating a monkey with an agar-agar culture heated to 60° C. If too large a dose of such a culture was given, however, the animal was enfeebled and remained susceptible.

CHAPTER XII.

TETRAGENUS.

THERE can sometimes be found in the normal saliva, more commonly in tuberculous sputum, and still more commonly in the cavities of tuberculosis pulmonalis, a large micrococcus grouped in fours and known as the Micrococcus tetragenus (Fig. 123). It was discovered by

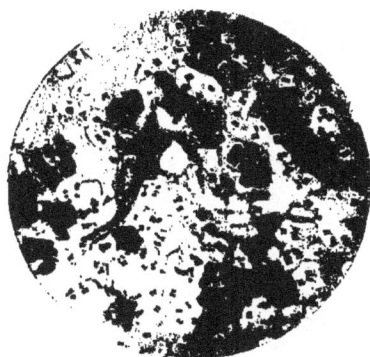

FIG. 123.—Micrococcus tetragenus in pus from a white mouse; × 615 (Heim).

Gaffky, and subsequently carefully studied by Koch and Gaffky. It sometimes occurs in the pus of acute abscesses, and may be of importance in connection with the pulmonary abscesses which so often complicate tuberculosis.

The cocci are rather large, measuring about 1 μ in diameter. In cultures they show no particular arrangement among themselves, but in the blood and tissues of animals they commonly appear arranged in groups of four surrounded by a transparent gelatinous capsule.

The organism stains well by ordinary methods, and

443

most beautifully by Gram's method, by which it can be best demonstrated in tissues.

Upon gelatin plates small white colonies are produced in from twenty-four to forty-eight hours. Under the microscope they are found to be spherical or elongate (lemon-shaped), finely granular, and lobulated like a raspberry or a mulberry. When superficial they form white, elevated, rather thick masses 1–2 mm. in diameter (Fig. 124).

In gelatin punctures a large white surface-growth

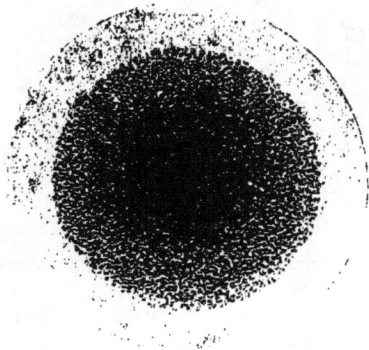

FIG. 124.—Micrococcus tetragenus: colony twenty-four hours old upon the surface of an agar-agar plate; × 100 (Heim).

takes place, but very scant development occurs in the puncture, where the small spherical colonies generally remain isolated.

Upon the surface of agar-agar spherical white colonies are produced. They may remain isolated or may become confluent.

Upon potato a luxuriant thick, white growth occurs.

The growth upon blood-serum is also abundant, especially at the temperature of the incubator. It has no distinctive peculiarities.

The introduction of tuberculous sputum or of a most minute quantity of a pure culture of this coccus into white mice generally causes a fatal septicemia.

The organisms are found in small numbers in the heart's blood, but are numerous in the spleen, lungs, liver, and kidneys.

House-mice and field-mice are comparatively immune; dogs and rabbits are also highly resistant. Guinea-pigs sometimes die from general infection, though sometimes local abscesses may be the only result of subcutaneous inoculation.

The tetragenococci are of no special importance in human pathology, but probably hasten the tissue-necrosis in tuberculosis pulmonalis, and may aid in the formation of abscesses of the lung and contribute to the production of the hectic fever.

CHAPTER XIII.

INFLUENZA.

NOTWITHSTANDING a large number of bacteriologic examinations conducted for the purpose of determining the cause of influenza, it was not until 1892, after the great epidemic, that there was found simultaneously by Canon and Pfeiffer a bacterium which conformed, at least in large part, to the requirements of specificity.

The observers mentioned found the same organism—one in the blood of influenza patients, the other in the purulent bronchial discharges.

The specific organisms (Fig. 125) are bacilli, very small in size, having about the same diameter as the bacillus

FIG. 125.—Bacillus influenzæ, from a gelatin culture; × 1000 (Itzerott and Niemann).

of mouse-septicemia, but only about half as long (0.2 by 0.5 μ). They are usually solitary, but may be united in chains of three or four elements. They stain rather poorly, except with such concentrated penetrating stains as carbol-fuchsin and Löffler's alkaline methylene blue, and even with these the bacilli stain more deeply at the ends than in the middle, so that they appear not a little like diplococci.

For the demonstration of the bacilli in the blood Canon recommends a rather complicated method. The blood is spread upon clean cover-glasses in the usual way, thoroughly dried, and then fixed by immersion in absolute alcohol for five minutes. The stain which seems best is Czenzynke's:

Concentrated aqueous solution of methylene blue,	40 ;
0.5 per cent. solution of eosin in 70 per cent. alcohol,	20 ;
Distilled water,	40.

The cover-glasses are immersed in this solution, and kept in the incubator for three to six hours, after which they are washed in water, dried, and then mounted in Canada balsam. By this method the erythrocytes are stained red, the leucocytes blue, and the bacillus, which is also blue, appears as a short rod or often as a dumb-bell.

Sometimes large numbers of the bacilli are present ; sometimes very few can be found after prolonged search. They are often enclosed within the leucocytes. It really is not necessary to pursue so tedious a staining method for demonstrating the bacilli, for they stain quite well by ordinary methods. They do not stain by Gram's method.

The bacillus is non-motile, and, so far as is known, does not form spores. Its resisting powers are very restricted, as it speedily succumbs to drying, and is certainly killed by an exposure to a temperature of 60° C. for five minutes. It will not grow at any temperature below 28° C.

The bacillus does not grow in gelatin or upon ordinary agar-agar. Upon glycerin agar-agar, after twenty-four hours in the incubator, minute colorless, transparent,

drop-like cultures may be seen along the line of inoculation. They do not look unlike condensed moisture, and
Kitasato makes a special point of the fact that the colonies never become confluent. The colonies may at times
be so small as to require a lens for their discovery.

In bouillon a scant development occurs, small whitish
particles appearing upon the surface, subsequently sinking to the bottom and causing a "woolly" deposit there.
While the growth is so delicate in these ordinary media,
the bacillus grows quite well upon culture-media contain-

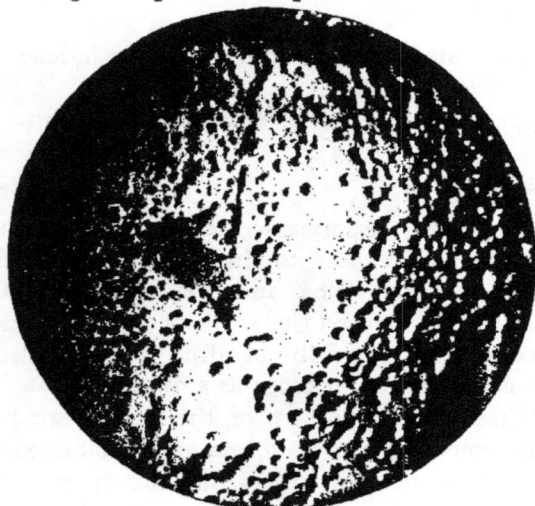

FIG. 126.—Bacillus of influenza; colonies on blood agar-agar; low magnifying
power (Pfeiffer).

ing hemoglobin or blood, and can be transferred from
culture to culture many times before it loses its vitality.

It cannot be positively proven that this bacillus is the
cause of influenza, but from the fact that the bacillus
can be found only in cases of influenza, that its presence
corresponds with the course of the disease in that it is
present as long as the purulent secretions last, and then
disappears, and that Pfeiffer was able to demonstrate its
presence in all cases of uncomplicated influenza, his conclusion that the bacillus is specific is certainly justifiable.

The bacillus is pathogenic for certain of the laboratory animals, the guinea-pig in particular being subject to fatal infection. The dose required to cause death of a guinea-pig varies considerably, in the immunization experiments of Deline and Kole[1] $\frac{1}{20}$ of a 24-hour old culture being fatal in twenty-four hours. These scholars found that the toxicity of the culture resides not in a soluble toxin, but in the bodies of the bacilli. The outcome of the researches, which were made most scientifically and

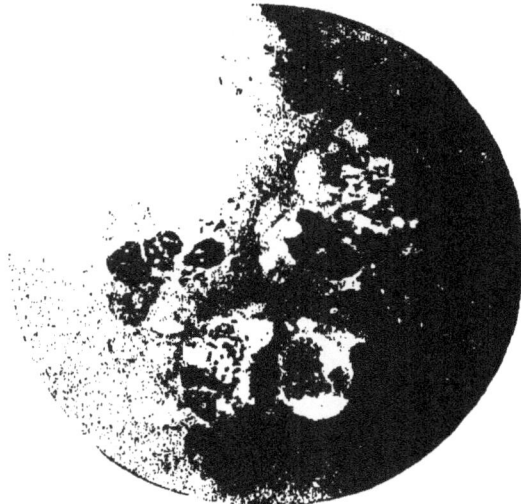

FIG. 127.—Bacillus of influenza; cover-glass preparation of sputum from a case of influenza, showing the bacilli in leukocytes; highly magnified (Pfeiffer).

painstakingly, was the total failure to produce immunity. Increasing doses of the cultures injected into the peritoneum resulted in enabling the animals to resist rather more than a fatal dose, but never enabled them to maintain vitality when large doses were administered. This discovery is in exact harmony with the familiar clinical observation that, instead of an individual being immune after an attack of influenza, he is as susceptible as before, if not more so.

[1] *Zeitschrift für Hygiene*, etc., Bd. xxiv., 1897, Heft. 2.

A. Catanni, Jr.[1] trephined rabbits and injected influenza toxin into their brains, at the same time trephining control-animals, into some of whose brains he injected water. The results were that animals thus receiving 0.5–1 mgr. of the living culture constantly died in twenty-four hours with all the nervous symptoms of the disease, dyspnea, paralysis beginning in the posterior extremities and extending over the whole body, clonic convulsions, stiffness of the neck, etc. Control-animals injected with a variety of pathogenic bacteria in the same manner never manifested similar symptoms. The virulence of the bacillus was also observed to increase rapidly when transplanted from brain to brain.

[1] *Zeitschrift für Hygiene,* etc., Bd. xxiii., 1896.

CHAPTER XIV.

MEASLES.

In 1892, Canon and Pielicke, after the investigation of fourteen cases of measles, reported the discovery of a specific bacillus in the blood in that disease.

The organism is quite variable in size, sometimes being quite small and resembling a diplococcus, sometimes larger, and occasionally quite long, so that one bacillus may be as long as the diameter of a red blood-corpuscle.

The discovery was made by means of a peculiar method of staining, as follows: The blood is spread in a very thin, even layer upon perfectly clean cover-glasses, and fixed by five to ten minutes' immersion in absolute alcohol. These glasses are then placed in a stain consisting of

Concentrated aqueous solution of methylene blue, 40;
0.25 per ct. solution of eosin in 70 per ct. alcohol, 20;
Distilled water,   40,

and stood in the incubator at 37° C. for from six to twenty-four hours. The bacilli do not all stain uniformly.

The discoverers of the bacillus claim to have made it grow several times in bouillon, but failed to induce a growth upon other media.

The bacilli do not stain by Gram's method; they seem to have motility; no spores were observed. They were found not only in the blood, but also in the secretions from the nose and eyes. They are said to persist throughout the whole course of the disease, even occasionally being found after the fever subsides.

Czajrowski asserts that the bacillus can be cultivated upon various albuminous media except gelatin and agar. On glycerin agar-agar, especially with the addition of hematogen, and on blood-serum, they should grow in three or four days with an appearance like that of dew-drops.   Under the microscope the colonies are structure-less.   Mice die of a septicemia after a subcutaneous in-oculation.

An interesting field for experimentation has been opened by Behla,[1] who seems to have successfully inoculated a sucking-pig with measles by introducing some of the nasal secretion from a case of measles into the nose, which had been prepared to receive it by scratching with a wire.

[1] *Centralbl. f. Bakt. u. Parasitenk.*, Oct. 24, 1896, Bd. xx., Nos. 16 and 17, p. 36.

D. MISCELLANEOUS.

CHAPTER I.

SYMPTOMATIC ANTHRAX.

"SYMPTOMATIC ANTHRAX," *charbon symptomatique*, *Rauschbrand*, "quarter-evil," and "black-leg" are the various names applied to a peculiar disease of cattle common during the summer season in the Bavarian Alps, Baden, Schleswig-Holstein, and some parts of the United States, characterized by the occurrence of irregular, emphysematous, crepitating subcutaneous pustules. Diseased areas are also found in the muscles, and are most common over the quarters, hence the name "quarter-evil." When incised the affected tissues have a dark color and contain a dark, bloody serum.

The micro-organismal nature of the disease had been suspected from an early date, but until the work of Faser and Bollinger the disease was confounded with anthrax. Still later, Arloing, Thomas, Cornevin, and Kitasato studied the disease, and succeeded in demonstrating the specific micro-organism, which Kitasato successfuly cultivated upon artificial media.

The bacillus which the results of these labors brought to light is a rather large individual (3–5 μ in length, 0.5–0.6 μ in breadth) with rounded ends. The bacilli are occasionally united in twos, but are never united in long chains (Fig. 128). They are actively motile (Thoinot and Masselin say scarcely at all motile) when examined in the hanging drop, but after a short time, perhaps because of the exposure to the oxygen required in the hanging-drop preparation, the movement is lost and the bacilli die. When stained by Löffler's method a considerable number of flagella can be demonstrated. Large

453

oval spores are found; by their presence they distort the bacilli in which they occur, causing them to assume a spindle shape (clostridium), or, when two are united and a spore occupies one of them, a drumstick shape. In-

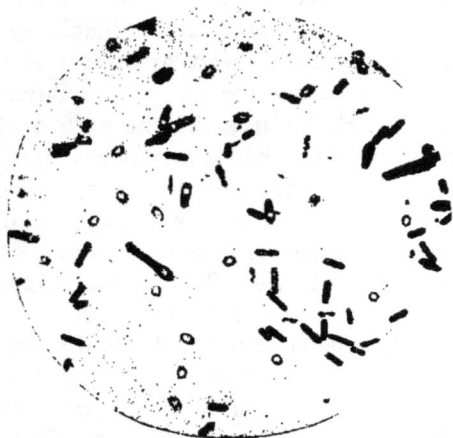

FIG. 128.—Bacillus of symptomatic anthrax, containing spores, from an agar-agar culture; × 1000 (Fränkel and Pfeiffer).

volution-forms are exceedingly common in old cultures, and are of enormous size and of granular appearance.

The bacillus can be stained with the ordinary aqueous solutions of the anilin dyes, but will not retain the color by Gram's method or Weigert's fibrin method. It can be colored in sections of tissue with Löffler's solution, and can be observed in the blood without staining shortly after death.

The spores, which can be stained by ordinary methods, are quite resistant to the action of heat and disinfectants, and withstand the effects of drying for a considerable length of time.

The bacillus of symptomatic anthrax (Fig. 129) is a strictly anaërobic, parasitic bacterium. It grows at temperatures above 18° C., but best at 37° C.

The artificial cultivation which was achieved by Kitasato is not more difficult than that of other anaërobic organisms. In gelatin containing 1 to 2 per cent. of glucose or 5 per cent. of glycerin the organism develops quite well, the exact appearance depending somewhat upon the method by which it was planted. If the bacteria are dispersed through the culture-medium, the little colonies will appear in the lower parts of the tube as nearly spherical or slightly irregular, clouded, liquefied areas containing bubbles of gas. If, on the other hand, the inoculation is made by a deep puncture, a stocking-shaped liquefaction forms along the whole lower part of the puncture, leads to considerable gas-production, and finally causes the liquefaction of all the gelatin except a thin superficial stratum. A peculiar acid odor is given off by the cultures.

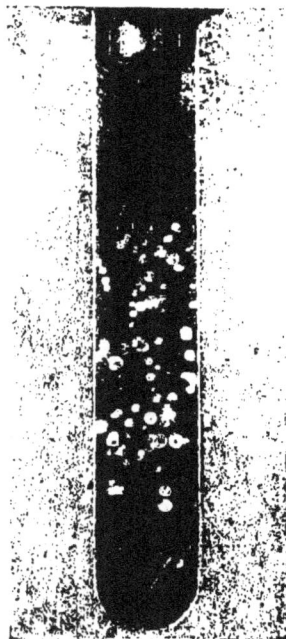

FIG. 129.—Bacillus of symptomatic anthrax : four-days-old culture in glucose-gelatin (Fränkel and Pfeiffer).

When the bacteria grow anaërobically in Esmarch tubes, the colonies are irregularly club-shaped or spherical, with a tangled mass of delicate projecting filaments visible upon microscopic examination.

In agar-agar the development is similar to that in gelatin. The gas-production is marked, the liquefaction of course absent, and the same acid odor pronounced.

The bacillus also develops quite well in bouillon, the bacillary masses sinking to the bottom in the form of whitish flakes, while the gas-bubbles collect at the top. In this medium the virulence is unfortunately soon lost.

Milk also seems to be a favorable culture-medium. The development of the bacilli is unaccompanied by coagulation.

The virulence of the organism is soon lost in all culture-media, but it is said that the virulence of the culture can be much increased by the addition to it of 20 per cent. of lactic acid.

When susceptible animals are inoculated with a minute portion of a pure culture in a little subcutaneous pocket, such as is described in connection with tetanus and malignant edema, the bacilli proceed to grow, produce the well-known affection, and lead to a certainly fatal outcome. Cows seem to be the most susceptible animals, especially those between six months and four years old; sheep and goats are also sometimes affected. Curiously enough, animals that are immune to malignant edema are seemingly more susceptible to Rauschbrand. Of the laboratory animals, the guinea-pig is most susceptible; swine, dogs, and rabbits are very slightly susceptible; horses, goats, and birds are immune.

The virulence of the bacillus is capable of ready attenuation by exposure to heat, by previous exposure of its spores to heat, or by drying combined with exposure to increased temperature. The inoculation of animals with the attenuated bacilli causes a very mild affection, followed by complete immunity to the virulent organisms. Upon this principle the "protective vaccination" is based. Kitt has, however, shown that almost the same method as that employed by Pasteur for vaccination against rabies may be employed against this bacillus, and that when muscular tissue from an animal dead of the disease is dried at a temperature of 32–35° C., and then exposed for six hours to a temperature of 100°–104° C., and a second portion is exposed in the same manner to a temperature of 90°–95° C., an emulsion of this tissue in distilled water, salt-solution, or bouillon, injected into the animals to be protected, will act in a manner resembling the pulverized spinal cords of the rabbits used in rabies, and give an almost perfect immunity. Roux and Chamberland have found that filtered cultures can also produce immunity when properly introduced into animals.

The immunity to symptomatic anthrax seems, however, to be one of degree, for Arloing, Cornevin, and Thomas found that when the bacillus was introduced into the animal body simultaneously with a 20 per cent. solution of lactic acid, either the virulence of the bacillus or the resistance of the tissues was so changed that natural immunity was destroyed and the bacteria allowed to develop and produce the disease. Roger found also that refractory animals, like the rabbit, mouse, pigeon, and chicken, could be made susceptible by the combined injection of the Rauschbrand bouillon, the Bacillus prodigiosus, Proteus vulgaris, and other harmless organisms.

When the guinea-pig is inoculated with the bacillus of symptomatic anthrax, it dies in from twenty-four to thirty-six hours. The post-mortem examination shows a bloody serum at the point of inoculation, and the muscles are dark red or black, like those of the "black-leg" of cattle. No changes are apparent in the internal organs. The bacilli are at first found near the point of inoculation in the inflammatory exudations only, but soon after death, being motile, they spread to all parts of the body.

The peculiarities of symptomatic anthrax point to the entrance of the bacteria into the animal body through wounds, but the occurrence of epidemics at certain geographical points, known technically as "Rauschbrand stations," suggests that infection may also take place through the respiratory and alimentary tracts.

At first thought, as Fränkel points out, one might imagine that an animal dead of quarter-evil and the discharges from its body might be harmless, as compared, for example, with the cadavers and discharges of anthrax, because of the purely anaërobic method of the growth of the bacillus of symptomatic anthrax and the rapidity of its death in the presence of oxygen. This is, however, untrue, for the rapid development of a permanent form in the resisting spores of the bacillus makes the pollution of the soil exceedingly dangerous for cows who subsequently browse upon it. That the spores are of great vitality is shown by the well-known laboratory method of keeping them on hand for experimental purposes, dried in the muscular tissue of a diseased animal.

Every precaution should be exerted to have the affected animals isolated, and their cadavers disinfected and destroyed or buried in such a manner that subsequent infection is impossible.

Statistical results of Guillod and Simon, based upon 3500 protective inoculations, show a distinct reduction of the death-rate from 5–20 per cent. in unprotected animals to 0.5–2 per cent. in protected animals.

CHAPTER II.

MALIGNANT EDEMA.

THE chief contaminating organism in the preparation of pure cultures of the tetanus bacillus is a large slender bacillus almost as large as that of anthrax, but with rounded ends and an individual motility accomplished by means of flagella attached to its ends and sides (Fig. 130). It is a strictly anaërobic bacterium, and was

FIG. 130.—Bacillus of malignant edema, from the body-juice of a guinea-pig inoculated with garden-earth; × 1000 (Fränkel and Pfeiffer).

originally described by Pasteur (1875) as the Vibrion septique. It grows well at the room-temperature, as well as at the temperature of the incubator, produces oval central spores, and, because of its association with a specific edema in certain animals, is known as the Bacillus œdema maligni.

The organism is widely distributed in nature, being almost always present in garden-earth. It is also found in various dusts, in the waste water from houses, and sometimes in the intestinal canals of animals.

When introduced beneath the skin this bacillus proves pathogenic for a large number of animals—mice, guinea-pigs, rabbits, horses, dogs, sheep, goats, pigs, calves, chickens, and pigeons. Cattle seem to be immune.

Günther points out that the simple inoculation of the bacillus upon an abraded surface is insufficient to produce the disease, because the oxygen which is, of course, abundant there is detrimental to its growth. When an experimental inoculation is performed, a small subcutaneous pocket should be made, and the bacilli introduced into it in such a manner as not to be in contact with the air.

If the inoculated animal be a mouse, guinea-pig, or rabbit, in about forty-eight hours it sickens and dies. The autopsy shows a general subcutaneous edema containing immense numbers of the bacilli. In the blood the bacilli are few or cannot be found, because of the oxygen which it contains. The great majority of them occupy the subcutaneous tissue, where very little oxygen is present and the conditions of growth are therefore good. If the animal is allowed to remain undisturbed for some time after death, the bacilli spread to the circulatory system and reach all the organs.

Brieger and Ehrlich have reported two cases of malignant edema in man. Both cases were typhoid-fever patients injected with musk, and developed the edema in consequence of impurity of the therapeutic agent. No case is reported, however, in which healthy men have been infected with the disease.

Cornevin declares that the passage of the bacillus through the white rat diminishes its virulence, and that the animals of various species that recover from this milder affection are subsequently immune to the virulent organisms.

The bacillus of malignant edema stains well with ordinary cold aqueous solutions of the anilin dyes, but not by Gram's method.

The organism is not a difficult one to secure in pure culture, as has been said, generally contaminating tetanus cultures and being much more easy to secure by itself than its congener. It is most easily obtained from the edematous tissues of guinea-pigs and rabbits inoculated with garden-earth.

The colonies which develop upon the surface of gelatin kept free of oxygen appear to the naked eye as small shining bodies with liquid grayish-white contents. They gradually increase in circumference, but do not change their appearance. Under the microscope they appear filled with a tangled mass of long filaments which under a high power exhibit individual movement. The edges of the

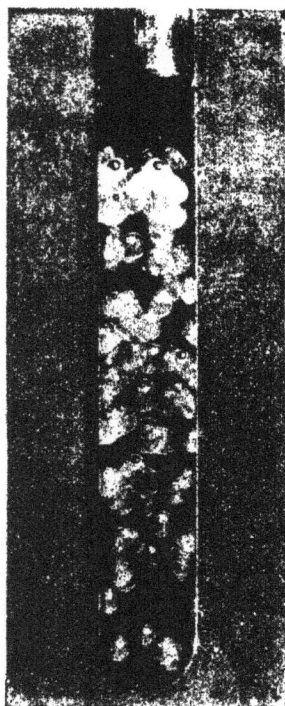

FIG. 131.—Bacillus of malignant edema growing in glucose gelatin (Fränkel and Pfeiffer).

colony have a fringed appearance, much like the hay or potato bacillus.

In gelatin tube-cultures the characteristic growth cannot be observed in a puncture, because of the air which remains in the path of the wire. The best preparation is made by heating the gelatin to expel the air it may contain, inoculating while still liquid, then replacing the air by hydrogen, and sealing the tube. In such a tube the bacilli develop near the bottom. The appearance of the growth is highly typical, as globular circumscribed areas of cloudy liquefaction result (Fig. 131), and may contain a small amount of gas. In gelatin to which a little grape-sugar has been added the gas-production is marked. The gas is partly inflammable, partly CO_2. A distinct odor accompanies the gas-production, and is especially noticeable in agar-agar cultures.

CHAPTER III.

BACILLUS AËROGENES CAPSULATUS.

THIS very interesting micro-organism was first described by Welch, and subsequently carefully studied by Welch and Nuttall,[1] and Welch and Flexner.[2] It was first secured from the body of a man dying suddenly of aneurysm with a peculiar condition of gaseous emphysema of the subcutaneous tissue and internal organs, and a copious formation of gas in the veins and arteries. The blood was thin and watery, of a lac-color, and everywhere contained large and small gas-bubbles. The blood-alteration was associated with a change in its coloring-matter, which dissolved out of the corpuscles and stained the tissues a deep red. The blood was found to contain many bacilli, which were also obtained from the various organs, especially in the neighborhood of the gas-bubbles. The bacilli were in nearly pure culture.

The bacillus is a large organism, measuring 3–5 μ in length, about the thickness of the anthrax bacillus, with ends slightly rounded, or, when joined, square (Fig. 132). It occurs chiefly in pairs and in irregular masses, but not in chains, in this particular differing very markedly from the anthrax bacillus. In culture-media the bacillus is usually straight, with slightly rounded ends. In old cultures the rods may be slightly bent, and involution-forms occur. When several bacilli are joined together the opposed ends are square-cut. The bacillus varies somewhat in size, especially in length, in different culture-media. It usually appears thicker and more vari-

[1] *Bull. of the Johns Hopkins Hospital*, July and Aug., 1892, vol. viii., No. 24.

[2] *Jour. of Exper. Med.*, vol. i., No. 1, Jan., 1896.

able in length in artificial cultures than in the blood of animals and of man. The bacilli occur singly, in pairs, in clumps, and sometimes in short chains. When united, an angle is often formed.

The bacillus is non-motile in both the ordinary hanging-drops and in anaërobic culture. No mention is made of the presence of flagella.

The organism stains well with the ordinary stains, and retains the color well in Gram's method. When stained with methylene blue a granular or vacuolated appearance

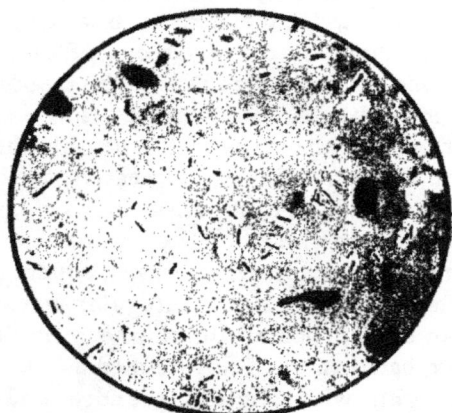

FIG. 132.—Bacillus aërogenes capsulatus (from photograph by Prof. Simon Flexner).

is sometimes observable, due to the presence of unstained dots in the protoplasm.

Usually in the body-fluids and often in cultures the bacilli are surrounded by distinct capsules—clear, unstained zones. To demonstrate this capsule to the best advantage, Welch and Nuttall devised the following special stain: a cover is thinly spread with the bacilli, dried, and fixed without over-heating. Upon the surface prepared, glacial acetic acid is dropped for a few moments, then allowed to drain off, and at once replaced by a strong aqueous solution of gentian violet, which is poured off and renewed several times until the acid has been replaced by the stain. The specimen is then examined in the coloring-solution, after soaking up the excess with filter paper, the thin layer of coloring fluid not interfering with a clear view of the bacteria and their capsules. After mounting in Canada balsam the capsules are not nearly so distinct. The width of the capsule varies from one-half to twice the thickness of the bacillus. Its outer margin is stained, leaving a clear zone immediately around the bacillus.

It was at first thought that the bacillus produced no spores, but Dunham[1] found that spores were produced upon blood-serum, and especially upon Löffler's blood-serum bouillon mixture. The spores resist desiccation and exposure to the air for ten months. They stain readily in hot solutions of fuchsin in anilin water, and are not decolorized by a moderate exposure to the action of 3 per cent. solution of hydrochloric acid in absolute alcohol. They are oval, and are usually situated near the middle of the bacillus, which is distended because of the large size of the spore and bulges at the sides.

The bacillus is *anaërobic.* It grows upon all culture-media, both at the room-temperature and at the temperature of incubation, best at the latter. The bacillus grows in ordinary neutral or alkaline gelatin, but better in gelatin containing glucose, in which the characteristic gas-production is marked. Soft gelatin, made with 5 instead of 10 per cent. of the crude gelatin, is said to be better than the ordinary medium.

There is no distinct liquefaction, but in 5 per cent. gelatin there is sometimes a softening that can be best demonstrated by tilting the tube and observing that the gas-bubbles change their position, as well as by noticing that the growth tends to sediment.

In making agar-agar cultures careful anaërobic precautions must be observed. The tubes should contain considerable of the medium, which should be boiled and freshly solidified before using. The implantation should be deeply made with a long wire. The growth takes

[1] *Bull. of the Johns Hopkins Hospital,* April, 1897, p. 68.

place slowly unless such tubes are placed in a Buchner's jar. The deeper colonies are the largest. Sometimes the growth only takes place within 10–12 mm. of the surface, at others within 3–4 cm. of it. After repeated cultivation the organism seems to become somewhat accustomed to the presence of oxygen, and will grow higher up in the tube than when freshly secured from animal tissue (see Fig. 133).

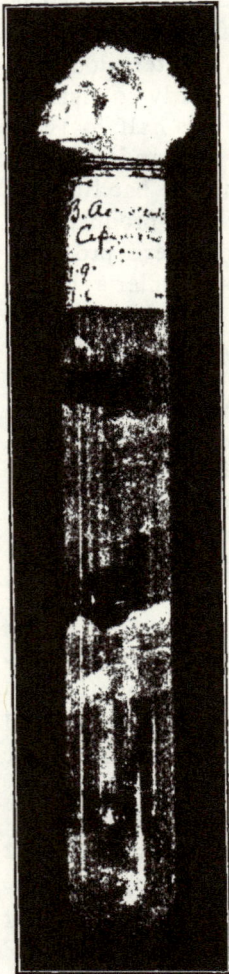

FIG. 133. — Bacillus aërogenes capsulatus, with gas-production (from photograph by Prof. Simon Flexner).

The colonies seen in the culture-media are grayish-white or brownish-white by transmitted light, and sometimes exhibit a central dark dot. At the end of twenty-four hours the larger colonies do not exceed 0.5–1.0 mm. in diameter, though they may subsequently attain a diameter of 2–3 mm. or more. Their first appearance is as little spheres or ovals, more or less flattened, with rather irregular contours, due to the presence of small projecting prongs, which are quite distinct under a lens. The colonies may appear as little irregular masses with projections.

After several days or weeks, single, well-separated colonies may attain a large size and be surrounded by projections, either in the form of little knobs or spikes or of fine branchings —hair-like or feathery. Their appearance has been compared to thistle-balls or powder-puffs and to thorn-apples. When the growth takes place in the puncture the feathery projections are continuous. Bubbles of gas make their appearance in plain agar as well as in sugar-agar, though, of course, less plentifully. They first appear in the line of growth; afterward throughout the agar, often at a distance from the actual growth. Any fluid collecting about the bubbles or at the surface of the agar-agar may be turbid from the presence of bacilli. The gas-production is more abundant at incubation- than at room-temperatures.

The agar-agar is not liquefied by the growth of the bacillus, but is often broken up into fragments and forced into the upper part of the tube by the excessive gas-production.

In its growth the bacillus produces acid in considerable amount.

In bouillon growth does not occur in tubes exposed to the air, but when the tubes are placed in Buchner's jars, or kept under anaërobic conditions, it occurs with abundant gas-formation, especially in glucose-bouillon, with the formation of a frothy layer on the surface. The growth is very rapid in its development, the bouillon becoming clouded in two to three hours. After a few days the bacilli sediment and the bouillon again becomes clear. The reaction of the bouillon becomes strongly acid.

In milk the growth is rapid and luxuriant under anaërobic conditions, but does not take place in cultures exposed to the air. The milk is coagulated in from twenty-four to forty-eight hours, the coagulum being either uniform or firm, retracted, and furrowed by gas-bubbles. When litmus has been added to the milk it becomes decolorized when the culture is kept without oxygen, but turns pink when it is exposed to the air.

The bacillus will also grow upon potato when the tubes are enclosed in an anaërobic apparatus. There is a copious gas-development in the fluid at the bottom and sides of the tube, so that the potato becomes surrounded by a froth. After complete absorption of the oxygen a thin, moist, grayish-white growth takes place upon the surface of the potato.

The vital resistance of the organism is not great. Its thermal death-point was found to be 58° C. after ten minutes' exposure. Cultures made by displacing the air with hydrogen are less vigorous than those in which the oxygen is absorbed from the air by pyrogallic acid. It was found that in the former class of cultures the bacillus generally died in three days, while in the absorption experiments it was kept alive at the body-temperature for one hundred and twenty-three days. It is said to live longer in plain than in sugar-agar. To keep the cultures alive it has been recommended to seal the agar-agar tube after two or three days' growth.

It is believed that the natural habitat of the bacterium is the soil, but there is reason to think that it occurs in the intestine at times, and it may occasionally be found upon the skin.

The pathogenic powers of the bacillus are limited, and while in some cases it seems to be the cause of a fatal outcome in infected cases, its power to do mischief in the body seems to depend upon the pre-existence of other depressing and devitalizing conditions predisposing to its growth.

Being anaërobic, the bacilli are unable to live in the circulating blood, but they grow in old clots and in cavities, such as the uterus, etc., where but little oxygen ever enters, and from such areas enter the blood and are distributed.

In support of these views Welch and Nuttall cite the result of inoculation into healthy and diseased rabbits. When a healthy rabbit is injected with 2½ c.cm. of a fresh sugar-bouillon into the ear-vein it generally recovers without any evident symptoms. One of their rabbits was pregnant, and at time of injection was carrying two dead embryos. After similar injection with but 1 c.cm. of the culture it died in twenty-one hours. It seems that the bacilli were first able to secure a foothold in the dead embryos, and there multiply sufficiently to bring about death later on.

After the death of the animal, when the blood is no longer oxygenated, the bacilli grow rapidly with a marked gas-production, which in some cases is said to have caused the bodies to swell to twice their normal size. The result of injection into guinea-pigs does not differ very much from that observed in rabbits. Gaseous phlegmons are sometimes produced.

Pigeons when inoculated subcutaneously in the pectoral region frequently succumb. Following the injection there is gas-production that causes the tissues of the chest to become emphysematous. The bird generally dies in from seven to twenty-four hours, but may live.

Intraperitoneal inoculation of animals sometimes causes fatal purulent peritonitis.

The infection as seen in man generally occurs from wounds into which dirt has been ground, as in the case of a compound, comminuted fracture of the humerus, with fatal infection, reported by Dunham, or in wounds and injuries in the neighborhood of the perineum.

Among the twenty-three cases reported by Welch and Flexner[1] we find wounds of the knee, leg, hip, and forearm, ulcer of the stomach, typhoid ulcerations of the intestine, strangulated hernia with operation, gastric and duodenal ulcer, perineal section, and aneurism, as conditions in which external or gastro-intestinal infection occurred.

Dobbin, P. Ernst, Graham Stewart and Baldwin, and Krönig have met cases of puerperal sepsis and sepsis following abortion caused by the bacillus, or in which it played an important *role.*

The symptoms following infection are quite uniform. There are usually redness and swelling of the wound, with rapid elevation of temperature and rapid pulse. The wound is usually more or less emphysematous, and discharges a thin, dirty, brownish, offensive fluid which con-

[1] *Jour. of Exper. Med.,* vol. I, No. I, Jan., 1896.

tains gas-bubbles and is sometimes frothy. Occasionally the patients recover, especially when the infected part is susceptible of amputation, but death is a more common outcome. After death the body begins to swell almost immediately; it may attain twice its normal size and be unrecognizable. Upon palpation a peculiar crepitation can be felt in the subcutaneous tissue nearly everywhere, and the presence of gas in the blood-vessels is easy of demonstration. The gas is inflammable, and as the bubbles ignite explosive sounds are heard.

At the autopsy the gas-bubbles are found in most of the internal organs, sometimes so numerously as to justify the German term "Schaumorgane" (frothy-organs). The liver especially is apt to show this frothy condition. When the tissues from such a case are hardened and examined microscopically it is found that the bubbles appear as open spaces in the tissue, the borders of which are lined with large numbers of the gas bacillus. There are also clumps of bacilli without gas-bubbles, but surrounded by tissue, whose nuclei show a disposition to fragment or disappear, and whose cells and fibers show signs of disintegration and fatty change. In discussing these changes Ernst[1] concluded that they were *ante-mortem* and due to the irritation caused by the bacillus. The gas-production he regarded as postmortem.

In the internal organs the bacillus is usually found in pure culture, but in the wound it is generally mixed with other bacteria. On this account it is difficult to estimate just how much of the damage before death is the result of the activity of the gas bacillus. That gas-production after death has nothing to do with pathogenesis during life is shown by injecting into the ear-vein of a rabbit a liquid culture of the gas bacillus, allowing about five minutes' time for the distribution of the bacilli throughout the circulation, and then killing the rabbit. In a few hours the rabbit will swell and his organs and tissues will be riddled with the gas-bubbles.

[1] Virchow's *Archiv*, Bd. 133, Heft ii.

At times, however, as in a case of Graham Stewart and Baldwin, there is no doubt that the bacillus produces gas in the tissues of the entire body during life. These observers, in a case of abortion with subsequent infection, found the patient "emphysematous from the top of her head to the soles of her feet" several hours before death.

In this case, in which the bacillus was found in pure culture, it would indeed be difficult to doubt that the fatal issue was due to the bacillus aërogenes capsulatus. Whether the fatal termination of the cases is due to the presence of gas in the vessels, or partly to that and partly to some toxic property it possesses, does not seem to have been worked out as yet. It would seem, however, to have a toxic property from the fact that the onset of the infection is first shown by the occurrence of chill, pyrexia, and rapid pulse, and from the change caused by the clumps of bacilli upon the surrounding cells of the tissues in which they occur.

CHAPTER IV.

BACILLUS PROTEUS VULGARIS (HAUSER).

THIS bacillus was first found by Hauser in decomposing animal infusions, generally in company with two closely allied forms, Proteus mirabilis and Proteus Zenkeri, which, as the experiments and observations of Sanfelice and others show, may be identical with or represent

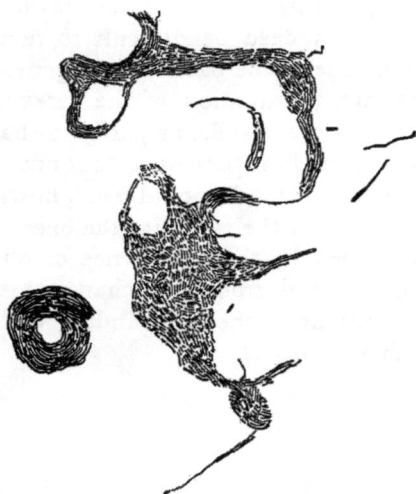

FIG. 134.—Swarming islands of proteus bacilli on the surface of gelatin; × 650 (Hauser).

attenuated forms of it. According to Kruse, it is quite probable that the old species called Bacterium termo was largely made up of the proteus.

The bacilli are very variable in size and shape—pleomorphic—and are named proteus from this peculiarity. Some forms differ very little from cocci, some are more like the colon bacteria in shape, others are found as very long filaments, and occasionally sporulina-forms are met with. True spirilla-forms are never found. All the forms mentioned may be met with in cultures of the same organism. The diameter of the bacillus is usually about 0.6 μ, but the length varies from 1.2 μ or less to 4 μ or more. No spores are formed. The organisms are actively motile. The long filaments frequently form loops and tangles. Flagella are present usually in large number; upon one of the longer bacilli as many as one hundred have been counted. Involution-forms are frequent in old cultures. The bacilli stain well by the ordinary methods. Gram's method is irregular in action, but usually fails to color the bacteria.

Upon gelatin plates a typical phenomenon is observed in connection with the development of the colonies, but for the most advantageous observation the gelatin used for making the cultures should contain only 5 per cent. of gelatin instead 10 per cent., as ordinarily used. Kruse[1] describes the phenomenon as follows: "at the temperature of the room rounded, saucer-shaped depressions, with a whitish central mass surrounded by a lighter zone, are quickly formed. Under low magnification the center of the growth is seen to be surrounded by radiations extending in all directions into the solid gelatin, and made up of chains of bacilli. Between the radiations and the granular center motile bacteria are seen in active motion. Upon the surface the colony extends as a thin patch, consisting of a layer of bacilli arranged in threads, sending numerous projections from the periphery. Occasionally filaments are found in the surroundings. Under certain conditions the wandering of the processes can be directly observed under the microscope. It depends not only upon the culture-medium, but, in part, upon the culture itself. Entire groups of bacilli or single threads, by gradual extension and circular movement, detach themselves from the colony and wander about upon the

[1] Flügge's *Mikroörganismen.*

plate. Often from the radiated central part of the colony peculiar zoöglea are formed, having a sausage- or screw-shape, or wound in spirals like a corkscrew. The younger colonies, which have not yet reached the surface of the gelatin, are more compact, rounded or nodular, later covered with hair, and then becoming radiated and like the superficial colonies."

When the culture-medium is more concentrated, or the culture one that has been frequently transplanted, the phenomenon is much less marked and sometimes does not take place at all.

Puncture-cultures in gelatin are not at all characteristic. They show a rapid stocking-like liquefaction of the gelatin, extending so as to take in the entire gelatin in the tube in a few days. Anaërobic cultures do not liquefy.

Upon agar-agar the bacillus grows with the production of a moist, thin, transparent, rapidly extending layer which probably rarely reaches the sides of the tube. Upon agar-agar plates the wandering of the colonies is also said to occur.

Upon potato the growth is in the form of a dirty-looking, smeary patch.

In culture-media containing either grape- or cane-sugar fermentation occurs both in the presence and in the absence of oxygen. Milk-sugar is not decomposed.

When grown in milk the medium is coagulated.

In its growth the bacillus usually produces a strong alkaline reaction. Indol and phenol are formed from the peptone of the culture-media. Nitrates are reduced to nitrites, and then partly reduced to NH_3. In most culture-media not containing sugar the bacillus produces a very disagreeable odor.

It is a question whether the Bacillus proteus is to be ranked among the pathogenic bacteria. Small doses of it are harmless for the laboratory animals; in large doses it produces abscesses. A toxic substance undoubtedly results from the metabolism of the organism, and is the cause of death in cases in which considerable quantities are injected into the peritoneal cavity or blood-vessels. The bacilli do not seem able to multiply in the animal body in health, but can do so when there has been previous injury to its tissues or when associated with pathogenic bacteria. In such cases, if it be enabled to grow in considerable quantity, its toxin may cause pronounced symptoms. By various observers the proteus has been secured in culture from cases of wound and puerperal infections, purulent peritonitis, endometritis, and pleurisy. When the local lesion in which it grows is small, as in endometritis, the danger of toxemia is slight, but when spread over large areas, as the peritoneum, may prove serious.

It is quite probable that in some of the cases in which blood-infection with the proteus has been found after death it did not exist previously, as the researches of Bordoni-Uffreduzzi have shown that the proteus quite regularly enters the tissues after death.

While thus apparently unable to keep up an independent existence in the tissues during life, and important in the body only in conjunction with other bacteria, the proteus seems able to grow abundantly in urine and to produce primary inflammation of the bladder when introduced spontaneously or experimentally into that viscus. The inflammatory process may extend from the bladder to the kidney, and so prove quite serious.

The Bacillus proteus has also been found in acute infectious jaundice and in acute febrile icterus, or Weil's disease.

CHAPTER V.

WHOOPING-COUGH.

It is only recently that the bacteriology of whooping-cough has begun to assume definiteness, and even yet there is no certainty that any of the various described bacteria play any specific part in its etiology. In all diseases of the respiratory apparatus the discharges are almost certain to be so contaminated with the nasal and oral bacteria as to make the isolation from them of a single probably specific organism a matter of difficulty, and its original recognition a matter of genius.

Of historical interest are the researches and observations of Deichler, Kurloff, Szemetzchenko, Cohn, Neumann, Ritter and Afanassiew. Those of Kurloff and Afanassiew are of especial importance because they opened the way for the recent studies of Koplik[1] and those of Czaplewski and Hensel.[2] Koplik and Czaplewski and Hensel worked entirely independently of each other, and while the bacterium studied by the former differs in several points from that of the latter, Czaplewski and Hensel have claimed to see in Koplik's work a confirmation of their own.

Koplik studied 16 cases of whooping-cough. The sputum was collected in sterile Petri dishes, in which it was allowed to stand for an hour or so in order that it should break up into mucous fragments.

When the clear viscid expectoration from uncomplicated cases of whooping-cough is allowed to stand for an

[1] *Centralbl. f. Bakt. u. Parasitenk.*, Sept. 15, 1897, xxii., Nos. 8 and 9, p. 222.

[2] *Deutsche med. Woch.*, 1897, No. 57, p. 586, and *Centralbl. f. Bakt. u. Parasitenk.*, Dec. 22, 1897, xxii., Nos. 22, 23, p. 641.

hour or so it separates into a fluid portion and a mass of whitish, opalescent, irregularly formed flakes or fragments. These were selected for study, and were transplanted by means of a platinum-wire hook to the culture-media. Czaplewski and Hensel used a rather better technique than this, and secured purity of the bacteria in the flakes by transferring them to a test-tube containing pepton solution and violently agitating the tube to wash off foreign bacteria. After washing, the flakes were sown upon culture-media.

Hydrocele-fluid was found most useful as a culture-fluid, but particles of sputum were planted upon all the culture-media, and attempts to cultivate bacteria from them were conducted both aërobically and anaërobically. In 13 out of the 16 cases the same bacillus (x) was isolated. The organism when stained and examined microscopically appeared as a remarkably short and delicate bacillus, shorter and more slender than the diphtheria bacillus, measuring about 0.8–1.7 μ in length and about 0.3–0.4 μ in breadth. When stained it appeared somewhat granular, and so resembled somewhat the diphtheria bacillus. Old cultures presented similar involution-forms to those seen in old cultures of the diphtheria bacillus. In general the bacillus resembles the organism found by Afanassiew[1] and others in cover-glass specimens of whooping-cough sputum, but differs in that spores were seen several times.

In pure cultures on coagulated hydrocele-fluid the bacillus forms a finely granular layer of pearl-white color.

On agar-agar the cultures are opaque, pearl-white, and occur as a thin layer.

The colonies upon agar-agar are whitish by reflected light, and straw-yellow or deeper olive-green by transmitted light. They are of an irregularly rounded shape and are granular.

In gelatin puncture-cultures the growth resembles that of the streptococcus, forming along the track of the wire

[1] *St. Petersburger med. Woch.*, 1887, Nos. 39–42.

a line of finely granular, non-liquefying colonies. Upon the surface of the gelatin the growth expands so·as to form the so-called "nail-growth."

The colonies upon gelatin have an irregularly circular form, appear white or straw-yellow by reflected light and olive-green by transmitted light, and are granular. They do not liquefy and do not grow to large colonies.

In bouillon after twenty-four hours there was a faint clouding of the liquid and subsequently a sedimentation of the bacteria in small clusters. After a week or so the surface of the medium is covered with a delicate pellicle, which grows thicker with the passage of time.

The bacillus grows quite well anaërobically. It is motile.

The bacillus is pathogenic for mice, but does not produce characteristic symptoms in any of the experiment-animals.

In discussing the results of Koplik's work, and comparing it with their own, which very shortly preceded it, Czaplewski and Hensel suggest that the bacillus is better described as a bacterium than as a bacillus. The finely granular ("fein punktiertes") appearance described by Koplik, in their observations seems to consist of a deeper staining at the poles of the cells. The growths on gelatin and on Löffler's blood-serum mixture correspond in every way. The agar-agar growths are similar, though a slight difference in color is noted, and is attributed to a difference in the quality of the medium used. The bouillon culture differs, the description of Czaplewski and Hensel being as follows: at the end of a day at 37° C. the bouillon is scarcely clouded. At the bottom of the tube is a sharply defined, lentil-like sediment, which arises in the form of slimy threads when the fluid is whirled about, and mixes with the fluid when energetically shaken. Neither bacillus grows on potato. Koplik's bacillus was also peculiar in that it was motile. Regarding Koplik's bacillus as identical with their own, Czaplewski and Hensel do not agree with him in believing it to be the same as that described by Afanassiew, and by comparison found the latter to be a much larger, shorter, more elongate bacillus. Czaplewski and Hensel's studies embraced 44 cases of whooping-cough, in which the bacillus was isolated 18 times; 5 cases of bronchitis, which subsequently developed whooping-cough, in all of which it was found; and 1 case of rhinitis and bronchitis which developed whooping-cough, and in which it was found on three different occasions.

From the preceding, it will be seen that many scholars have labored to detect the specific organism of this disease. At present several agree upon the presence of a certain bacillus in the expectorated matter; but none of them have yet succeeded in producing the disease or any modification of it in the lower animals. The specificity is, therefore, a matter of much doubt, and rests solely upon the constancy of the presence of the micro-organism in the sputum.

INDEX.

31

32

Registered Telegraph Address, - - *"HIRSCHFELD, LONDON"*

CATALOGUE OF
HENRY KIMPTON'S
STANDARD
MEDICAL PUBLICATIONS.
82, HIGH HOLBORN,
LONDON, W.C.

The American Text=Books of Dentistry.
IN CONTRIBUTIONS BY EMINENT AUTHORITIES.

PROSTHETIC DENTISTRY. Edited by CHARLES J. ESSIG, M.D., D.D.S., Professor of Mechanical Dentistry and Metallurgy, Department of Dentistry, University of Pennsylvania, Philadelphia. In one royal octavo volume of 760 pages, with 983 engravings. Cloth, gilt top. 26s net.

OPERATIVE DENTISTRY. Edited by EDWARD C. KIRK, D.D.S., Professor of Clinical Dentistry, University of Pennsylvania, Department of Dentistry. In one royal octavo volume of 702 pages. Cloth, gilt top. Price 26s net.

A TEXT-BOOK OF DENTAL PATHOLOGY AND THERAPEUTICS, including Pharmacology; being a Treatise on the Principles and Practice of Dental Medicine for Students and Practitioners. By HENRY H. BURCHARD, M.D., D.D.S., Special Lecturer on Dental Pathology and Therapeutics in the Philadelphia Dental College. One volume, royal 8vo, 287 pages, with 388 engravings and 2 coloured plates. Cloth, gilt top. Price 22s net.

"The tendency of Essig's book will be to raise the standard of mechanical dentistry, not merely in name, but in fact. No one can rise from the reading of the pages here presented without a feeling of increased respect for the basis of his profession—prosthetic dentistry. No more thorough production will be found either in this country or in any country where dentistry is understood as a part of civilisation."—*The International Dental Journal.*

A

Reviews continued from previous page :—

"This work is very much the best in its line in our literature. It is written and edited by masters in their art. It is up-to-date in every particular. It is a practical course on prosthetics which any student can take up during or after college."—*Dominion Dental Journal.*

"The editor and his collaborators are to be congratulated upon having presented all that can be regarded as really essential in such an acceptable manner. As a specimen of book-making, the publishers have added another to their list of text-books *par excellence.* The illustrations, typography, paper, and press-work are beyond criticism. As a text-book on Prosthetic Dentistry it is a decided step in advance of anything that has appeared on that subject."—*The Dental Cosmos.*

"The appearance of this magnificent work marks an important era in dental literature and in dental art, elevating, as it does, the mechanical side of dentistry to the same scientific plane which has been reached in the operative. This volume will unquestionably take merited rank as the standard text-book on Prosthetic Dentistry. The several contributors have been chosen with wonderful discrimination for the chapters which they have written. The illustrations are especially worthy of note, far surpassing anything that has heretofore been seen.

"As the third of the series of *American Text-Books of Dentistry*, Dr. Burchard's work has a high standard already set for it, but we confidently believe that all expectations will be met. The author is widely and favourably known, not only as a teacher, but as an illuminating writer on a broad range of dental subjects, and in fact many chapters in *Kirk's Operative Dentistry* and *Essig's Prosthetic Dentistry* are from his pen. It will be noted that these three works cover the principal fields of dental instruction, and moreover that Dr. Burchard has cleverly selected subjects for his new work which are naturally cognate and therefore best taught in conjunction. He has approached them from the conviction that the rational practice of dentistry must be laid upon the same basis which underlies scientific medicine and surgery—namely, accurate pathology. Accordingly he discusses every dental disease from this standpoint, bringing to bear a full acquaintance with dental literature and collateral sciences, and he is thus enabled to complete a system of dental medicine by furnishing rational recommendations as to treatment in ample detail. The section on Dental Pharmacology, though condensed, is the result of equally careful thought, and will answer every need.

"The series of illustrations is rich, not only in number but in instructiveness. An exceptionally large proportion of the engravings are from the author's own drawings, and being thus produced in immediate connection with his singularly clear text they possess unique directness in aiding its assimilation."

THE AMERICAN SYSTEM OF PRACTICAL MEDICINE. In Contributions by Eminent Clinicians. Edited by ALFRED LEE LOOMIS, M.D., L.L.D., late Professor of Pathology and Practical Medicine in the New York University, and WILLIAM GILMAN THOMSON, M.D., Professor of Materia Medica, Therapeutics, and Clinical Medicine in the New York University. In four handsome imperial octavo volumes, containing from 900 to 1,000 pages each, fully illustrated in colours and in black. In Roxburgh Binding, gilt top. Price per volume, £1 5s net.

Vol. I.—Infectious Diseases.

Vol. II.—Diseases of the Respiratory and Circulatory Systems, and of the Blood, Kidneys, Bladder, and Prostate Gland.

Vol. III.—Diseases of the Digestive System, of the Liver, Spleen, Pancreas, Thyroid and other Glands. Addison's Disease, Drug Habits, Infectious Diseases Common to Man and Animals.

Vol. IV.—Diseases of the Nervous System and of the Muscles. Gout, Rheumatism, Diabetes, and Rachitis. Diseases of Doubtful Origin, Insolation, etc.

The phenomenal growth which has of late characterised medical science so rapidly renders obsolete existing works that there is an admitted necessity for a comprehensive System of Medicine which shall collect and embody in convenient form the present aspect of the science in its most advanced condition as viewed from the standpoint of the experience of the clinicians and teachers who are acknowledged leaders in professional thought and in practical work. A reference to the list of contributors will show the generous rivalry with which the most distinguished men—from the East and the West, from the North and the South, from all the prominent centres of education, and from all the hospitals which afford special opportunity for study and practice, have united in bringing together their aggregate of specialised knowledge.

The design of The American System of Medicine is to present a thoroughly practical work of ready reference for the practitioner of general medicine. Extended historical statements and discussions of mooted theories have been omitted, but each author has sought to present the results of his personal experience and to combine them with the views of other acknowledged authorities. In conformity with the practical character of The System, it will contain no general articles upon Hygiene, Bacteriology, Pathology, or Symptomatology, but these subjects will be separately presented in connection with each disease, thus facilitating consultation, by making each article a complete practical treatise in itself. Much original research and investigation have been undertaken by the authors expressly for this work, the results of which the reader will find both in the text and in the illustrations. The latter have been made a special feature of those articles which admit of such elucidation. Minute details are given in each practical subject, such as the examination of the blood in malaria and in anæmia, the examination of the sputa, the physical diagnosis of the chest, the localisation of disease of the brain, spinal chord, etc. Particular attention has everywhere been bestowed to give full directions for treatment, original prescriptions, formulas, diagrams, charts, and tables being inserted wherever their admission seemed desirable. Much care has been devoted to the preparation of the indexes, as upon their completeness depend very greatly the convenience and utility of The System.

The American System of Medicine is therefore a work of which the American profession may reasonably feel proud, in which the practitioner will find a safe and trustworthy counsellor in the daily responsibilities of practice, and for which the publishers confidently anticipate a circulation unexampled in the annals of medical literature. That these expectations are fully justified will appear from the list of contributors.

"In the style and substance of its contributions, as well as in the handsome manner in which the work has been produced, it leaves nothing to be desired. . . . The work is, indeed, one of a high class, and will be all the more welcome as the selection of writers has evidently been made with great judiciousness, so that their contributions fully represent the high position which American medicine may well claim to have attained."—*Lancet.*

"This System of Medicine will prove one of the most important contributions of the kind ever made to medical literature."—*Scottish Medical and Surgical Journal.*

"It is impossible to speak in detail of the twenty-eight articles contained in this, the first, volume, but there can be no question as to their value, and we warmly recommend the first instalment of a work which bids fair to do honour to the fair name of America in the world of medicine."—*British Medical Journal.*

AULDE.—The Pocket Pharmacy, with Therapeutic Index : A *Resumé* of the Clinical Applications of Remedies adapted to the Pocket Case, for the Treatment of Emergencies and Acute Diseases. By JOHN AULDE, M.D., Member of the American Medical Association. Crown 8vo. Price 7s 6d.

BROWN.—The Animal Alkaloids, Cadaveric and Vital; or the Ptomaines and Leucomaines Chemically, Physiologically, and Pathologically considered in Relation to Scientific Medicine. By A. M. BROWN, M.D. With an Introduction by Professor ARMAND GAUTIER, of the Faculté de Médicine of Paris, and Member of the Académie de Médicine and of the Académie des Sciences. Second edition. Price 7s 6d.

"The book is brief, well-written, and easily studied. . . As a first edition of an advanced and most important subject, it promises a distinguished career."—Benjamin Ward Richardson, in *The Asclepiad*.

"The pages in which these views are elaborated are full of interest."—*Lancet*.

"As among the notable books, mention must be made of Dr. Brown's 'Animal Alkaloids,' which stands out as the most original work of the year; one opening up a new line of thought and investigation."—*The Medical Press and Circular*.

BUSHONG.—Modern Gynæcology, a Treatise on Diseases of Women, comprising the results of the latest investigations and treatment in this branch of Medical Science. By CHARLES H. BUSHONG, M.D., Assistant Gynæcologist to the Demilt Dispensary, New York, etc. One volume, 8vo, illustrated, 380 pages. Price 7s 6d net.

"This work will prove a very useful addition to the library of every medical man who has much to do with the treatment of the diseases peculiar to women. It is evidently the outcome of the author's personal experience in dealing with this class of disease, and therefore possesses the advantage of being thoroughly practical. Another point in its favour is that it deals with its subject from the general practitioner's point of view rather than that of the specialist. The symptoms of each disease are clearly stated, and the methods of treatment are those which have been well tried and approved by experience. In inflammatory affections of the generative organs, Dr. Bushong is a thorough believer in the benefits of complete rest, copious hot water injections, and of plugs of cotton wool well soaked in glycerine. To the use of the bromides and ergot, he adds that of hydrastus canadensus and salix nigra. He also speaks highly of hamamelis. We have read the book with pleasure, and feel justified in recommending it to our readers."—*Quarterly Medical Journal*.

CERNA.—Notes on the Newer Remedies, their Therapeutic Applications and Modes of Administration. By DAVIE CERNA, M.D., Ph.D., Demonstrator of and Lecturer on Experimental Therapeutics in the University of Pennsylvania. Forming a small octavo volume of 253 pages. Price 5s net. *Second Edition, Revised and greatly Enlarged.*

The work takes up in alphabetical order all the Newer Remedies, giving their physical properties, solubility, therapeutic applications, administration, and chemical formula.

In this way it forms a very valuable addition to the various works on Therapeutics now in existence.

Chemists are so multiplying compounds that if each compound is to be thoroughly studied, investigations must be carried far enough to determine the practical importance of the new agents.

"The volume is a useful one, and should have a large distribution."—*Journal of the American Medical Association*.

"These 'Notes' will be found very useful to practitioners who take an interest in the many newer remedies of the present day."—*Edinburgh Medical Journal*.

CHAPMAN. — Medical Jurisprudence and Toxicology. By HENRY C. CHAPMAN, M.D., Professor of Institutes of Medicine and Medical Jurisprudence in the Jefferson Medical College of Philadelphia: Member of the College of Physicians of Philadelphia, of the Academy of Natural Sciences of Philadelphia, of the American Philosophical Society, and of the Zoological Society of Philadelphia. 232 pages, with 36 illustrations, some of which are in colours. Price 5s net.

For many years there has been a demand from members of the medical and legal professions for a medium-sized work on this most important branch of medicine. The necessarily prescribed limits of the work permit the consideration only of those parts of this extensive subject which the experience of the author as coroner's physician of the city of Philadelphia for a period of six years leads him to regard as the most material for practical purposes.

Particular attention is drawn to the illustrations, many being produced in colours, thus conveying to the layman a far clearer idea of the more intricate cases.

"The salient points are clearly defined, and ascertained facts are laid down with a clearness that is unequivocal."—*St. Louis Medical and Surgical Journal.*

"The presentation is always thorough, the text is liberally interspersed with illustrations, and the style of the author is at once pleasing and interesting."—*Therapeutic Gazette.*

"One that is not overloaded with an unnecessary detail of a large amount of literature on the subject, requiring hours of research for the essential points in the decision of a question; that contains the most lucid symptomatology of questionable conditions, tests of poisons, and the readiest means of making them —such is the new book before us."—*The Sanitarian.*

COBLENTZ.—Handbook of Pharmacy, embracing the theory
and practice of Pharmacy and the art of dispensing. For students of Pharmacy and Medicine, Practical Pharmacists, and Physicians. By VIRGIL COBLENTZ, Ph.G., Phil.D., F.C.S., etc., Professor of Pharmacy and Pharmaceutical Chemistry, and Director of the Pharmaceutical Laboratory in the College of Pharmacy of the City of New York; Fellow of the Chemical Societies of London and Berlin, of the Society of Chemical Industry, etc., etc. Second edition, revised and enlarged, 572 pages, with 437 illustrations. Price 18s net.

DA COSTA.—A Manual of Surgery, General and Operative.
By JOHN CHALMERS DA COSTA, M.D., Demonstrator of Surgery, Jefferson Medical College, Philadelphia; Chief Assistant Surgeon, Jefferson Medical College Hospital; Surgical Registrar, Philadelphia Hospital, &c. Second edition in preparation.

A new manual of the Principles and Practice of Surgery, intended to meet the demands of students and working practitioners for a medium-sized work which will embody all the newer methods of procedure detailed in the larger text-books. The work has been written in a concise, practical manner, and especial attention has been given to the most recent methods of treatment. Illustrations are freely used to elucidate the text.

DAVIS.—A Manual of Practical Obstetrics. By EDWARD J.
DAVIS, A.M., M.D., Clinical Lecturer on Obstetrics in the Jefferson Medical College, Professor of Obstetrics and diseases of children in the Philadelphia Polyclinic, &c., &c., with 140 illustrations, several of which are coloured, 298 pages. Price 6s net.

DENCH.—Diseases of the Ear. A Text-book for Practitioners
and Students of Medicine. By EDWARD BRADFORD DENCH, Ph.B., M.D., Professor of Otology in the Bellevue Hospital Medical College; Aural Surgeon to the New York Eye and Ear Infirmary, &c., 8vo, 645 pages. With 8 coloured plates and 152 illustrations in the Text. Cloth. 21s net.

"This is a work of excellence, and well adapted, as its name implies, for both the practitioner and the student. The chapters on anatomy and physiology are complete, and the facts are presented with a clearness that must certainly aid the student. . . . The work is a valuable addition to otological literature, and will prove of great service to every practitioner and student."—*Cincinnati Lancet-Clinic.*

"The present volume combines all the good points of the most recent works, as well as descriptions of the various manipulative procedures, for the benefit of those not familiar with the subject. In a careful reading of this work one is impressed with the numerous excellent illustrations; with the details of the writer's personal experience, and the care with which he has selected cases for operation. The neglect of this care has been the great cause for which operations on the middle ear have been decried. . . . After a careful examination of the volume, we consider it not only the best work, but also the most practical text-book in the English language."—*Medical and Surgical Reporter.*

"Dr. Dench, although still a young man, has attained prominence as an aurist, rivalling . . . other leaders in this country, and has written what is probably, all things considered, the best American text-book on this subject to-day. . . . The work is up-to-date in every respect. It is written in a clear and interesting style, and the print is all that could be desired."—*Indiana Medical Journal.*

The name of the author is so well known in connection with advanced aural surgery, that one approaches this volume with feelings of the greatest anticipation. feelings which are truly satisfied, for there exists but one or two works on Aural Surgery which can compare with it, and they are all of slightly older issue. This volume is by far the most scientific work of its kind. It is complete, full of detail, and exhibits at the same time the knowledge and skill of the writer, and his aptitude in teaching the same. . . .

The portion of the work devoted to Anatomy and Physiology is exceptionally clearly rendered, the plates being excellent as well as numerous."—*Treatment.*

DEXTER.—The Anatomy of the Peritonæum. By FRANKLIN

DEXTER, M.D., Assistant Demonstrator of Anatomy, College of Physicians and Surgeons (Columbia University), New York. With 38 full-page illustrations in colours. Price 6s net.

DORLAND—A Manual of Obstetrics. By W. A. NEWMAN

DORLAND, A.M., M.D., Assistant Demonstrator of Obstetrics, University of Pennsylvania ; Instructor in Gynæcology in the Philadelphia Polyclinic ; one of the consulting Obstetricians to the South-Eastern Dispensary for Women ; Fellow of the American Academy of Medicine. With 163 illustrations in the text, and 6 full-page plates. 760 pages. Price 12s net.

"Among the many recent manuals of midwifery—and truly their name is legion —the work now under review deserves more than a passing notice. In all its parts the book shows evidence of great care and up-to-dateness, a remark which applies even to the somewhat recondite matters of fœtal disease and deformity, often very inadequately discussed in obstetric text-books. By the help of paragraphing, italicising, and numbering, the information is made easy of access to the busy practitioner, and the diagnostic tables, of which there are many, will doubtless serve a useful end. The illustrations are plentiful and good."—*Scottish Medical and Surgical Journal.*

DRUMMOND. — Diseases of Brain and Spinal Cord;

Their Diagnosis, Pathology, and Treatment. By DAVID DRUMMOND, M.A., M.D., T.C.D., et Dunelm, Physician to the Infirmary, Newcastle-on-Tyne. 8vo, 300 pages, with 50 illustrations. 10s 6d.

FICK.—Diseases of the Eye and Ophthalmoscopy. A Hand-

book for Physicians and Students. By Dr. EUGENE FICK, University of Zurich. Authorised Translation by A. B. HALE, M.D., Assistant to the Eye Department, Post-Graduate Medical School, and Consulting Oculist to Charity Hospital, Chicago ; late Vol. Assistant, Imperial Eye Clinic, University of Kiel. With a Glossary and 157 illustrations, many of which are in colours. Octavo. 21s net.

Fick represents the ambitious Zurich school, a middle ground between the German and French schools. This book takes an entirely unoccupied place in German literature. It is compact, thorough, and exhaustive, has no padding in the way of statistics or unnecessary pathology. Its physics is clearer and more orderly than in the majority of books, while its arrangement is far superior and more logical. The treatment is modern, simply and plainly given. Disputed and special operations do not occupy an unequal amount of space. The translator will also assume the *role* of editor, and adapt the text, when necessary, to American and English methods, and has added sections on skiascopy, etc. Dr. Fick has contributed some special notes for this edition.

"The volume before us has been written by the author, because he is of opinion that the best text-books of ophthalmology are too exhaustive, and he has endeavoured to supply the student with a compact treatise, in which pathological statements and hypotheses, as well as authorities, should be referred to, only so far as they may be necessary to illustrate diseased conditions, and which might prove supplementary and complementary to the clinical study of diseases. The translation, we may say at once, is creditable to Dr. Hale of Chicago. It reads easily, and is, as a rule, satisfactory. . . .

"The treatise is divided into two parts, the first dealing with the methods of examination, including the means of determining the acuteness of vision and errors of refraction, the sense of light and of colour, the field of vision, and the tests for binocular vision and for strabismus, and giving also an account of the objective methods of examination, such as keratoscopy, oblique illumination, and the use of the ophthalmoscope. The second part is devoted to the diseases of the eye, which are considered in the usual topographical order, each being preceded by a short account of the histology of the part. The observations made by the author are, as a rule, those of an unprejudiced mind, and although they might, in some instances, have been extended with advantage, yet they are sufficiently intelligible. . . .

"The book is a valuable one, and represents truthfully and well the present state of ophthalmic science and practice."—*Lancet.*

FROTHINGHAM.—A Guide to the Bacteriological Laboratory. By Langdon Frothingham, M.D. Illustrated. Price 4s net.

The technical methods involved in bacteria culture, methods of staining, and microscopical study are fully described and arranged as simply and concisely as possible. The book is especially intended for use in laboratory work.

GARRIGUES.—Diseases of Women. By Henry J. Garrigues, A.M., M.D., Professor of Obstetrics in the New York Post-Graduate Medical School and Hospital ; Gynæcologist to St. Mark's Hospital, and to the German Dispensary, etc., New York City. In one very handsome octavo volume of about 700 pages, illustrated by numerous wood-cuts and coloured plates. Price, cloth, 21s net.

A practical work on Gynæcology for the use of students and practitioners, written in a terse and concise manner. The importance of a thorough knowledge of the anatomy of the female pelvic organs has been fully recognised by the author, and considerable space has been devoted to the subject. The chapters on Operations and on Treatment are thoroughly modern, and are based upon the large hospital and private practice of the author. The text is elucidated by a large number of illustrations and coloured plates, many of them being original, and forming a complete atlas for studying *embryology* and the *anatomy* of the *female genitalia*, besides exemplifying, whenever needed, morbid conditions, instruments, apparatus, and operations.

Development of the Female Genitals—Anatomy of the Female Pelvic Organs—
Physiology—Puberty—Menstruation and Ovulation—Copulation—Fecundation—
The Climacteric—Etiology in General—Examinations in General—Treatment in
General—Abnormal Menstruation and Metrorrhagia—Leucorrhea—Diseases of
the Vulva—Diseases of the Perineum—Diseases of the Vagina—Diseases of the
Uterus—Diseases of the Fallopian Tubes—Diseases of the Ovaries—Diseases of
the Pelvis—Sterility.

*The reception accorded to this work has been most flattering. In the short period
which has elapsed since its issue, it has been adopted and recommended as a text-
book by more than sixty of the Medical Schools and Universities of the United
States and Canada.*

" One of the best text-books for students and practitioners which has been published in the English language ; it is condensed, clear, and comprehensive. The
profound learning and great clinical experience of the distinguished author finds
expression in this book in a most attractive and instructive form. Young practitioners, to whom experienced consultants may not be available, will find in this
book invaluable counsel and help."

THAD. A. REAMY, M.D., LL.D.

*Professor of Clinical Gynæcology, Medical College of Ohio ; Gynæcologist
to the Good Samaritan and Cincinnati Hospitals.*

GOULEY.—Diseases of the Urinary Apparatus, Phlegmasic
Affections. By John W. S. Gouley, M.D., Surgeon to Bellevue
Hospital. 355 pages. Price 7s 6d.

HARE.—Practical Diagnosis. The use of Symptoms in the
Diagnosis of Disease. By Hobart Amory Hare, M.D., Professor of Therapeutics and Materia Medica in the Jefferson Medical College of Philadelphia,
Laureate of the Medical Society of London, of the Royal Academy in
Belgium, etc., etc. Second Edition revised and enlarged. In one octavo
volume of 605 pages, with 201 engravings, and 13 coloured plates. Price, 21s
net.

" The rapidity with which a second edition has followed the first in little over
a year, is the best possible proof of the success of this book. Dr. Amory Hare has
the gift of making whatever he writes interesting, and those unacquainted with
his work could not have a better introduction than this volume. It will prove of
most value to those recently qualified, but useful and suggestive to almost every-
one. The general arrangement is to take the various parts of the body one after
another, and describe the abnormalities of signs and symptoms associated with
each. Separate chapters are devoted to the Face and Head, Hands and Arms,
Feet and Legs, and so on, with special chapters interpolated on subjects that need
fuller treatment, such as Hemiplegia and Convulsions. The illustrations are well
chosen, and good in themselves. In the chapter on the Face and Head, there are
in close juxtaposition excellent figures of a mouth-breather with post-nasal
growths, a cretin, an acromegalic, a patient with myxœdema, syphilitic ptosis,
and exophthalmic goitre, with short, pithy descriptions of the conditions represented. The grouping is unlike that which is ordinarily employed, and is therefore striking. In the chapter on the Hands and Arms there are some good photographs and skiagams of gout and rheumatoid arthritis, and progressive muscular
atrophy. Dr. Hare's large clinical experience and knowledge of students come
out well in the third chapter on the Feet and Legs, where the usual difficulties of
the different forms of paralysis are made as clear as possible by good plates and
tables. . . .

" The description of the diseases of the eye is very full and good, and the difficult subject of diplopia is well treated. There is a long and elaborate chapter on
the skin, giving practically every abnormality met with, and good coloured

diagrams are supplied of the skin areas, corresponding to the different nerve roots as mapped out by Thorburn, Starr, and Head. The chapter on the Thorax and its Viscera could not be better done; it is everywhere obvious that the statements made are the result of careful thought and experience. The latest methods of diagnosis in abdominal disease, 'the gastrodiaphane of Einhorn,' and Törck's gyromele all find their proper place. There is a good clinical account of the abnormalities of blood and urine, of forms of vomiting, and types of sputa. The index, forty-six pages in length, is excellent. DR. HARE IS TO BE CONGRATU-LATED ON HAVING WRITTEN A MOST STIMULATING AND SUGGESTIVE BOOK."—*Lancet*.

"No better criterion of the value of this handsome and beautifully illustrated volume can be given than the fact that the first edition, which was published in August, 1896, was so rapidly exhausted that a second edition had to be issued last September. The book was written as a guide to bedside practice, and that the profession needed such a book is proved by the welcome given to the first edition. A striking feature of the book is the wealth of illustration, more especially of the appearances, attitudes and deformities characteristic of certain diseases. For example, pictures of acromegaly, exophthalmic goitre, spastic paraplegia, paralysis agitans, pseudo-hypertrophic paralysis, hysterical spasm, the ape hand in progressive muscular atrophy, tabetic ulcer, etc., will enable the reader to recognise these diseases at a glance. Beautiful coloured pictures of the eye-ground in health and in certain medical diseases are given. . . .

"Much can be learnt at a glance from the coloured charts of localisation of cortical centres, and from the equally beautifully executed diagram showing course of motor fibres from cerebrum and cord to the periphery. The matter is worthy of the illustrations, and greater praise cannot be given to it. It is made very readily accessible by a very full index, which fills more than forty pages."—*Quarterly Medical Journal*.

"The warm appreciation of Dr. Hare's treatise, which we expressed in our columns only a few months since, has been fully justified by the rapidity with which a second edition has followed upon the first. The profession were in need of a book dealing with diagnosis from the standpoint of the symptoms, and this gap has now been satisfactorily filled. The second edition is no mere replica of the first, but the author has made a substantial addition of material in every part. We notice also several new engravings. However, it is high testimony to the care bestowed upon the former edition that the only important alterations in the latter are of the nature of addition and not of revision. We congratulate the author on a book that has been of value to a very varied class of readers, both students and practitioners."—*Practitioner*.

HARE.—A Text-book of Practical Therapeutics, with especial reference to the Application of Remedial Measures to Disease, and their Employment upon a Rational Basis. By HOBART AMORY HARE, M.D., B.Sc., Professor of Therapeutics and Materia Medica in the Jefferson Medical College of Philadelphia ; Physician to the Jefferson Medical College Hospital; Laureate of the Royal Academy of Medicine in Belgium, of the Medical Society of London ; Corresponding Fellow of the Sociedad Espanola de Higiene of Madrid ; Member of the Association of American Physicians ; Author of "A Text-Book of Practical Diagnosis," etc. Sixth Edition, enlarged, thoroughly revised and largely rewritten, in one royal octavo volume of 758 pages. Price, 21s net.

"The fact that this work has passed through five editions in seven years, and that a sixth is now called for, is sufficient evidence that not only has a want been supplied, but that the author has been successful in his endeavours to carry out his intention of producing a work on therapeutics which should teach a distinct practical application of remedial agents in the treatment of disease, and their employment upon a rational basis.

"The book is divided into four parts. Part I. is concerned with general therapeutical considerations, modes of administering drugs, dosage, strength and reliability of drugs, classification of drugs, etc.

B

" Part II., which occupies the main part of the work, is simply headed ' Drugs,' and contains a full description of the various mechanical agents included under that term, together with their therapeutic measures. THIS PORTION IS WORTHY OF THE HIGHEST PRAISE. After a technical description of the drug, its source and preparation, follows its physiological action on the various systems of the body ; then the modes of its elimination ; next any peculiar properties, such as an antiseptic action or toxic changes from prolonged use, etc. Then we find the symptoms of poisoning, and the measures to be adopted should such an occurrence present itself. The therapeutics are very plainly and fully considered, and conclude with ' untoward effects' (if any). Then follow the methods of administration, the doses being given according to the two systems in vogue (apothecaries and metric), and finally are the ' contraindications' for the use of each drug.

"In Part III. are described remedial measures other than drugs, and a description of the methods employed in preparing foods for the sick.

" Part IV. commences with a consideration of the various diseases, purely from a therapeutic point of view. Here also will be found a large amount of useful information. In conclusion, various tables are given—namely, ' doses of medicines,' ' tables of relative weights and measures in the metric and apothecaries' systems' ; index of drugs and remedial measures, and index of diseases and remedies.

" WE CAN THOROUGHLY RECOMMEND THIS BOOK TO PRACTITIONERS AND STUDENTS."—Lancet.

" WE STRONGLY RECOMMEND THE BOOK AS A USEFUL AID TO THE PRACTICAL WORK OF THE PROFESSION."—Scottish Medical and Surgical Journal.

" THE WORK CAN BE STRONGLY RECOMMENDED TO ENGLISH PRACTITIONERS, TO WHOM, PERHAPS, IT IS NOT SO WELL KNOWN AS IT UNDOUBTEDLY DESERVES TO BE."—Quarterly Medical Journal.

HAYNES.—A Manual of Anatomy by IRVING S. HAYNES, Ph.D., M.D., Adjunct Professor and Demonstrator of Anatomy in the Medical Department of the New York University, visiting Surgeon to the Harlein Hospital, etc., etc. With 134 half-tone illustrations and 42 diagrams. 680 pages. Price 12s net.

HEMMETER.—Diseases of the Stomach. Their special Pathology, Diagnosis and Treatment, with sections on Anatomy, Physiology, Analysis of Stomach contents, Dietetics, Surgery of the Stomach, etc. In three parts. By JOHN C. HEMMETER, M.B., M.D., Philos.D., Clinical Professor of Medicine at the Baltimore Medical College ; Consultant to the Maryland General Hospital, etc. With many illustrations, a number of which are in original colours, and a lithograph frontispiece. 1 volume, royal 8vo, 788 pages. Price 30s net.

HIRT.—The Diseases of the Nervous System. A Text-Book for Physicians and Students. By Dr. LUDWIG HIRT, Professor at the University of Breslau. Translated, with permission of the Author, by August Hoch, M.D., assisted by Frank R. Smith, A.M. (Cantab.), M.D., Assistant Physicians to the Johns Hopkins Hospital. With an Introduction by William Osler, M.D., F.R.C.P., Professor of Medicine in the Johns Hopkins University, and Physician-in-Chief to the Johns Hopkins Hospital, Baltimore. 8vo, 671 pages. With 178 illustrations. Cloth. 21s net.

HOLT.—The Diseases of Infancy and Childhood. For the Use of Students and Practitioners of Medicine. By L. EMMETT HOLT, A.M., M.D., Professor of Diseases of Children in the New York Polyclinic ; Attending Physician to the Babies' Hospital and to the Nursery and Child's Hospital, New York ; Consulting Physician to the New York Infant Asylum and to the Hospital for Ruptured and Crippled. 1 volume of 1134 pages, with 7 full-page coloured plates and 203 illustrations. Half-morocco gilt. 25s net.

"This is in every way an admirable volume, and we are genuinely pleased to congratulate Dr. Holt on his work. Its very size led us to expect something of the nature of a dictionary—a mere book of reference—but we have found it conspicuously free of the stock-in-trade of the wholesale compiler. It is a monument of labour, and labour not of collation, but the ripe fruit of the many-sided practical experience of the author himself. It is a book that we can confidently recommend to every practitioner as the best we know in this department of medicine, and full of interest and useful suggestiveness from cover to cover. And when to excellence of matter and style are linked good printing, good paper, and good binding, we have a most acceptable volume. To the pathologist also there is a special attraction in the large amount of space devoted to the morbid anatomy of infantile disease, a subject that receives sparse illustration in existing text-books ; lesions are fully described, and by means of numerous drawings, photographs, and coloured plates, brought more within the range of those whose duties withdraw them from the *post-mortem* room to the bedside. The coloured plate of acute meningitis is a masterpiece of its kind, and represents most exactly what we so often see in the deadhouse.

"Detailed attention is very properly devoted to the question of nutrition, with its derangements and associated diseases, and great stress laid upon diet and hygiene, 'since in this rather than in drug-giving lies the secret of success, certainly in all disorders of digestion and nutrition,' and there is no more promising field for therapeutic activity than the prevention of disease in children. The experience of our large Children's Hospitals goes far to show that there are two chief factors in the causation of infantile disease—bad feeding and squalor. The former we can only hope to remedy by the better instruction of ignorant mothers, and this by medical men whose therapeutic range is not entirely limited to grey powder and circumcision. We notice with pleasure a praiseworthy absence of the numerous formulæ of food-stuffs that make most text-books unreadable ; while the graphic chart method brings the essentials of composition readily to recognition. We quite agree with Dr. Holt that artificial feeding, as at present ignorantly practised, is the most fertile cause of infantile disease, and fully endorse his experience that 'it is exceedingly rare to find a healthy child who has been reared in a tenement house, and who has been artificially fed from birth.'

"A most instructive chapter is that on the 'Peculiarities of Disease in Children,' while another of no less value is devoted to a discussion of Rickets, with copious illustrations of the incident bony deformities. We should have said that the antero-posterior curvature of the lower third of the tibia was much more frequent than the author suggests, and not necessarily associated with bow legs ; indeed, it is the common and usually the only curvature in those children who, while kept off their legs, have been nursed on their mother's lap, with the leg supported in such a way as to incur the bending strain of the full weight of the foot. The carbo-hydrate phantom, too, is relegated to the subservient position it really occupies in the ætiology of the disease. The handling of the system diseases, one and all, leaves but little to be desired. True to the intention expressed on the title-page, the author caters at once both for the student and the practitioner; the general principles of treatment are so explained as to be most helpful to the uninitiated, while many practical hints of the highest value are to be found on every page. We miss many old friends—to wit, the fallacies and misstatements handed from author to author—and we welcome many new ones that are usually conspicuous by their absence. On the whole, the chapters on diseases of the lungs attract us most in this portion of the book. The statistics of pneumonia and broncho-pneumonia point to a much greater frequency of pneumonia in infancy than is generally imagined to be the case. In the first twelve months of life the highly bronchial texture of the lung favours the peribronchial variety; but after this period, as the vesicular element becomes relatively more abundant, we find at first a tendency to a mixed process, and after the third year a great preponderance of the croupous type. Thus it is that though the pneumococcus is the infective agent in almost every case of pneumonia and primary broncho-pneumonia, the anatomical distinction is maintained. In secondary broncho-pneumonia, however, there is nearly always a mixed infection, and with the familiar streptococcus are often found Friedlander's bacillus and staphylococci, of the specific germ of influenza, diphtheria, pneumonia, or tuberculosis. Dr. Holt is certainly much to be congratulated on his discussions of the bacterial agencies at work in the production of disease. We have said enough to convey the very high opinion we have formed of the whole volume, and we confidently expect that it will rank in the estimation of the profession as one of the best of many good books that have come to us from across the Atlantic."— *Practitioner.*

HYDE AND MONTGOMERY.—A Manual of Syphilis and the Venereal Diseases. By JAMES NEVINS HYDE, M.D., Professor of Skin and Venereal Diseases, Rush Medical College, Chicago, and FRANK H. MONTGOMERY, M.D., Lecturer on Dermatology and Genito-Urinary Diseases, Rush Medical College, Chicago. Profusely illustrated. Price 12s net.

This Manual is intended as a thoroughly practical guide, and represents the latest knowledge of the Venereal Diseases which are included under the heads of Syphilis, and Gonorrhœa and its complications, with very complete instructions for their diagnosis and carefully prepared instructions for their treatment, cure, and alleviation.

The illustrations (some of which are coloured) have been selected with the greatest possible care, and with the view of elucidating the text.

" We can commend this Manual to the student as a help to him in his study of venereal diseases."—*Liverpool Med.-Chi. Journal.*

" The work may safely be recommended, being modern in spirit and concise and complete."—*The Physician and Surgeon.*

INGALS.—Diseases of the Chest, Throat, and Nasal Cavaties, including Physical Diagnosis and Diseases of the Lungs, Heart, and Aorta, Laryngology and Diseases of the Pharynx, Larynx, Nose, Thyroid Gland, and Œsophagus. By E. FLETCHER INGALS, A.M., M.D., Professor of Laryngology and Practice of Medicine, Rush Medical College ; Professor of Diseases of the Throat and Chest, North-Western University Woman's Medical School ; Professor of Laryngology and Rhinology, Chicago Polyclinic ; Laryngologist to the St. Joseph's Hospital and to the Presbyterian Hospital, etc. ; Fellow of the American Laryngological Association and American Climatological Association ; Member of the American Medical Association, Illinois State Medical Society, Chicago Medical Society, Chicago Pathological Society, etc., etc. Third Edition, revised. With 240 illustrations, in one volume. Price 21s net.

KELLY.—Operative Gynecology. By HOWARD A. KELLY, A.B., M.D., Fellow of the American Gynecological Society ; Professor of Gynecology and Obstetrics in the Johns Hopkins University, and Gynecologist and Obstetrician to the Johns Hopkins Hospital, Baltimore ; formerly Associate Professor of Obstetrics in the University of Pennsylvania ; Corresponding Member of the Société Obstétricale et Gynécologique de Paris, and of the Gesellschaft Für Geburtshulfe Zu Leipzig. In two royal octavo volumes, with 24 plates and 590 illustrations. Handsomely bound in half-morocco. Gilt tops. Vol. I., 580 pages ; Vol. II., 573 pages. Price £3 3s net.

The author's aim in preparing this book has been to place in the hands of his friends who have followed his gynecological work, and before the medical public, a summary of the various gynecological operations that he has found best in his own practice. He does not undertake to present a digest of the literature of the subject, and the work is not burdened by numerous references. His claims to originality lie in his special researches in connection with the operation for suspension of the uterus, and in the investigation of vesical and ureteral diseases. THE WORK COVERS QUITE FULLY THE GENERAL FIELD OF GYNECOLOGICAL SURGERY, AND IS ENRICHED BY MORE THAN FIVE HUNDRED ORIGINAL ILLUSTRATIONS, WHICH FROM A SCIENTIFIC AS WELL AS FROM AN ARTISTIC STANDPOINT ARE EQUALLED BY THOSE OF NO OTHER WORK EXTANT. Expense and labour have not been spared in the preparation of the drawings for the illustrations, or in their reproduction for the books, of which, by their accuracy in detail and clearness in delineation, they form a very important and valuable part. The work does not appeal to the gynecological surgeon only, but is one which will be found of inestimable value to the general practitioner and to the surgeon, whose practices bring them in contact with gynecological cases. Dr. Kelly has had a long and successful career as a gynecological surgeon, and his experience has fitted him pre-eminently for the preparation of a work such as he has written.

KIMPTON'S POCKET MEDICAL LEXICON ; or, Dictionary of Terms and Words used in Medicine and Surgery.

By JOHN M. KEATING, M.D., editor of "Cyclopædia of Diseases of Children," etc., author of the "New Pronouncing Dictionary of Medicine"; and HENRY HAMILTON, author of "A New Translation of Virgil's Æneid into English Verse"; co-author of a "New Pronouncing Dictionary of Medicine." A new and revised edition. 32mo, 282 pages. Price, cloth, 2s 6d net.

This new and comprehensive work of reference is the outcome of a demand for a more modern handbook of its class than those at present on the market, which, dating as they do from 1855 to 1884, are of but trifling use to the student by their not containing the hundreds of new words now used in current literature, especially those relating to Electricity and Bacteriology.

"Remarkably accurate in terminology, accentuation, and definition."—*Journal of American Medical Association.*

"Brief, yet complete . . . it contains the very latest nomenclature in even the newest departments of medicine."—*New York Medical Record.*

KING'S MANUAL OF OBSTETRICS.—New (7th) Edition. A Manual of Obstetrics.

By A. F. KING, M.D., Professor of Obstetrics and Diseases of Women in the Medical Department of the Columbian University, Washington, D.C., and in the University of Vermont, etc. Seventh Edition, revised and enlarged. In one demy 8vo volume of 574 pages, with 223 illustrations. Cloth, 10s 6d net.

"Prof. King's Manual is now so well known that the appearance of the sixth edition calls only for a congratulatory note from the reviewer. A large number of additional illustrations have been introduced into the work, which is quite worthy of the high place it has attained in the undergraduate mind."—*Edinburgh Medical Journal.*

"For clearness of diction it is not excelled by any book of similar nature, and by its system of captions and italics it is abundantly suited to the needs of the medical student. THE BOOK IS UNDOUBTEDLY THE BEST MANUAL OF OBSTETRICS EXTANT IN ENGLISH."—*The Philadelphia Polyclinic.*

"Prof. King's Manual has had a remarkably successful career, passing rapidly from one edition to another. It is just such a work as the obstetrician turns to in time of need with the assurance that he will in a moment refresh his memory on the subject. A vast amount of knowledge is expressed in small space."—*The Ohio Medical Journal.*

"This is undoubtedly the best manual of obstetrics. Six editions in thirteen years show not only a demand for a book of this kind, but that this particular one meets the requirements for popularity, being clear, concise, and practical. The present edition has been carefully revised, and a number of additions and modifications have been introduced to bring the book to date. It is well illustrated, well arranged; in short, a modern manual."—*The Chicago Medical Recorder*.

"This popular manual now appears in the sixth edition. Published originally in 1882, and designed particularly for the students attending the author's lectures on obstetrics, the work maintains much of its peculiar character. It cannot be regarded as more than it professes to be—a manual for students and junior practitioners. There is no straining after abstruse problems, no elaborate arguments, nor portentous bibliographies. But, so far as it goes, it is an excellent and reliable guide to the junior student of midwifery. The language employed is clear and simple, and there is a healthy dogmatism about the methods of practice recommended which suggests the sort of teacher a student loves to listen to. Chapter vi., on Fecundation, is a most valuable one. We do not know of any work of similar size which treats of the early physiology of pregnancy with equal lucidity. Young practitioners will derive much help from chapter viii. on the Diseases of Pregnancy. Palpation of the abdomen for diagnosis of the position of the fœtus is clearly described at page 193. It would be well if more attention were given to this mode of examination. It is now taught as an important clinical method in America and on the Continent, but we know that our English students are less conversant with it than is desirable. The mechanism of labour is well described in accordance with generally accepted beliefs. The chapter on Symphysiotomy will be welcomed by practitioners desirous of knowing the most recent ideas from America regarding this reviving procedure. Chapter xxii. on Pelvic Deformities is terse, but clear and practical. We notice that Dr. King adopts the now favoured treatment of puerperal eclampsia in America by hypodermic injections of veratrum viride. There is a useful chapter on the Jurisprudence of Midwifery, containing many valuable points of information. We welcome this new edition, which GIVES A VERY EXCELLENT RESUME OF THE MAIN FACTS OF OBSTETRICAL THEORY AND PRACTICE, AND IS LIKELY TO PROVE AS FULLY POPULAR AS ITS PREDECESSORS."—*British Gynæcological Journal*.

LOCKWOOD.—Manual of the Practice of Medicine. By
GEORGE ROE LOCKWOOD, M.D., Professor of Practice in the Woman's Medical College and in the New York Infirmary; attending Physician to the Coloured Hospital and to the City (late Charity) Hospital; Pathologist to the French Hospital, etc. 935 pages, with 75 illustrations in text, and 22 coloured and half-tone plates. Price, 12s net.

This manual presents the essential facts and Principles of the Practice of Medicine in a concise and available form.

LONG.—A Syllabus of Gynæcology, arranged in conformity with
the American Text-Book of Gynecology. By J. W. LONG, M.D., Professor of Diseases of Women and Children, Medical College of Virginia, etc. Cloth (interleaved). Price 4s net.

"Based upon the teaching and methods laid down in the larger work, this will not only be useful as a supplementary volume, but to those who do not already possess the text-book it will also have an independent value as an aid to the practitioner in gynecological work, and to the student as a guide in the lecture-room, as the subject is presented in a manner at once systematic, clear, succinct, and practical."

McFARLAND. — Text-Book upon the Pathogenic Bacteria.
For Students of Medicine and Physicians. By JOSEPH McFARLAND, M.D. Demonstrator of Pathological Histology and Lecturer on Bacteriology in the Medical Department of the University of Pennsylvania; Fellow of the College of Physicians of Philadelphia; Pathologist to the Rush Hospital for Consumption and Allied Diseases. New edition in preparation.

"In a work of moderate size, the author has succeeded admirably in presenting the essential details of bacteriological technics, together with a judiciously chosen summary of our present knowledge of pathogenic bacteria. As indicated in the preface, the work is intended as an elementary text-book for students of medicine, but Part II., or Specific Diseases and their Bacteria, will readily commend itself to a large class of practitioners who recognise the value of acquaintance with the behaviour of the bacterial causes of disease, even without a technical knowledge of bacteriology. It is no unfavourable reflection on the scientific character of the treatise, moreover, to mention the fitness of this second part for the use of the non-professional readers who may be interested in the science or in its bearing on matters of vital general interest.

"In the Introduction the author has sketched briefly, but in a sufficiently complete and very interesting way, the history of bacteriology. The chapter on Immunity and Susceptibility is a more than usually successful attempt to briefly outline the present status of this very occult study, and in the discussion of the various theories presented, the author has not given undue prominence to any of the tenets. Tuberculosis is considered at comparative length, and all the more important relations of this subject have received attention in the practical way best adapted to the class of readers to which the book is addressed.

"Numerous photographic plates illustrate the text in the description of the various bacterial species. Of these photographs, many are very characteristic. The author has adhered with considerable uniformity to an easy and correct style of diction, which is so often lacking in the treatment of very technical subjects. The work, we think, should have a wide circulation among English-speaking students of medicine."—*New York Medical Journal.*

MAISCH'S Materia Medica.—Sixth Edition.—A Manual of Organic Materia Medica : Being a Guide to Materia Medica of the Vegetable and Animal Kingdoms. For the use of Students, Druggists, Pharmacists, and Physicians. By JOHN M. MAISCH, Phar.D., Professor of Materia Medica and Botany of the Philadelphia College of Pharmacy. New (sixth) edition, thoroughly revised by H. C. C. MAISCH, Ph.G. In one very handsome 12mo volume of 509 pages, with 285 engravings. Cloth. 10s 6d net.

"New matter has been added, and the whole work has received careful revision, so as to conform to the New United States Pharmacopœia. The great value of the work is the simplicity of style and the accuracy of each description. It considers each article of the vegetable and animal pharmacopœia, and, where important, sections on antidotes, etc., are added. Several useful tables are incorporated."—*Virginia Medical Monthly.*

"The best hand-book upon pharmacognosy of any published in this country. The revision brings the work up to date, and is in accord with its previous high standard."—*The Boston Medical and Surgical Journal.*

"We can add nothing to our previous commendatory notices of this standard text-book of materia medica. It is a work of such well-tried merit that it stands in no danger of being superseded."—*American Druggist and Pharmaceutical Record.*

OSLER.—Lectures on the Diagnosis of Abdominal Tumours.

Delivered before the Post-Graduate Class, Johns Hopkins University, By WILLIAM OSLER, M.D., Professor of Medicine, Johns Hopkins University : Physician-in-Chief to Johns Hopkins Hospital, Baltimore, M.D., Small 8vo. Illustrated. Cloth. 6s net.

"The volume before us contains six lectures delivered before the post-graduate course at the Johns Hopkins University, which have already appeared in the pages of the 'New York Medical Journal.' The first two are devoted to the stomach, the first dealing with tumours formed by the dilated stomach itself, almost always associated with a nodular mass at the pylorus. Amongst the special points to which he calls attention in reference to diagnosis may be mentioned the two kinds of movement that are observable—namely, a peristalsis that can be seen in the walls of the stomach, which occurs from left to right ; and, secondly, the development of irregular protuberances of the stomach wall, generally near the greater curvature, and often synchronous with the above-mentioned peristalsis. Another point of importance is the gurgling of gas through the pylorus, which can sometimes be felt. Inflation constitutes a most valuable aid to diagnosis, and is best effected by administering half a drachm of bicarbonate of soda in solution, followed by a similar quantity of tartaric acid, also in solution. In a few cases a tumour may be formed by a contracted stomach, as in œsophageal obstruction, or from cirrhosis or diffuse cancer of the stomach walls. The second lecture is devoted to nodular and massive tumours of the stomach, including thereby instances of thickening and induration round an old ulcer. In none of his cases was a tumour situated at the cardiac orifice or on the posterior wall. Tumours of the liver form the subject of the third lecture, cases of abscess, syphilis, and cancer being described, whilst dilated gall bladder and cancer of the gall bladder are considered in the fourth. The diagnosis of the latter condition is not always easy, but the following points would be helpful : Two-thirds of the patients are women, and in seven-eighths of the cases there is an association with gall stones, so that a history of colic and previous attacks of jaundice should be sought for. Rapid emaciation and the development of cachexia within three or four months would favour cancer ; chills and fevers would be against it ; ascites is often present, but jaundice is not necessary till the disease spreads to the walls of the duct. The fifth lecture deals with tumours of the intestine, omentum, and pancreas, and some miscellaneous cases of obscure origin, whilst the last lecture is devoted to tumours of the kidney, dealing with movable kidney, which is so common that he says they are never without an example in the wards, intermittent hydronephrosis, sarcoma of the kidney, including a very interesting case in which the tumour was successfully extirpated, and tuberculosis. The lectures are entirely *confined to a consideration of cases that had been under treatment during the preceding twelve months, and we may congratulate* Dr. OSLER both on the wealth of his material and on the excellent use he has made of it. THE WHOLE SET CONSTITUTES A MOST EXCELLENT PIECE OF CLINICAL WORK, AND WE BELIEVE THAT NO PHYSICIAN COULD FAIL TO DERIVE BENEFIT FROM A CAREFUL PERUSAL of these lectures, which, we may add, are profusely illustrated with photographs and diagrams."—*British Medical Journal.*

PELLEW.—Manual of Practical Medical and Physiological Chemistry. By CHARLES E. PELLEW, E.M. Demonstrator of Physics and Chemistry in the College of Physicians and Surgeons (Medical Department of Columbia College), New York. Honorary Assistant in Chemistry at the School of Mines, Columbia College, etc. With illustrations, 330 pages. Price 15s.

PHELPS.—Traumatic Injuries of the Brain and its Membranes. With a Special Study of Pistol-Shot Wounds of the Head in their Medico-Legal and Surgical Relations. By CHARLES PHELPS, M.D., Surgeon to Bellevue and St. Vincent's Hospitals. 8vo, 596 pages, with 49 illustrations. Cloth. £1 1s net.

RAYMOND.—A Manual of Physiology. By JOSEPH H. RAYMOND, A.M., M.D., Professor of Physiology and Hygiene, and Lecturer on Gynecology in the Long Island College Hospital; Director of Physiology in the Hoagland Laboratory ; formerly Lecturer on Physiology and Hygiene in the Brooklyn Normal School for Physical Education ; Ex-Vice-President of the American Public Health Association ; Ex-Health Commissioner, City of Brooklyn, etc. Illustrated. Cloth. Price 6s net.

In this manual the author has endeavoured to put into a concrete and available form the results of twenty years' experience as a teacher of physiology to medical students, and has produced a work for the student and practitioner, representing in a concise form the existing state of physiology and its methods of investigation, based upon comparative and pathological anatomy, clinical medicine, physic, and chemistry, as well as upon experimental research.

SAUNDER'S POCKET MEDICAL FORMULARY. By WM. M. POWELL, M.D., Attending Physician to the Mercer House for Invalid Women at Atlantic City. Containing 1,750 Formulæ, selected from several hundred of the best known authorities. Forming a handsome and convenient pocket companion of nearly 300 printed pages, with blank leaves for Additions; with an Appendix containing Posological Table, Formulæ and Doses for Hypodermatic Medication, Poisons and their Antidotes, Diameters of the Female Pelvis and Fœtal Head, Obstetrical Table, Diet List for Various Diseases, Materials and Drugs used in Antiseptic Surgery, Treatment of Asphyxia from Drowning, Surgical Remembrancer, Tables of Incompatibles, Eruptive Fevers, Weights and Measures, etc. Fourth Edition, revised and greatly enlarged. Handsomely bound in morocco, with side index, wallet and flap. Price 7s 6d net.

A concise, clear, and correct record of the many hundreds of famous formulæ which are found scattered through the works of the most eminent physicians and surgeons of the world. The work is helpful to the student and practitioner alike, as through it they become acquainted with numerous formulæ which are not found in text-books, but have been collected from among the rising generation of the profession, college professors, and hospital physicians and surgeons.

"This volume contains a collection of prescriptions arranged under the head of various diseases which they are designed to benefit. The diseases are classified in alphabetical order, and the volume is supplied with a thumb-nail index, which renders consultation the more easy. The prescriptions given appear to have been selected with judgment from a large number of sources, and this handbook will doubtless often be useful in indicating how an unfamiliar drug may best be prescribed. It will also be of use sometimes in suggesting new lines of treatment, for there is no doubt that we are all rather disposed to fall into habits in the matter of drug prescribing."—*British Medical Journal.*

"Designed to be of immense help to the general practitioner in the exercise of his daily calling."—*Boston Medical and Surgical Journal.*

"An excellent pocket companion, containing the most satisfactory and rational formulæ used by the leading medical men of Europe and America, introducing in the many prescriptions contained therein a considerable number of the more important recently-discovered drugs."—*Southern Practitioner.*

SIMON'S CLINICAL DIAGNOSIS. A Manual of Clinical Diagnosis by means of Microscopic and Chemical Methods. For Students, Hospital Physicians and Practitioners. By CHARLES E. SIMON, M.D., Late Assistant Resident Physician, Johns Hopkins Hospital, Baltimore. In one very handsome octavo volume of 563 pages, with 133 Illustrations on wood, and 14 full-page coloured plates. Second Edition, revised and enlarged. Cloth, price, 16s net.

"The author sets forth the methods most satisfactory and most approved for determining pathological conditions by chemical and microscopical examinations. Without other special training the work will be a guide to the attaining of the essential facts which only chemistry and the microscope can reveal."—*The North American Practitioner.*

"This is a very much-needed book. It tells the meaning of the clinical chemistry and results of microscopical examination of a case, and without their aid it is impossible to master a diagnostic study of many diseases told by the various secretions and excretions. A most excellent arrangement consists in the Differential Table of the More Important Diseases, or of the fluid, secretion or excretion, under consideration—the table being at the end of each subject discussed. Another excellence of the book consists in the full detail of the technique as to mode of securing, preparing, and examining specimens. There are so many practical, helpful points in this book that we must add it to the library which we regard as essential for the practitioner in his daily round of duties.—*The Va. Med. Semi-Monthly.*

"There is little need in the present day to dwell on the value and importance of the assistance given to clinical diagnosis, and therefore to treatment, by a thorough microscopical and chemical examination of the products of disease or of the blood and the various excretions. So important is it that within the past decade many a work has been published devoted solely to this one branch of clinical investigation, and there is no medical school where instruction upon it of a systematic kind is not to some extent imparted. Nevertheless, this necessary extension of the field of observation is in itself so wide and comprehensive that it becomes more and more difficult for the practitioner to follow. It needs a special department and a staff of highly-trained experts to carry it out to the full; and it is this class of work which is being so well undertaken in this country by the Clinical Research Association. The author of the volume before us has enjoyed at the Johns Hopkins Hospital, Baltimore, ample opportunities for the study of the subject, and his treatise is in every respect excellent. COVERING PRACTICALLY THE SAME GROUND AS THE WELL-KNOWN WORK OF PROFESSOR VON JAKSCH, THE BOOK CONTAINS IN SOME SECTIONS EVEN MORE INFORMATION THAN DOES THAT VOLUME. It is evident, too, that the author has himself largely confirmed the statements which he makes, and occasionally he feels bound to differ from the somewhat too dogmatic teaching that has dominated parts of the subject. WE HAVE, AFTER A CAREFUL REVIEW OF THE CONTENTS OF THE BOOK, NO HESITATION IN COMMENDING IT AS ONE OF THE BEST AND MOST COMPENDIOUS MANUALS FOR THE CLINICAL LABORATORY THAT HAS APPEARED. The subject-matter is arranged on a very systematic plan, the text is not burdened by references to literature, and the descriptions of apparatus as well as the instructions for the performance of tests are clear and concise. Perhaps the best section is that devoted to the urine, occupying about one-half of the volume, but the sections on the blood and on the gastric juice and gastric contents are little, if at all, inferior in scope and fulness. In his preface Dr. SIMON pleads for a more thorough recognition of these studies in places of instruction, and urges the younger members of the profession to pursue them with diligence. As he says, 'It is inconceivable that a physician can rationally diagnose and treat diseases of the stomach, intestines, kidneys and liver, etc., without laboratory facilities.' Whether his suggestion that physicians might usefully employ a laboratory assistant to enable them to carry out this duty will ever be realised, time, with its advance of knowledge, can alone show."—*Lancet.*

"The sciences of chemistry and microscopy, as applied to medicine, are year by year becoming of great importance; and while both form part of every medical curriculum in the preliminary stages, it is rare to find a medical school in which they are taught purely from the point of view of their clinical application. Too often they are learned by the student only to be forgotten as soon as he commences the 'professional' part of his studies. That the time has come when this state of things should be altered, and a separate study made of these sciences in their application to diagnosis, will impress all who read Dr. SIMON's volume.

"It has evidently been the author's aim in this work to present to students and practitioners not only the facts of physical science which are of practical importance, but also the reasons which have led up to that union of empirical deduction and scientific reasoning of which the modern science of diagnosis largely consists. Consequently, we find in the volume precise descriptions for the examination of the various fluids, secretions, and exudates of the body, both in health and disease. In every case a description of the normal material precedes the pathological considerations, which latter are in turn followed by a detailed account of the methods and apparatus used in examination. Following the directions given, no worker ought to find any insuperable difficulty in learning to recognise, say, the presence of tubercle bacillus in sputum, or of the diphtheria bacillus in membranous exudate.

"The volume is most appropriately illustrated both by coloured plates and by woodcuts in the text. WE HEARTILY WELCOME THE APPEARANCE OF THE WORK, WHICH WE FEEL SURE WILL FIND A PERMANENT PLACE IN THE WORKING LITERATURE OF THE PROFESSION, AND WILL ADEQUATELY SUPPLY A WELL-RECOGNISED DEFICIENCY."— *British Medical Journal*.

STARR.—Diets for Infants and Children in Health and in Disease. By LOUIS STARR, M.D., Editor of "An American Text-Book of the Diseases of Children." 230 blanks (pocket-book size), perforated and neatly bound in flexible morocco. Price 6s net.

"The first series of blanks are prepared for the first seven months of infant life. Each blank indicates the ingredients, but not the quantities, of the food, the latter directions being left for the physician. After the seventh month, modifications being less necessary, the diet-lists are printed in full. Formulæ for the preparation of diluents and foods are appended."

"We recommend every one who has occasion to treat infants and children to obtain a copy."—*St. Louis Med. and Surg. Journal*.

"The work on the whole will commend itself highly to the practitioner."— *Archives of Pediatrics*.

STEVENS.—A Manual of Practice of Medicine. By A. A. STEVENS, A.M., M.D., Instructor of Physical Diagnosis in the University of Pennsylvania, and Demonstrator of Pathology in the Women's Medical College of Philadelphia. Specially intended for students preparing for graduation and hospital examinations, and includes the following sections : General Diseases, Diseases of the Digestive Organs, Diseases of the Respiratory System, Diseases of the Circulatory System, Diseases of the Nervous System, Diseases of the Blood, Diseases of the Kidneys, and Diseases of the Skin. Each section is prefaced by a chapter on General Symptomatology. Third Edition. Post 8vo, 502 pages. Numerous illustrations and selected formulæ. Price 6s net.

"Contributions to the science of medicine have poured in so rapidly during the last quarter of a century that it is well-nigh impossible for the student, with the limited time at his disposal, to master elaborate treatises or to cull from them that knowledge which is absolutely essential. From an extended experience in teaching, the author has been enabled, by classification, to group allied symptoms, and by the judicious elimination of theories and redundant explanations to bring within a comparatively small compass a complete outline of the practice of medicine."

TAYLOR.—A Practical Treatise on Sexual Disorders of the Male and Female. By ROBERT W. TAYLOR, A.M., M.D., Clinical Professor of Venereal Diseases at the College of Physicians and Surgeons (Columbia College), New York ; Surgeon to Bellevue Hospital, and Consulting Surgeon to the City (Charity) Hospital, New York. In one octavo volume of 451 pages. With 73 illustrations and 8 plates in colour and monotone. Price 12s net.

"The branch of surgery with which this work deals is one about which very little is said in most text-books of surgery, and yet its importance is by no means small, for whether we consider the frequency with which such cases present themselves, or the amount of unhappiness which results from them, it is very obvious that they are worthy of the most careful attention of the surgeon. Dr. Taylor deals in the first place with the anatomy and physiology of the male sexual apparatus, and it is interesting to note that he is inclined to accept the result of the researches of Professor George S. Huntingdon, who asserts that the vesiculæ seminales never contain semen, and that they do not act as places of storage of this fluid, but they provide a special form of mucus to dilute and carry on the semen. Impotence and sterility in the male are thoroughly considered, and a chapter is devoted to the mental effects of sexual disorders. With regard to sterility in the male, the author thinks that probably in one case in six of unfruitful marriages this is the cause. The second half of the book deals with sexual disorders in the female. The final chapter treats of a peculiar new growth of the vulva, three examples of which Dr. Taylor has seen. He has already written on this condition in the *American Journal of the Medical Sciences*. In some respects it resembled a tertiary syphilitic condition, but potassium iodide seemed to have no effect upon it, and microscopically it appeared to be inflammatory. THE VOLUME IS A TRUSTWORTHY TREATISE ON A DIFFICULT SUBJECT."—*Lancet.*

THAYER.—Lectures on the Malarial Fevers. By WILLIAM
SYDNEY THAYER, M.D., Associate Professor of Medicine in the Johns Hopkins University. Small 8vo, 326 pages. With 19 charts, and 3 lithographic plates showing the Parasite of Tertian, Quartan, and Æstivo-Autumnal Fevers. Cloth, 12s net.

THOMAS.—Abortion and its Treatment: From a Standpoint
of Practical Experience. By T. GAILLARD THOMAS, M.D., Emeritus Prof. of Obstetrics and Gynæcology. Crown 8vo, 5s.

THOMPSON.—Practical Dietetics, with Special Reference to
Diet in Disease. By W. GILMAN THOMPSON, M.D., Professor of Materia Medica, Therapeutics, and Clinical Medicine in the University of the City of New York ; Visiting Physician to the Presbyterian and Bellevue Hospitals, New York. Large 8vo, 830 pages, illustrated. Cloth. Price 21s net.

"We quite agree with the author that the subjects which are so fully discussed in this volume are frequently dismissed in brief and indefinite phrases by the writers upon the theory and practice of medicine. . . . The fact that the author has written a successful book is due not only to his knowledge as a chemist and his studies as a physiologist, but as well as to the fact that he is a practical physician. . . . On the whole, the book shows that the author has industriously collected the best opinions upon the subject, that he has drawn from the results of his own experience, that he has endeavoured to bring the findings of the laboratory into practical relations with the observations of the consulting-room, and, finally, to produce a book of value to the practising physician. We believe that he has succeeded admirably in presenting a useful and readable book."—*The American Journal of the Medical Sciences.*

"The book will be of great assistance to the practitioner in the dietetic treatment of diseases that are influenced by proper feeding to the trained nurse in hospital and private nursing, and as a guide in the administration of proper food to infants and invalids in the home."—*College and Clinical Record.*

"It is a great pleasure to welcome Dr. Thompson's work on dietetics. For a long time we have longed for a book giving detailed and accurate information as to foods, their nutritive values, and their appropriate uses in disease. Other books have appeared, written by English and Continental writers, but they have not been suited to American needs. . . . The book is encyclopedic in its completeness. . . . We recommend it most heartily. It fills a place in medicine more important even than therapeutics, and one which has been too much neglected."—*University Medical Magazine.*

"Fewer subjects in medicine present greater difficulties to an author than that of dietetics ; and Dr. Thompson has done the profession a service in collecting so much information on this subject, and presenting it in so systematic and attractive a manner."—*Boston Medical and Surgical Journal.*

"The work is so complete, and has been so systematically prepared, that it is almost impossible to find a condition in which some benefit cannot be obtained by suitable diet."—*Ohio Medical Journal.*

TILLMANNS.—A Text-Book of General Surgery. By Dr.
HERMANN TILLMANNS, Professor in the University of Leipsig. Edited by LEWIS A. STIMSON, M.D., Professor of Surgery in the New York University. 8vo. Cloth, £1 1s net, per vol.

Vol. I.—The Principles of Surgery and Surgical Pathology. General Rules governing Operations and the Application of Dressings. Translated from the Third German Edition by JOHN ROGERS, M.D., and BENJAMIN T. TILTON, M.D. With 447 Illustrations.

Vol. II.—Regional Surgery. Translated from the Fourth German Edition by BENJAMIN T. TILTON, M.D., New York. With 417 Illustrations.

Vol. III.—Regional Surgery. With 517 Illustrations.

Dr. Hermann Tillmanns, Professor of Surgery in the University of Leipsig, possesses as a teacher those rare qualities which enable him to instruct the student step by step, beginning by the laying of a firm, broad foundation, upon which is built the solid surgical structure. It was on account of these exceptional qualities of the author that his work was selected as the best for the use of students, and at the same time well adapted to the needs of the practitioner.

Surgery, as presented in the present volumes, is a translation of his works on General Surgery and Surgical Pathology, and on Regional Surgery. Of the latter there are two volumes.

Volume I., General Surgery and Surgical Pathology, is largely devoted to the exposition of the essential principles which underlie a solid surgical structure. This applies not only to general surgical operations, but also to all surgical conditions. The work covers the entire field of general surgery and of surgical diseases, dealing not so much with special operations as with the conditions which should govern them—general directions for their performance, after-treatment, and the etiology, pathology, and treatment of the various surgical diseases.

Volume II., Regional Surgery, is devoted to the surgery of the head, neck, thorax, and spine and spinal cord; including, in the *first division*, injuries and diseases of the scalp, of the cranial bones, of the brain and its adnexa, of the face, of the nose and nasal fossæ, of the jaws, of the mouth, fauces, and pharynx, of the ear, and of the salivary glands. The *second division* includes injuries and surgical diseases of the neck, of the larynx and trachea, and of the œsophagus. The *third division* covers injuries and diseases of the thorax and of the heart; and the *fourth division* treats of the surgery of the spine and spinal cord, including deformities, fractures, gunshot injuries, tumours, etc.

Volume III., Regional Surgery, is devoted to the surgery of the abdomen, the upper extremity, and the lower extremity; including in the first section injuries and diseases of the abdominal wall, of the peritoneal cavity, the surgery of the liver, gall bladder, pancreas, spleen, stomach, and intestinal canal (with the exception of the rectum and anus), injuries and diseases of the rectum and anus, hernia, surgery of the kidney and ureter, injuries and diseases of the male bladder, of the urethra and penis, of the scrotum, testicle, epididymis, spermatic cord, and seminal vesicles, of the prostate and Cowper's glands, surgery of the female genito-urinary organs, and injuries and diseases of the pelvis. The second section includes injuries and diseases in the region of the shoulder, of the upper arm and the elbow joint, of the forearm and the wrist, and of the hand and the fingers. The third section includes injuries and diseases of the hip-joint and the thigh, of the knee-joint and the leg, and of the ankle and the foot.

The list of subjects is so full that it includes even the great surgical rarities, and the descriptions are sufficiently complete to save the reader from the necessity of consulting other works to obtain the knowledge necessary to understand and to treat.

"The translators are to be congratulated on their selection of this work as a medium through which to bring the current views of German surgeons before the English-reading medical public. Written by an acknowledged master of his art, accepted in the country of its production as a standard text-book, and bearing the imprimatur of a fourth edition within five years of its publication, it is admirably calculated to reflect the opinions and practice of the surgeons of to-day in Germany. . . .

"The first volume consists of the ' Principles of Surgery and Surgical Pathology,' and constitutes one of the best expositions of these subjects at present available. . . . The work before us (vol. ii.) deals with the Regional Surgery of the head, neck, thorax, and spine, and after a careful survey of it we do not hesitate to say that it would be difficult to find a more satisfactory presentation of the modern aspects of scientific surgery than its pages afford. . . . It is sufficient praise to the publisher to say that the paper, type, and illustrations are worthy of the text."— *Scottish Medical and Surgical Journal.*

Essentials of Physiology. By H. A. HARE, M.D., Prof. Therapeutics in Jefferson Med. College, Philadelphia. Numerous illustrations. Fourth edition, revised and enlarged, containing a series of handsome plate illustrations taken from the celebrated "Icomes Nervorum Capitis" of Arnold. 192 pages. Cloth, 4s net, post free.

Essentials of Diagnosis. By SOLOMON SOLIS COHEN, M.D., Prof. of Clin. Med. in the Philadelphia Polyclinic; and AUGUSTUS A. ESHNER, M.D., Instructor in Clin. Med. in Jefferson Med. Coll. 382 pages. 55 illustrations, some of which are coloured, and a frontispiece. 6s net, post free.

Essentials of Obstetrics. By EASTERLY ASHTON, M.D., Obstetrician to Philadelphia Hospital. 75 illustrations. Third edition, thoroughly revised and enlarged. 252 pages. Cloth, 4s net, post free.

Essentials of Gynæcology. By EDWIN B. CRAIGIN, M.D., Gynæcologist to Roosevelt Hosp., New York. 200 pages. 62 fine illustrations. Third edition. Cloth, 4s net, post free.

Essentials of Diseases of the Skin, including the Syphilodermata. By HENRY W. STELWAGON, M.D., Ph.D., Lect. on Dermatology in Jefferson Med. Coll. Third edition, revised and enlarged, with 71 letterpress cuts and 15 half-tone illustrations. 270 pages. Cloth, 4s net, post free.

Essentials of Refraction and Diseases of the Eye. By EDWARD JACKSON, A.M., M.D., Prof. Dis. of Eye in the Philadelphia Polyclinic ; and

Essentials of Diseases of the Nose and Throat. By E. BALDWIN GLEASON, M.D., Surg. to Nose, Throat, and Ear Dept., Northern Dispensary, Philadelphia. Second edition, revised. 290 pages, 124 illustrations. Cloth, 4s net, post free.

The above two volumes in one.

Essentials of Diseases of Children. By WILLIAM M. POWELL, M.D. Second Edition. 222 pages. Cloth, 4s net, post free.

Essentials of Medical Electricity. By D. D. STEWART, M.D., Chief of Neurological Clinic in the Jefferson Med. Coll. ; and E. S. LAWRANCE, M.D. 65 illustrations. Cloth, 4s net, post free.

Essentials of Practice of Medicine. By HENRY MORRIS, M.D., late Demonstrator, Jefferson Med. Coll. Philadelphia. Numerous illustrations and a coloured frontispiece. Third Edition. Cloth, 6s net, post free.

Essentials of Surgery. By EDWARD MARTIN, A.M., M.D., Clin. Prof. of Genito-Urinary Dis., Instructor in Operative Surg., and Lect. on Minor Surg., Univ. of Pennsylvania. Illustrated. Fifth Edition, revised and enlarged, 338 pages. Cloth, 4s net, post free.

Essentials of Pathology and Morbid Anatomy. By C. E. ARMAND SEMPLE, B.A., M.B., Phys. to the Bloomsbury Dispensary, etc. 160 pages, 46 illustrations. Cloth, 3s 6d net, post free.

Essentials of Forensic Medicine, Toxicology, and Hygiene. By C. E. ARMAND SEMPLE, B.A., M.B. 196 pages, 130 illustrations. Cloth, 4s 6d net, post free.

Essentials of Bacteriology. By M. V. BALL, M.D. Bacteriologist to St. Agnes' Hospital, Philadelphia. Third Edition, revised, with 81 illustrations, some in colours, and five plates. 218 pages. Cloth, 4s net, post free.

Essentials of Nervous Diseases and Insanity. By JOHN C. SHAW, M.D., Clin. Prof. of the Dis. of the Mind and Nervous System, Long Island Coll. Hosp. Med. School. 48 original illustrations. Second Edition. Cloth, 4s net, post free.

Essentials of Diseases of the Ear. By E. B. GLEASON, S. B.-M.D., Clin. Prof. of Otology, Philadelphia Med. Coll. Cloth, 4s net. 147 pages, fully illustrated.

Essentials of Anatomy. By FRED J. BROCKWAY, M.D., Assist-Demonstrator of Anatomy Coll. of Phys. and Surg., New York, and A. O'MALLEY, M.D., Instructor in Surg., New York Polyclinic. Second Edition. With full page plates, 376 pages. Cloth, 6s net, post free.

The following Catalogues can be had Post Free on application.

Catalogue 110 :

MIDWIFERY AND DISEASES OF WOMEN AND CHILDREN
URINARY AND VENEREAL DISEASES, SYPHILIS
SKIN DISEASES
FEVERS
CHOLERA
GOUT, RHEUMATISM, LIVER, STOMACH, RECTUM, AND DROPSY
SPINAL DISEASES

Catalogue 111 :

ANATOMY, PHYSIOLOGY
MEDICINE, PATHOLOGY
SMALL POX, VACCINATION, CHOLERA, FEVERS, &c.
MIDWIFERY, DISEASES OF WOMEN
CUTANEOUS DISEASES AND VARIOUS SUBJECTS

Catalogue 112 :

DISEASES OF THE CHEST, LUNGS, HEART, BLOOD, ASTHMA,
    CONSUMPTION
DISEASES OF THE BONES AND JOINTS, DEFORMITIES, CLUBFOOT,
    FRACTURES
INDIGESTION, DIET, FOOD, HEALTH
CANCER, TUMOURS, ULCERS, WOUNDS
DISEASES OF INDIA, EAST & WEST INDIES & TROPICAL CLIMATES
SURGERY
VALUABLE WORKS AND SETS

Catalogue 113 :

INSANITY, DISEASES OF THE BRAIN AND NERVOUS SYSTEM
DISEASES OF THE EYE

Catalogue 114 :

DENTISTRY

Catalogue 115 :

VACCINATION AND SMALL POX

Catalogue 116 :

DISEASES OF THE EAR AND THROAT

Printed by Cowan & Co., Limited, Perth.

www.ingramcontent.com/pod-product-compliance
Lightning Source LLC
Chambersburg PA
CBHW020856210326
41598CB00018B/1685